Principles and Applications of Fermentation Technology

Scrivener Publishing
100 Cummings Center, Suite 541J
Beverly, MA 01915-6106

Publishers at Scrivener
Martin Scrivener (martin@scrivenerpublishing.com)
Phillip Carmical (pcarmical@scrivenerpublishing.com)

Principles and Applications of Fermentation Technology

Edited by

Arindam Kuila and Vinay Sharma

Scrivener Publishing

WILEY

This edition first published 2018 by John Wiley & Sons, Inc., 111 River Street, Hoboken, NJ 07030, USA and Scrivener Publishing LLC, 100 Cummings Center, Suite 541J, Beverly, MA 01915, USA

For more information about Scrivener publications please visit www.scrivenerpublishing.com.

Wiley Global Headquarters
111 River Street, Hoboken, NJ 07030, USA

For details of our global editorial offices, customer services, and more information about Wiley products visit us at www.wiley.com.

Library of Congress Cataloging-in-Publication Data

ISBN: 978-1-119-46026-8

Cover image: Pixabay.Com
Cover design by Russell Richardson

Set in size of 11pt and Minion Pro by Exeter Premedia Services Private Ltd., Chennai, India

Printed in Singapore

10 9 8 7 6 5 4 3 2 1

Contents

Part I

PRINCIPLES OF FERMENTATION TECHNOLOGY

1

Fermentation Technology: Current Status and Future Prospects

Ritika Joshi, Vinay Sharma and Arindam Kuila*

Bioscience & Biotechnology Department, Banasthali University, Rajasthan, India

Abstract

This chapter deals with the current status and future prospects of the fermentation technology (FT). It discusses the different types of fermentation processes (solid-state and submerged fermentation) as well as the different types of enzyme and antibiotics production by FT. In addition, various industrial applications (enzyme production, organic acid production, biofuel production, etc.) of solid-state fermentation are also discussed. Also discussed are the future prospects of FT with regard to enhanced value product development.

Keywords: Fermentation technology, solid-state fermentation, enzyme production, biofuel production

1.1 Introduction

Fermentation technology is defined as field that involves the use of microbial enzymes for production of compounds that have application within the energy production, material, pharmaceutical industries, chemical, and food industries [1].

It appears naturally in various foods. The human beings are using it from the ancient times for preservation and organoleptic properties of food. It is a well-established technology of the ancient time used for food preservation, production of bread, beer, vinegar, yogurt, cheese, and wine. From time to time, it has got refined and diversified [2].

**Corresponding author*: arindammcb@gmail.com

Arindam Kuila and Vinay Sharma (eds.) Principles and Applications of Fermentation Technology, (3–14) © 2018 Scrivener Publishing LLC

It is the biological process in which various microorganisms such as yeast, bacteria, and fungi are involved in the conversion of complex substrate into simple compounds which are useful to humans (enzymes production, metabolites, biomass, recombinant technology, and bio-transformation product) on industrial scale. Organic acid and alcohol are the main products of fermentation. In this process, there is liberation of secondary metabolites like antibiotics, enzymes, and growth factors [3, 4].

They acquire biological activity so they are also known as bioactive compounds. These compounds contain plant and food constituents in small amount which are very nutritional. Various bioactive compounds consist of secondary metabolites, for example phenolic compounds, growth factors, food pigments, antibiotics, mycotoxins, and alkaloids [5, 6]. The constituent of phenolic compounds are flavonoids, tannins, and phenolic acids. Flavanones, flavonols, flavones, anthocyanidins, and iso-flavones are some major classes of flavonoids. Flavonoid comprises largest collection of plant phenolics where most of them are naturally occurring compounds [7].

According to their diverse perspectives, food and beverage are used in modern industrial fermentation processes. On the bases on different parameters such as environmental parameters and organisms required for fermentation, these techniques have become more advanced.

Generally, bioreactor is required in the middle of this process which can be arranged on the basis of their feeding of the batch, continuous and fed-batch fermentation, immobilization process. In the presence of the available amount of oxygen, mixing of substrate take place in single and mixed culture in submerged fermentation (SmF) [8].

1.2 Types of Fermentation Processes

1.2.1 Solid-State Fermentation

Solid-state (or substrate) fermentation (SSF) are define as fermentation that place in solid supporting, non-specific, natural state, and low moisture content. In this process, substrates such as nutrient rich waste can be reused. Bran, bagasses, and paper pulp are the solid substrates used in SSF. Since the process is slow the fermentation of substrate takes long time. So, the discharge of the nutrients is in controlled manner. It requires less moisture content so it is the best fermentation technology used for fungi and microorganism. However, this process is not applicable for bacteria because this fermentation cannot be used for organism that requires high water condition [9].

1.2.2 Submerged Fermentation

In SmF, microorganism required a controlled atmosphere for proficient manufacture of good quality end products; attain optimum productivity and high yield.

Batch, fed-batch, or continuous modes are used in industrial bioreactors for the production of different type of microorganism in broad range [8].

For the manufacture of alcoholic beverages (whisky, beer, brandy, rum, and wine), preservatives or acidifiers (lactic acids, citric, and vinegar) are used in food industry and for flavor enhancers (monosodium glutamate) or sweeteners (aspartate) amino acid are used in submerged batch cultivation.

In this part, there are different ways of submerged cultivation using microorganisms in bioreactors. Here we have discussed briefly about typical features and advantages and faults of each fermentation methods are displayed. Lastly, the production of microorganism in liquid medium in various type of food industrial product has been determined as the most important application for continuous, batch, and fed-batch cultivation.

1.2.2.1 Batch Cultivation

Batch culture is a closed system which works under aseptic condition. In these cultivations, inoculums, nutrients, and medium are mixed in the bioreactor in which the volume of the culture broth remains constant.

1.2.2.2 Substrates Used for Fermentation

It is very important to select a good substrate as the product of fermentation extremely varies. This technique is used for optimization of every substrate. This is mainly due to the cause that microorganism reacts in different way in every substrate.

The rate of consumption of different nutrient vary in every substrate, and so that their productivity. Some commonly used substrates in SSF are rice straw, vegetable waste, wheat bran, fruit bagasse, synthetic media, and paper pulp. Liquid media, molasses, waste water, vegetable juices, and soluble sugar are common substrates used in SmF to extract bioactive compounds.

Enzymes [10], antioxidants [11], antibiotics [12], biosurfactants [13], and pigments [14] are variety of bioactive compounds which are extracted using fermentation.

1.3 Enzymes

Enzyme cultivation is the most important technique for the manufacturing of different enzymes.

When fermentation on appropriate substrates is done, both fungus and bacterial microbes are required for the precious collection of enzyme. Enzyme production can be together performed by submerged and SSF. Bacterial enzyme production commonly implies SmF method because it requires high water potential [15]. In fungus, where less water potential is required, SSF method is applied [16].

In the world, 75% of the industries are using SmF for the production of enzymes. The major reason of using SSF is that it does not support genetically modified organisms (gmo) to the extent to which SmF does, so we prefer SmF rather than SSF.

One more reason of using SmF is that it has lack of paraphernalia as related to the cultivation of variety of enzymes using SSF. The microorganism is dissimilar in SmF and SSF by the detailed metabolism display that's way this is highly critical process. Here, influx of nutrients and efflux of waste substance is carried out in different metabolic parameters of cultivation. Some small variation from the particular parameters will affect the undesirable product.

1.3.1 Bacterial Enzymes

Cellulose, amylase, xylanase, and L-asparaginase are some well know enzymes produced from bacteria. Previously we have thought that SmF is one of the best ways to produce enzyme from bacteria. Current studies have shown that for bacterial enzyme production SSF is more capable than SmF. The most important explanation can be given by metabolic differences. In SmF system, lowering of enzyme activity and production efficiency is done by gathering of different intermediate metabolites.

1.3.2 Fungal Enzymes

Numerous genus of fungus, *Aspergillus*, has been isolated from this process which is industrially important for the production of enzyme. This fungus has been a well-known model of microorganism for the production of fungus enzyme [17]. *Aspergillus* is one of the largest sources of fungal enzyme. The common difference between SSF and SmF are straight lying on the productivity of the fungus [17]. Using SmF, phytase is extracted from *Thermoascusaurantícus* [18].

1.4 Antibiotics

The most important extract from microorganism using fermentation technology is antibiotics. It is a bioactive compound. Penicillin from *Penicillium notatum* is the first antibiotic produced from fermentation. It was completed in 1940s using SSF and SmF but today *P. chrysogenum* isolates are higher yielding producers [19]. Aminocillins, Carbapencins, Monobactams, Cephalosporins and Penicillins together they are known as β-lactam antibiotics [19]. Some other antibiotics like Tetracyclin, Streptomycin, Cyclosporin, Cephalosporin and Surfactin are manufactured from this process. *Streptomyces clavuligerus, Nocardialactamdurans,* and *Streptomyces cattleya* produces Cephamycin C from sunflower cake and cotton-de-oiled cake in which wheat raw is supplemented in SSF system as substrates for manufacturing Cephamycin C. In SSF, penicillin was produced by actinomycetes and fungi in mixed cultures.

In current time, the growth of proper substrates has led to the widespread use of SSF more than SmF. On the other hand, some results show that several microbial stains are extra suitable to SSF and others are more suitable for SmF. Thus, this technology is determined on the bases of microorganism that is being used for production. Recently, it has been studied that several antibiotics produced through SSF are more constant and high in quantity than SmF.

This is associated to minor production of bioactive substance that are intermediary compounds in SSF. However, the characteristics of the substrate material and their quality make SSF implementation limited. Due to this property, it is compulsory to check the production ability of different substrates earlier than optimization of the fermentation process.

Typically, in the beginning of batch cultivation, the bioreactors are filled with sterilized medium and the quantity of viable cell is known which is inoculated in the bioreactor. It is beneficial for the construction of biomass (Baker's yeasts) and primary metabolites (lactic acid, citric acid, acetic acid or ethanol production).

In food industries, organic acids used as preservatives or acidifiers(lactic acids, citric acids, and acetic acids), alcoholic beverages (wine, beer, and distilled spirits i.e. brandy, whisky, and rum) and sweeteners (e.g., aspartate) or amino acids used as flavoring agents (e.g., monosodium glutamate)are the various product manufactured by submerged batch cultivation.

Fermentation of whisky is taken as a good example, the manufacturing of distilled spirits are made from wood or stainless steel and it is made in simple cylindrical vessels known as wash backs.

Even it is very difficult to clean it but they used it, mainly in malt whisky distilleries. In this process, wort is pumped and cooled to 20 °C and inoculated with the yeast cells.

It has been found that manufacturingof citric acid has reached 1.8×10^6 tons in 2010 and about 90% of this is synthesized by the fungus *Aspergillus niger* from sugar containing material like sugarcane, corn, and sugar beet and food industry consumed 60% of it. We can follow surface liquid fermentation, SSF, and submerged liquid fermentation for the production of citric acid in industrial scale, however, the end predominates [24].

1.5 Fed-Batch Cultivation

In fed batch cultivation, one or more nutrients are added aseptically, it is a semi-open system and the culture is supplemented step-by-step into the bioreactor at the same time the volume of the liquid culture in the bioreactor increase within this time.

The increase in productivity, enhanced yield by controlled sequential addition of nutrients, ability to achieve higher cell densities, and prolonged product synthesis are the main advantages of fed-batch over batch cultures.

Immobilized Cell Technology Active Biocatalyst also known as enzyme or microbial cell has increased the productivity of bioprocesses and it is managed through controlled contact with high concentration. Through cell immobilization or recycling by feeding strategies in high density cultures [20]. Cell immobilization mostly studied in the food and gas-liquid mass. It is done in three phase bioreactor; it requires all three phases in competent mass transfer. These bioreactor aims in the region where main process amplification can be managed through the improvement of gas-liquid mass transfer [21].

Fundamental difference between SmF and SSF

Submerged fermentation	**Solid-state (substrate) fermentation**
Water cultivation medium (~95%).	Water cultivation medium is low (40–80%).
Liquid–gas are the two phase of the system.	Solid–liquid–gas are three phase of system.
Homogeneous.	Heterogeneous.
Low nutrient content, water soluble.	High nutrient content, water insoluble.
Oxygen transfer: gas–liquid.	Oxygen transfer: liquid–solid and gas–liquid.
Microorganism growth: liquid medium.	Microorganism growth: medium surface.
Only oxygen is transfer, process is not limited.	Oxygen, heat, and nutrient transfer is limited.
Product: soluble in the liquid phase.	Product: high concentration.

1.6 Application of SSF

1.6.1 Enzyme Production

In SSF, agriculture industrial substrates are considered the most excellent for enzyme production.

The expenditure of enzyme production by SmF is high as compared to SSF.

Approximately, all well-known microbial enzymes are produced through this process. According to research study, large amount of work has been done on the enzyme production of industrial importance like cellulases, lipase, proteases, glucoamylases, amylases, ligninases, xylanases, pectinases, and peroxidases. Thermostable enzyme xylanase by thermophilic *Bacillus licheniformis* has been produced from this process. Enzymes produced from this process are more thermo-stable than SmF process. It has 22- folds higher in SSF system than in SmF system.

The bacterial strain extracted from open xylan agar plate are characterized as xylanase produced from *Bacillus pumilus* from both the processes (submerged and SSF fermentation) [22]. *Rhizopus oligosporus* is used to produce acid protease from rice bran and during its production no toxin effect occurred in SSF.

1.6.2 Organic Acids

Gallic acid, citric acid, fumaric acid, kojic acid, and lactic acid are various acid produced by SSF. Wheat bran, de-oiled rice bran, sugarcane, carob pods, coffee husk, kiwi fruit peels, pineapple wastes, grape pomace, and apple are some agro-industrial wastes which are very resourceful substrates for production of citric acid in SSF. For the production of citric acid from *Aspergillus*, pine apple waste was used as substrate [23]. Sugarcane bagasse impregnated with glucose and $CaCO_3$ for the production of lactic acid from *Rhizopus oryzae*is used.

1.6.3 Secondary Metabolites

Fungus produce secondary metabolite, gibberellic acid, in its stationary phase. Gibberellic acid production increases in SSF system. Gathering of gibberellic acid was 1.626 times greater in SSF than SmF using *Gibberellafujikuroi* in the production of gibberellic acid in which wheat bran is used as substrate.

1.6.4 Antibiotic

Cephamycin C, Cyclosporin A, Penicillin, Neomycin, Iturin, and Cephalosporins are some common antibiotics produced from SSF. Penicillin is produced from *Penicillum chrysogenum* in which wheat bran and sugarcane bagasse are used as substrate under high moisture content (70%). *Nocardia lactamdurans, Streptomycesclavuligerus*, and *Streptomyces cattleya* produces Cephamycin C. In SSF, antibiotic penicillin is produced from Actinomycetes and fungi through mixed cultures.

1.6.5 Biofuel

Today, ethanol is the most extensively used biofuel. Even though it is very easier to produce ethanol using SmF, it is preferred because of low water requirement, little volumes of fermentation mash, end product protection is inhibited and less liquid water disposal, it decreases pollution problem and it is most commonly used for ethanol production because of abundant availability. *Saccharomyces cerevisiae* is used for ethanol production in SSF of apple pomace supplemented with ammonium sulfate in controlled fermentation. Sweet potato, rice starch, wheat flour, potato starch, and sweet sorghum are commonly used substrate.

1.6.6 Biocontrol Agents

On the bases of different mode of action, fungal agent has greater potential to act as biocontrol agents. To control mosquitoes Liagenidium *giganteum* is used as fungal agent. It works by encysting on their larvae. Here they use larvae as a substrate for growth.

1.6.7 Vitamin

Nicotinic acid, vitamin B12, thiamine, riboflavin, and vitamins B6 are the water soluble enzyme produced on SSF with the help of different species of *Rhizophus* and *Klebsiella*, which is well-known producer of vitamin B12.

1.7 Future Perspectives

In food industries, processing microbial enzymes are extensively used as gift to fermentation technology. Yet, it is essential to make this kind of enzyme for the future development. In recent years, various new industrial and analytical applications have been drawn out for the manufacture of new products.

Fermentation technology needs evolution and enhancement for the food and beverage industries. It aim is to humanizing higher yield and production amount by means of construction, new models, bacterial strain, and process monitoring. In these areas, they have developed some modern ideas that could show the mode of cost-effectively attractive solutions.

In SSF, the area of modern instrumentation and sensor development is commendation of process monitoring is very important.

The modern technology characterized so far include different sensor of technologies like infrared spectrometry, magnetic resonance imaging, x-rays, image analysis, and respirometry. The chief drawback is high cost, so for large-scale applications this technique is unsuitable. Algae and micro/macro algae derived food production is one of the best bioreactor design for development of large-scale photo-bioreactors and phytocultures (seaweed). The use of properly controlled ultra-sonication in bioprocesses is another potential approach to enhance the metabolic productivity.

Sono-bioreactor performance (mass transfer enhancement), their function (e.g., cross-membrane ion fluxes, stimulated sterol synthesis, altered cell morphology, and increased enzyme activity) and biocatalysts (cells and enzymes) are advantageous effects of ultrasound which can be exploited.

Its prospective in the field of food fermentation for genetic engineering is indisputable. On the basis of understanding of their diet and human gastro-intestinal microbiota, food fermentation has improved the nutritional status by the balanced choice of food-fermenting microbes. In this respect, food fermentation has attributed beneficial towards health and regarded as an extension of the food digestion.

References

1. Singh, V., Haque, S., Niwas, R., Srivastava, A., Pasupuleti, M., Tripathi, C.K.M., Strategies for fermentation medium optimization: an in-depth review. *Front. Microbiol.*, 7, 2087, 2017.
2. Motarjemi, Y., Impact of small scale fermentation technology on food safety in developing countries. *Int. J. Food Microbiol.*, 75(3), 213–29, 2002.
3. Subramaniyam, R., Vimala, R., Solid state and submerged fermentation for the production of bioactive substances: a comparative study. *Int. J. Secur. Net.*, 3, 480, 2012.
4. Machado, C.M., Oishi, B.O, Pandey, A., Socco, C.R., Kinetics of *Gibberellafujikori*growth andGibberellic acid production by solid state fermentation in a packed-bed column bioreactor. *Biotechnol. Prog.*, 20, 1449, 2004.
5. Martins, S., Mussatto, S.I., Martinez-Avila, G., Montanez-Saenz, J., Aguilar, C.N., Teixeira, J.A., Bioactive phenolic compounds: production and extraction by solid-state fermentation. a review. *Biotechnol. Adv.*, 29, 373, 2011.
6. Nigam, P.S., Pandey, A., Solid-state fermentation technology for bioconversion of biomass and agricultural residues. *Biotechnol. Agro-Ind. Res. Util.*, 197, 221, 2009.
7. Harborne, J.B., Baxter, H., Moss, G.P., *Phytochemical dictionary: handbook of bioactive compounds from plants, 2nd ed.* London: Taylor & Francis, 1999.
8. Inui, M., Vertes, A. A., Yukawa, H., Advanced fermentation technologies, in: *Biomass to biofuels*, A.A. Vertes, N. Qureshi, H.P. Blashek, H. Yukawa (Eds.), 311–330. Oxford, UK: Blackwell Publishing, Ltd., 2010.
9. Babu, K.R., Satyanarayana, T., Production of bacterial enzymes by solid state fermentation. *J. Sci. Ind. Res.*, 55, 464–467, 1996.
10. Kokila, R., Mrudula, S., Optimization of culture conditions for amylase production by thermohilic *Bacillus* sp. in submerged fermentation. *Asian J. Microbiol. Biotechnol. Environ. Sci.*, 12, 653, 2010.
11. Tafulo, P.K.R., Queiros, R.B., Delerue-Matos, C.M., Ferreira, M.G., Control and comparison of the antioxidant capacity of beers. *Food Res. J.*, 43, 1702, 2010.

12. Maragkoudakis, P.A., Mountzouris, K.C., Psyrras, D., Cremonese, S., Fischer, J., Cantor, M.D., Tsakalidou, E., Functional properties of novel protective lactic acidbacteria and application in raw chicken meat against *Listeria monocytogenes* and *Salmonella enteritidis*. *Int. J. Food Microbiol.*, 130, 219, 2009.
13. Pritchard, S.R., Phillips, M., Kailasapathy, K., Identification of bioactive peptides in commercial cheddar cheese. *Food Res. J.*, 43, 1545, 2010.
14. Dharmaraj, S. Askokkumar, B., Dhevendran, K., Food-grade pigments from *Streptomyces* sp.isolated from the marine sponge *Callyspongiadiffusa*. *Food Res. Int.*, 42, 487–492, 2009.
15. Chahal, D.S., Foundations of biochemical engineering kinetics and thermodynamics in biological systems, in: H.W. Blanch, E.T. Papontsakis, G. Stephanopoulas (Eds.), *ACS symposium series,* Washington:American Chemical Society, 1983.
16. Troller, J.A., Christian, J.H.B., *Water activity and food.* London: Academic Press, 1978.
17. Holker, U., Hofer, M., Lenz, J., Biotechnological advantages of laboratory-scale solidstate fermentation with fungi. *Appl. Microbiol. Biotechnol.*, 64, 175–186, 2004.
18. Nampoothiri, K.M., Tomes, G.J., Roopesh, K., Szakacs, G., Nagy, V., Soccol, C.R., Pandey, A., Thermostable phytase production by *Thermoascusaurantiacus* in submerged fermentation. *Appl. Biochem. Biotechnol.*, 118(1–3), 205–214, 2004.
19. Balakrishnan, K., Pandey, A., Production of biologically active secondary metabolites in solid state fermentation. *J. Sci. Ind. Res.*, 55, 365, 1996.
20. Bumbak, F., Cook, S., Zachleder, V., Hauser, S., Kovar, K., Best practices in heterotrophic high-cell-density microalgal processes: achievements, potential and possible limitations. *Appl. Microbiol. Biotechnol.*, 91, 31–46, 2011.
21. Suresh, S., Srivastava, V.C., Mishra, I.M., Critical analysis of engineering aspects of shaken flask bioreactors. *Crit. Rev. Biotechnol.*, 29, 255–278, 2009.
22. Kapilan, R., Arasaratnam, V., Paddy husk as support for solid state fermentation to produce xylanase from *Bacillus pumilus*. *Rice Sci.*, 18 (1), 36–45, 2011.
23. Oliveira, F.C., Freire, D.M.G., Castilho, L.R., Production of poly(3-hydroxybutyrate) by solid-state fermentation with *Ralstoniaeutropha*. *Biotechnol. Lett.*, 26, 24, 2004.
24. Soccol, C.R., Vandenberghe, L.P.S., Rodrigues, C., Pandey, A., New perspectives for citric acid production and application. *Food Technol. Biotechnol.*, 44, 141–149, 2006.

2

Modeling and Kinetics of Fermentation Technology

Biva Ghosh[1], Debalina Bhattacharya[2] and Mainak Mukhopadhyay[1]*

[1]Department of Biotechnology, JIS University, Kolkata, West Bengal, India
[2]Department of Biochemistry, University of Calcutta, Kolkata, West Bengal, India

Abstract

Fermentation is a biochemical process of microorganism for the production of different valuable products such as enzymes, hormones, biofuels, etc. Fermentation process generally includes batch fermentation, feb-batch fermentation, and continuous culture. For enzyme production submerged and solid state fermentation process is involved. Microorganisms utilize the nutrients present in the substrates for their growth and product synthesis. Change in chemical or physical environment highly effects the product formation and its quality and yield. These changes effect the growth and product synthesis kinetics leading to different quality and yield of products. Thus, to ensure that the product formation is high quality and high yield, fermentation process has to be monitored properly. Mathematical calculation and statistical analysis is needed to track the fermentation process and monitor this process for best results. This enhances the product quality as well as leads to high yield. Many researchers has also developed strategies for the production of zero waste or to reuse the waste produced from one system to produce value added products of other system and leads to no waste technology. But all these strategies depend on the mathematical calculation, observation and statistical analysis, kinetics of product formation and monitoring. Different microorganisms have different growth kinetics and needs different modeling for high yield. It also enhances the economic value of product and economic status of the country. Thus this chapter focuses on the modeling and kinetics involved in high yield and high quality product formation from fermentation system.

Keywords: Modeling, kinetics, statistical analysis, mathematical calculations

**Corresponding author*: m.mukhopadhyay85@gmail.com

Arindam Kuila and Vinay Sharma (eds.) Principles and Applications of Fermentation Technology, (15–44) © 2018 Scrivener Publishing LLC

2.1 Introduction

Fermentation is a biochemical process of microorganism for the production of variable products. Different organisms need different conditions to produce some specific products. Some of the variables such as biomass resource, type of microorganism, growth rate, agitation speed, substrate composition, reaction time and pH of the culture medium, simultaneous sacharification, and fermentation (SSF) are factors needed to be optimized for efficient production of product. Thus, optimization with modeling and kinetics solves the problem [1]. Kinetics is the analysis of the interpretation of observations and factors influencing the fermentation process. Such analysis can be explained by mainly three approaches: phenomenological, thermodynamic, and kinetic [2]. Though modeling and kinetics are differentially explained by different authors but the main interpretation remains same. Modeling and kinetics of the system is best interpreted by mathematical representations. Mathematical modeling is the representation of the essential aspects of reality with the help of function, symbols, and numbers. Manipulations and conversion of mathematical expression according to the need of the system help to create an optimized model of fermentation system for a particular product formation. It helps to estimate the convenience and cost of product formation in reality before performing the experiment in reality [1]. Modeling the kinetics of fermentation process helps to process-control and research efforts and thus, is considered as one of the most important aspect in fermentation process study. It effectively reduces the cost of production and increases quantity and quality of product formation. Modeling of the fermentation process not only includes kinetics of the cell system but also includes the condition of the bioreactor's performance [3]. Thus, modeling has two parts microbial kinetics and bioreactor's performance [3]. Now, as fermentation process involves many factors such as temperature, aeration, substrate, biomass, etc. on which products formation depends. Absence of perfects sensors for quantification of product formation and substrate and biomass leads to low productivity and manually optimizing the system is a tedious job. Thus, to increase the productivity, other factors affecting fermentation process needs to be controlled which leads to need of more man power and increase the cost of production [4]. Thus, to minimize the cost, the fermentation systems need to be automated. Thus, modeling and kinetics of the fermentation process using computerization is also an interesting topic discussed in this chapter. Fundamental aspects and need of modeling are explained in this chapter. This chapter helps to better understand the generalized notion of the application of modeling and use of kinetics for increasing productivity of fermentation process in recent days.

2.2 Modeling

Models consist of relationship between the system and the variables that affects the system. A system can be any equipment of unit operation such as bioreactors, a single cell, a microbial culture, an immobilized cell, HPLC etc. A system is affected by different variable of interest such as time, temperature, rate of reaction etc. Changing the variables, effects, the system or the surrounding environment. Thus, modeling of a system optimizes the conditions for better performance of the system. In case of fermentation there are many variables such as feed rate, pH, the rate and mode of agitation, inoculum quality, temperature, costs of production system, etc. which affects the system and surrounding environment [5]. Modeling can be done by using mathematical expression or non-mathematical by experimental methods. Mathematical modeling is best as it estimates the outcome of the system without actually performing the experiment. Whereas, in case of non-mathematical experimental methods is tedious as it takes long time and recurring of experimental methods and are also non-predictable [6]. Mathematical modeling is cost effective as it predicts the outcome before-hand thus, decrease the cost of system's modeling. Modeling of a system is a cyclic process which involves many aspects which needs to monitored. Some of the aspects are biological, physicochemical, technological constrains, literature study, database together forms data from which assumptions are derived. Further combination of experiment with these assumptions leads to model formation. More analysis is done to improve the model and produce an optimized model for the specific system [5].

As modeling is cyclic process consisting many steps of optimization thus, to start a model formation we can consider any simple components such as set of results from a batch culture (Figure 2.1). Changing the components of culture media and observing the rate of cell growth and extracellular component production with respect to times is also a small example of modeling the culture system [6]. Changing the parameter which is involved in the system leads to modeling the system. Mathematical modeling is the best method of modeling as discussed earlier in this chapter. Now, this mathematical expression when combined with the power of automatization form dynamic model of system. In this 21st century automatization is achieved with the help of computer system. Nowadays, many software has developed which can easily analyze data and interpret it. Many more sophisticated sensors have developed which can precisely sense the production of required components in the system. Modeling of fermentation system using computers has enhanced the productivity [4, 6].

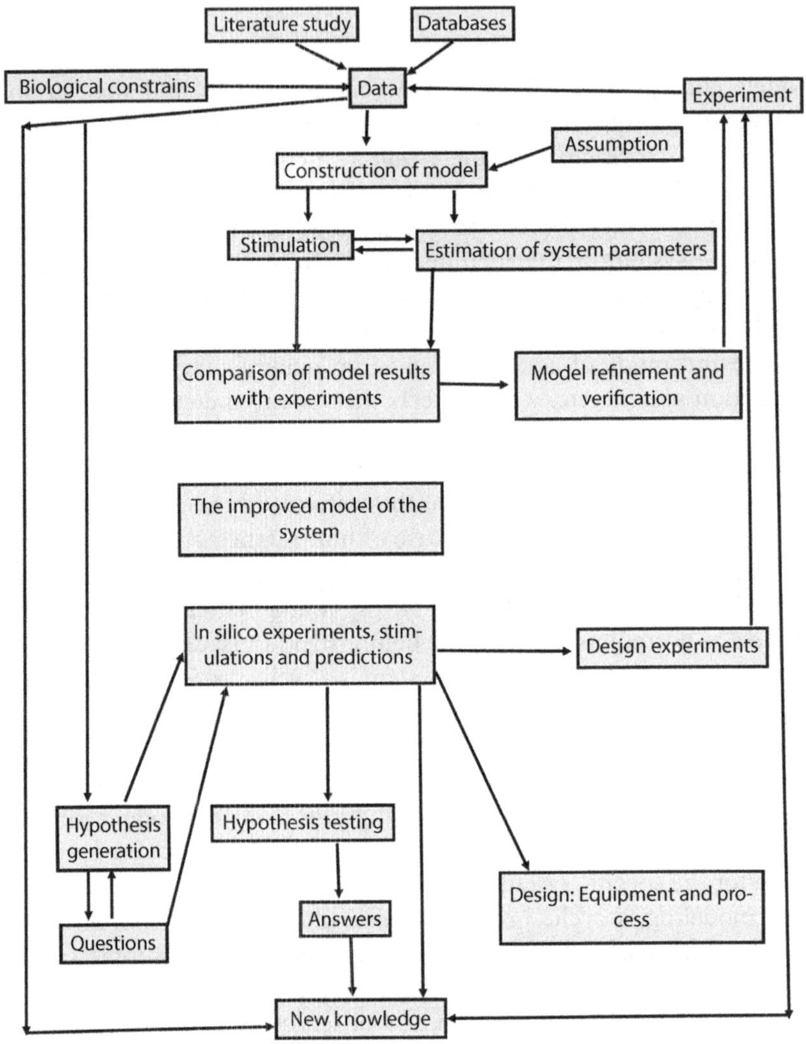

Figure 2.1 A flowchart describing the cyclic nature of modeling process.

2.2.1 Importance of Modeling

Fermentation is biochemical process which involves conversion of different compound into industrially valuable compounds. Fermentation system is an innovative piece of instrument which makes the fermentation process simpler and easier to produce complex compounds in a simple process.

Fermentation process is affected by many parameters such as composition of media, pH, temperature, aeration, feed rate, mode of agitation, inoculation quantity etc. Change in these parameters affects the fermentation process. Thus, optimization and monitoring of the system increases the production rate [6].

With the advancement of technology, such as improved measurement, instrumentation, information technology, molecular biology and high-throughput techniques enormous data of quantitative and qualitative in fermentation processing, and biotechnology engineering is produced. These data are analyzed, looked for relations and connectivity among them using various software. Once the relation and connectivity is found a model is developed [4]. As modeling is a cyclic process, construction of hypothesis as a first step towards construction of model is the best method. Modeling thus, provides predictive information regarding the action of fermentation system. It prior to perform an experiment predicts the outcomes or results. In this ways we can choose a perfect model or construct a new model with the existing model according to our needs regarding the product formation. It also reduces the labor or manpower cost and automation of the system provides error less analysis leading to minimum loss [7]. In this way, a cost effect but high yielding fermentation system is generated. By using model based terminologies, it also acts as a communicating language among scientist and engineers of different backgrounds. It acts as a universal language for communication regarding a fermentation system. It helps to predict and decide the next experiment precisely without hassle of repeating experiments. Model automatically measure and monitors factors and sometimes highlights factors which are consider as less importance but are actually highly important to the fermentation system. These applications of model signifies the importance of modeling a system [8].

For constructing a model, the components of modeling need to be understood. The knowledge of the parameters of modeling helps to predict the system. Constructing a model is precise when it is tested by its ability to predict the outcome of the system reaction by a set of independent experiments which consist of different forms of experiment including parameters involved in the fermentation system [9]. In constructing a model, experimental error and physical constrain should also be taken care of. Experimental errors may include omitting data with high degree of error. Thus, the model should consist of replicate of experiments, sampling and analysis. Physical constrains includes technical, biological, chemical and physical, upper and lower limits of the range of values of the system variables and parameters which needs to be taken care of [5].

2.2.2 Components of Modeling

Components of modeling include control volume, variables, parameters, and the equations (Figure 2.2). Other than this, assumption and hypothesis are also indispensable part of the fermentation modeling system [10].

2.2.2.1 Control Volume

Control region or volume is one of the most important components of modeling. Control regions is the space in the system where all the variable (concentration, pH, temperature, pressure etc.) chosen for the system are kept uniform. It is need not to be necessary that the concentration in the control region to be constant with time. Rather, concentration may vary or may remain constant with time but, any change occurring in the control region remains uniforms with time [5]. This means that the concentration of the compounds for example in the system remains uniform with time in the control region. As in most real system is heterogeneous thus, control region is mostly considered as an imaginary space of the system by the modelers. In case of a heterogeneous system more than one control region is considered depending on the bulk of homogeneity. Control region can be best with an example such as in bioreactor where the concentration of

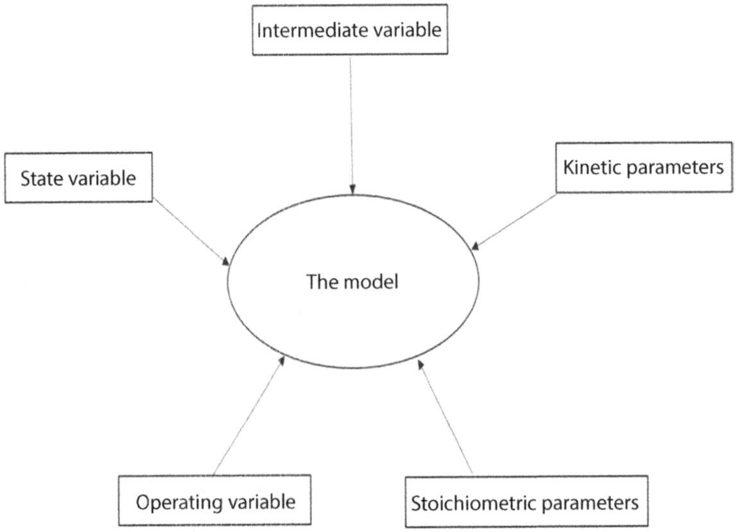

Figure 2.2 The components of modeling.

a compound in uniform in the whole system than the bulk liquid is the control region and has single control region. But if in a bioreactor, the concentration of compounds is divided by the impeller in two or more than two halves, but the concentration of compounds in each region remains uniform than the system consist of more than two control region. The bulk liquid is the control region (Figure 2.3). There may be exchange of matter, energy or momentum with the control regions. The volume of the control region may vary or remain constant. The control regions can be finite or infinitesimal. Control regions has some boundaries that can be defined as: phase boundaries across which no exchange occurs, phase boundaries across which an exchange of mass and energy takes place and geometrically defined boundaries in a single phase within which the exchange takes place by bulk flow or molecular diffusion [5]. Choosing of control volume is a crucial step in modeling process for the success of the model. Though the process seems to be easy, but many factors and variable are needs to be considered which make the process complicated. Thus, it can be interpreted that to construct a model first the system to be designed should be assumed and then the consideration of what should be the mode of operation or activity which further help to decide whether the system will be steady or unsteady that is whether the system properties should change or not. This heterogeneity of the system further decides whether the control region will be finite or infinitesimal.

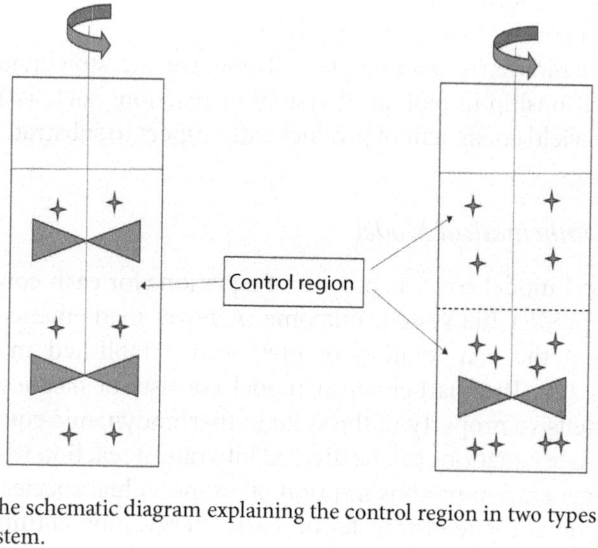

Figure 2.3 The schematic diagram explaining the control region in two types of bioreactor system.

2.2.2.2 *Variables*

Variables are the component of the system whose change in the system affects the system. Variable are of three types: state, operating, and intermediate variables [7].

1. State variables – It defines the state of the process and for every extensive property of the system one variable is present. For example, viable cell concentration (X_v), non-viable cell concentration (X_d) etc. [5].
2. Operating variables – These are the variable whose values can be set by the operator of the process. For example, dilution rate (D), volumetric feed flow rate (F) etc. [5].
3. Intermediate variables – It is defined as volumetric rate variables which can also be defined under state variable [5].

2.2.2.3 *Parameters*

Parameter are a set of constrains or measurable factors which limits or boundaries the scope of a particular process [2].

1. Kinetic parameter – The kinetic rate expression constants for the system is defined as kinetic parameters. Such as μ_{max} is maximum specific growth rate per hour, K_S is the saturation constant kg per m^3 etc. [2, 11]
2. Stoichiometric parameters – These are the stoichiometric relationship in biological system or reaction. Such as $Y_{P/S}$ is the yield coefficient of product with respect to substrate [12].

2.2.2.4 *Mathematical Model*

Mathematical model consists of a set of equations for each control model which can predict the system outcome. A novel mathematical model is derived from the combination of previously established mathematical expressions [1]. The mathematical model consists of balance equations for each extensive property of the system, thermodynamic equations, rate equations. Rate equations can be divided into rate of reaction which defines the rate of generation or consumption of an individual species within the control region and rate of transfer of mass, energy, momentum across the boundaries of the control region [13].

2.2.2.4.1 Mass Balance Equation

Balance equations are needed for every extensive property of interest in every control region. Extensive properties are those that are additive over the whole system such as mass and energy whereas concentration and temperature are intrusive properties of the system [14]. Other than this each and every balance equation are linearly independent that means no balance equation is formed by the addition or combination other equations [5]. Such as:

Rate of accumulation or rate of deletion = (rate of input to control region) – (rate of output to control region) (Eq. 2.1)

Input and output can be defined by the rate of mass transfer and reaction phenomenon as:

- Generation (input) and consumption (output) due to reaction within the control region.
- Transfer occurs across the phase boundaries
- Bulk flow across the boundaries of control region
- Diffusion across the boundaries of control region

Extensive properties of the control region can accumulate or deplete which can be measured by numeric value or magnitude of input and output of the control region [15]. Here input is considered as positive and output as negative term. Accumulations and depletions are the rate of extensive properties change in the control region with respect to time [16]. If the total of input term is larger than those of the output term, then the extensive properties are accumulating in the control region and if the total of output term is larger than the input then the depletion of extensive properties occurs the control region (Eq. 2.1) [5].

2.2.2.5 Automatization

From 18th century with the invention of computers, steady increase in the use of computers in different sectors has occurred. Automatization using computers has crept into every sectors of industries replacing the power of manpower. It has also lead to more error free and precise process. Automatization has already well prospered in industries such as oil industries, metallurgy, chemical industries etc. whereas it took long time to prosper in fermentation industries [8]. The reason behind this are: lack of proper sensor for product, substrate and biomass; absence of reliable process model for process control analysis; investment for computers in

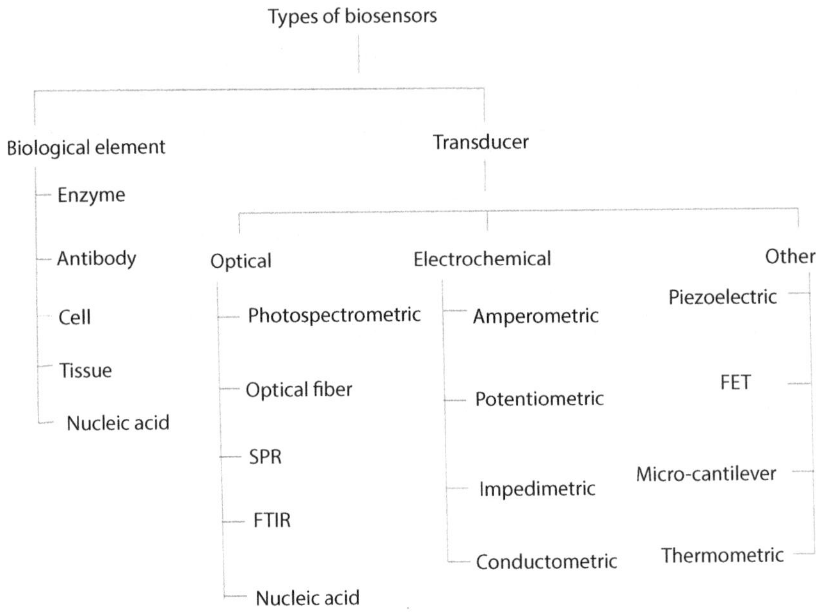

Figure 2.4 The type of sensors.

case of fermentation field were costlier with respect to other industries as fermentation were small scale production earlier. But now fermentation industries are growing rapidly and are large scale industry as well as now better sensors and fermentation models are present to facilitate the fermentation process (Figure 2.4) [8].

2.2.2.5.1 Some Fundamental Component of Computer-Control Fermentation

1. Fermentor – Fermentor is vessel with controlled condition of aeration, agitation, temperature, pH in which microbes are grown for fermentation process. It consists of an input and output port. A sensor could be attached with the output port for controlling the input of the substrates with respective to the output of the product. In this way other factor of a fermentor which is controlled manually could be automatized and the whole process could be tracked and sensed in the computer system [4].
2. Computer – Computers are the digital machines which could performs the task given to them automatically by performing a set of operation in accordance with predetermined set

Table 2.1 Some fermentation control software in recent days.

Sl.no.	Software	Description	Reference
1	Matlab	analyzing data, developing algorithms, or creating models	[17]
2	Minifor	includes all the electronics for visualizing and regulation of 6 parameters (temperature, pH, DO, air flow rate, agitation and parameter 'X')	[18]
4	Process control software (PCS)	for completely automatic control, data acquisition and real-time visualization of parameters	[19]
5	FNet	ready to use software for MINIFOR fermentor and bioreactor	[20]
6	SIAM	industrial fermentation software with unlimited possibilities (e.g.: redox potential, CARBOMETER and other instruments	[21]
7	MINI-4-GAS software	an extension of SIAM for automatic gas-mixing	[22]

of variables and programs assigned to them. In case of fermentation system, a computer should consist of programs and software which could analyze the generated data and reproduce it as an understandable format (Table 2.1). Thus, a powerful computer with more storage capacity and high speed of performance could be best suited for automatization of fermentation system [4].

2.2.2.5.2 Interphase Between Computer and Fermentor

A computer system is accompanied with input and output port which is used to transfer data and control signals to and fro between computer and fermentor. There are mainly two types of ports that are parallel port and serial port. A parallel port transfer bits of data simultaneously whereas serial port transfers bits of data one at a time. Serial port is used to communicate between process operator console and process computer. Serial port is useful in managing data traffic that exists between the computer and terminal [23]. As in case of serial port single link is enough to transfer all data bits even in long distance connection but in parallel ports as many links are needed as many bits of data are to be transferred. Parallel lines are mainly used for conjunction between computer and process system. They are used as input and output port for data transfer and for senses and control lines. Parallel port can be used to introduce switch in connection with computer without using expensive interface. But a buffering is needed in case the system doesn't crash due to overload. As well as some control valves are needed to handle the high voltage or current. In case of analogue signal, analogue digital converter is needed to convert the singles in digital. On the other hand, a digital to analogue converter is also needed to send signals from computer to the control process. Now, when computer is connected to the whole fermentation system then, a question arises that when and how should the input and output port work and how much bits of data should pass through input and output port. Here, software is need which could analyze the collected data in the memory and logically decide depending on the set programs and control the input output port and take care of the proper addressing of the data transmission [23]. Other than this some floating sensors such as pH meter, spectrometer etc. need to be attached with the system and should be connected with computer so the data generated could be analyzed and compared by the computer in the memory disk and the system become completely automatized [4].

2.2.2.5.3 Set Point Control and Direct Digital Control

Analogue controller controls the process actions in fermentation plants that are not computerized. Thus, this process system is equipped with analogous regulatory mechanism which keeps the variables controlled at a set point. As this control points are set manually, thus if the control points are not constant then they have to be changed repeatedly by manually. This procedure of manual setting is called set point control. As discussed in earlier section computers can also perform controlling functions, thus, by comparing the data produced by sensors with the rated value point inside the memory of the computer, it can decide a combination or differential action for the fermentation system. This procedure of direct interaction of fermentation system with the computers is called direct digital control [4].

2.3 Kinetics of Modeling

Study of fermentation system includes growth of the microorganism, substrate utilization, product formation with respect to time. Thus, kinetic analysis approaches are: thermodynamic, phenomenological, and kinetic [2].

2.3.1 Thermodynamic

Thermodynamic approach was first used by Calam *et al.* (1951). It is the measurement of fermentation rate by calculating the activation energy of rate determining step in the reactions involved including all the metabolite functions [2].

2.3.2 Phenomenological

Phenomenological approach is the measurement related to phenomenon such as growth rate, rate of product formations etc. Gaden (1955) first classified it on the basis of specific reaction rate as a comparison between rate per unit weight of cellular tissue and utilization of substrate. Further Maxon (1955) classified it as comparison between growth rate and rate of product formation. This was fist approach toward kinetic study of fermentation. Later Gaden (1958) further classified it into cell propagation, direct metabolic product and indirect metabolic product [2].

2.3.3 Kinetic

Kinetic approach was first approached by Luedeking (1958) in a homofermentative lactic acid production by *Lactobacillus delbrueclcii.* He found the formation of lactic acid depends both on growth and non-growth phase by a mathematical expression

$$dC/dt = x\, dM/dt + y\, M \qquad \text{(Eq. 2.2)}$$

where C is product concentration, M is concentration of cell mass, t is time and x and y are parameters which are functions of pH in this system. Deindoerfer and Humphrey (1959) further modified the above reaction as:

$$x = 0 \text{ (non-growth)}$$

$$dC/dt = y\, M = -z\, dN/dt \qquad \text{(Eq. 2.3)}$$

$$y = 0 \text{ (growth)}$$

$$dC/dt = x\, dM/dt = -z\, dN/dt \qquad \text{(Eq. 2.4)}$$

where N is the concentration of substrate or limiting nutrient concentration. Thus, lactic acid fermentation can be defined by two simple mathematical expressions [2].

With the increasing knowledge in the recent days, immense knowledge regarding the metabolic reactions, product formations, cell growth, cell death, the changes in their morphology, etc. are generated. These also has effect on the fermentation system. Thus, illustration with the help of mathematical modeling can be done in the biological system. Hence, each biological reaction is considered as a single reaction system. Here, cell maintenance or endogenous respiration is considered as basic cell activities. Other assumption includes all microbial cells are of same physiology (shape, size, etc.) and treat the whole living cells in the microbial culture as one uniform biomass. Collection of dead cells are also considered as another one uniform biomass [24].

2.3.3.1 *Volumetric Rate and Specific Rate*

Volumetric rate is defined by the following equation:

$$\text{volumetric rate} \equiv \frac{\text{Amount of a compound produced or consumed}}{(\text{unit volume})(\text{unit time})} \qquad \text{(Eq. 2.5)}$$

Where the amount of a compound produced per unit volume is the concentration [5]. Any microbial activity encompasses, which is expressed as volumetric rates, depends on the viable biomass x_v of the control region. But in certain unusual cases where the product formed is found in the media after the cell's death, then the volumetric rate depends on the dead biomass of the control region [25]. Though volumetric rates help to approximate the process or design experiment, but it doesn't allow any inferences regarding comparison of the performance between same or different microorganisms. For this type of inferences specific rates are used.

Specific rates depend on growth of microorganism, product formation, and substrate consumption [5]. It is defined by the following mathematical expression:

$$\text{specific rate} \equiv \frac{\text{volumetric rate}}{\text{biomass concentration}} \qquad \text{(Eq. 2.6)}$$

Thus, combining Equations 2.5 and 2.6 we can rewrite specific rate as:

$$\text{specific rate} \equiv \frac{\text{amount of compounds produced or consumed}}{(\text{unit volume})(\text{unit time})(\text{biomass concentration})} \quad \text{(Eq. 2.7)}$$

Thus, volumetric growth rate can be expressed as:

$$r_x \equiv \text{volumetric growth rate} \equiv \frac{\text{amount of biomass formed}}{(\text{unit volume})(\text{unit time})} \quad \text{(Eq. 2.8)}$$

the unit of volumetric growth rate is kg m^{-3} h^{-1}.

Thus, specific growth rate can be written as:

$$\mu \equiv \text{specific growth rate} \equiv \frac{\text{volumetric growth rate } (r_x)}{\text{concentation of viable biomass } (x_v)} \quad \text{(Eq. 2.9)}$$

the unit of specific growth rate is h^{-1}.

Similarly, in case of volumetric death rate can be defined as:

$$r_d \equiv \text{volumetric death rate} \equiv \frac{\text{amount of dead cells}}{(\text{unit volume})(\text{unit time})} \quad \text{(Eq. 2.10)}$$

the unit of volumetric death rate is kg m^{-3} h^{-1}.

Thus, the specific death rate can be written as:

$$k_d \equiv \text{specific death rate} \equiv \frac{\text{volumetric death rate } (r_d)}{\text{concentration of viable biomass } (x_v)} \quad \text{(Eq. 2.11)}$$

The unit of specific death rate is h^{-1}. In case of specific death rate concentration of viable biomass is again used as living cell contributes to dead cells but once a cell is dead it doesn't die again to contribute to the process [5].

Now, in case of product formation,

$$r_p \equiv \text{volumetric product formation rate}$$
$$\equiv \frac{\text{amount of product formed}}{\text{(unit volume)(unit time)}} \qquad \text{(Eq. 2.12)}$$

The unit of volumetric product formation rate can be written as kg m^{-3} h^{-1}.

Thus, specific product formation rate can be written as:

$$q_p \equiv \text{specific product formation rate}$$
$$\equiv \frac{\text{volumetric product formation rate } (r_p)}{\text{concentation of viable cells } (x_v)} \qquad \text{(Eq. 2.13)}$$

The unit of specific product formation rate is h^{-1}. Here, as formation of product is generally done by live cells thus, specific product formation rate is found with respect to concentration of viable cells. Though, in very few cases concentration of live cells is replaced by dead cells such as, in case of product formation from autolysis of cells [5].

Now, in case of substrate consumption,

$$r_s \equiv \text{volumetric substrate consumption rate}$$
$$\equiv \frac{\text{amount of substrate consumed}}{\text{(unit volume)(unit time)}} \qquad \text{(Eq. 2.14)}$$

The unit of volumetric substrate consumption rate is kg m^{-3} h^{-1}.

Thus, specific substrate consumption rate can be written as:

$$q_s \equiv \text{specific substrate consumption rate}$$
$$\equiv \frac{\text{volumetric substrate consumption rate } (r_s)}{\text{concentration of viable cells } (x_v)} \qquad \text{(Eq. 2.15)}$$

The unit of specific substrate consumption rate is h^{-1}. In this case as only living has the ability to use substrate thus specific substrate consumption rate is defined with respect to concentration of living cells [5].

2.3.3.2 Rate Expression for Microbial Culture

The knowledge of these basic forms of kinetic rate expression is needed to understand the kinetic expression in the balance equations of batch, fed-batch, continuous culture, and solid state in bioreactor [13, 26, 27].

The basic need for the growth of the microorganism includes: a viable inoculum, carbon source, energy source, essential nutrient, and physico-chemical conditions. When all these requirements for growth are provided the rate of increasing viable cells will be proportional to viable biomass concentration. The specific growth rate can be divided into two ways: substrate dependant and substrate independent [3].

2.3.3.2.1 Monod Kinetic for Growth of Microbial Cell

As in a cell's metabolism multiple biochemical reactions are occurring which includes enzymes and even production of these enzymes are also included in cell's metabolism. Thus, growth is assumed as result of hundreds of enzyme-catalyzed reactions [12]. Michaelis–Menten expression is the simplest form expression relating enzymatic reaction rate to the rate limiting substrate concentration which is written as follows [5]:

$$v = \frac{kES}{K_M + S} \qquad \text{(Eq. 2.16)}$$

where v = the reaction velocity; k = the rate contant; E = the total amount of enzyme;

K_M = the Michaelis–Menten constant; S = the substrate concentration. kE is maximum when $S >> K_M$ and the enzyme reaction proceeds. It is written as v_m.

If an enzyme reaction following Michaelis–Menten type with a rate controlling step for growth and it is assumed that the concentration of the rate controlling enzyme is proportional to the viable cells concentration, whereas the concentration of substrate of the rate controlling step is proportional to the limiting substrate concentration in the nutrient media then, the expression can be written as:

From Equation 2.9.

$$r_x = \mu x_v \qquad \text{(Eq. 2.17)}$$

Thus, the relationship between the specific growth rate and limiting substrate concentration as proposed by Monod is:

$$\mu = \mu_{max} \frac{S}{(K_S + S)} \quad \text{(Eq. 2.18)}$$

where, r_x is the volumetric rate of cell growth, kg m^{-3} h^{-1};
μ_{max} is the maximum specific growth rate, h^{-1};
S is limiting substrate concentration, kg m^{-3};
K_S is the saturation constant, kg m^{-3};
x_v is the viable cell concentration, kg m^{-3}

when $S >> K_S$ then, K_S can be ignored then,

$$\mu = \mu_{max} \quad \text{(Eq. 2.19)}$$

which is a Monod's zero order asymptote for specific growth rate with respect to substrate concentration.

Similarly, when $S << K_S$ then, S can be ignored the reaction become first order reaction as follows:

$$\mu = \frac{\mu_{max}}{K_S} S \quad \text{(Eq. 2.20)}$$

Thus, according to Monod's expression the specific rate is a function of the substrate concentration. In transition from zero order to Monod equation specific growth rate can be defined as critical substrate concentration. Now, a critical substrate concentration (S_{crit}) using Monod's expression can be defined as:

$$\mu = 0.99\ \mu_{max} \quad \text{(Eq. 2.21)}$$

Now, combining Equations 2.18 and 2.20 we get,

$$S_{crit} = 99\ K_S \quad \text{(Eq. 2.22)}$$

Here, Equation 2.22 depends on the arbitrary definition of critical substrate concentration. Now, if the Equation 2.21 is defined as 90% instead of 99% then, $S_{crit} = 90\ K_S$.

When substrate concentration (S) will be equal to K_S then the Monod's expression for specific growth rate can be rewritten from Equation 2.18 as:

$$\mu = \frac{\mu_{max}}{2} \qquad \text{(Eq. 2.23)}$$

Thus, Monod saturation constant K_S is also called as the critical substrate concentration. It is used to design medium concentration when a particular substrate becomes growth limiting.

Though Monod's equations are simple and easy to understand and work, but it still has some disadvantages. The equation doesn't work for complicated situations such as when the intracellular substrate concentration is reduced due to fast cellular growth even though enough amount of nutrients is still present in the media, then, the correlation of specific growth rate and substrate concentration does not occur according to Monod's equation. To solve such problems logistic equations are alternative solution.

In case of logistic equation, the expression for specific growth rate is written as:

$$\mu = \mu_{max}\left(1 - \frac{x_v}{x_{vm}}\right) \qquad \text{(Eq. 2.24)}$$

and for microbial cell growth rate:

$$r_x = \mu_{max}\left(1 - \frac{x_v}{x_{vm}}\right) \qquad \text{(Eq. 2.25)}$$

where, μ_{max} is the maximum specific growth rate h^{-1} and x_{vm} is the maximum viable biomass concentration.

Some other growth rate expression such as:

Tessier model:

$$\mu = \mu_{max}(1 - e^{-s/K_S}) \qquad \text{(Eq. 2.26)}$$

Moser model:

$$\mu = \mu_{max}(1 + K_S S^{-\lambda})^{-1} \qquad \text{(Eq. 2.27)}$$

where λ is constant

Contois model:

$$\mu = \mu_{max} \frac{S}{Bx + S} \quad \text{(Eq. 2.28)}$$

where B is a constant and Bx is apparent Monod constant which is proportional to biomass concentration x.

2.3.3.2.2 Growth Inhibition

In the above section microbial growth rate is discussed but inhibition factors of microbial growth are also an important factor in fermentation system. Some of the reasons for growth inhibition of microbial growth are substrate inhibition, antibiotics, poisons, accumulation of product which is poisonous above a particle threshold concentration. Some of the expression used for growth inhibition estimations are [5]:

For substrate inhibition:

$$\mu = \frac{\mu_{max} S}{K_S + S + \left(\frac{S}{k_i}\right)^2} \quad \text{(Eq. 2.29)}$$

where, k_i = inhibition coefficient

Other inhibition substance:

$$\mu = \frac{\mu_{max} S}{K_S + S}(1 - k_i I) \quad \text{(Eq. 2.30)}$$

$$\mu = \frac{\mu_{max} S}{K_S + S} \cdot \frac{k_i}{k_i + I} \quad \text{(Eq. 2.31)}$$

$$\mu = \frac{\mu_{max} S}{K_S + S} exp(-k_i I) \quad \text{(Eq. 2.32)}$$

where I = inhibitor concentration and k_i = inhibition constant.

2.3.3.2.3 Autolysis

Autolysis or natural cell death is a phenomenon which occurs in the fermentation system when some cells become non-viable. In such cases, cells loose its cell wall's integrity and autolysis of cell occurs. The cell death rate with respect to viable cells as first order expression can be expressed as [5]:

$$r_d = k_d x_v \quad \text{(Eq. 2.33)}$$

where, r_d = volumetric rate of conversion to non-viable form (kg m^{-3} h^{-1})
x_v = concentration of viable cells (kg m^{-3})
k_d = rate constant (h^{-1}).

The Equation 2.33 can be defined as the rate of conversion of viable cells to non-viable cells is assumed to be directly proportional to the concentration of viable cells. It should be noted that only the viable cells contribute to the death rate because once the cell is dead it cannot contribute to death rate. But dead cell does contribute to volumetric rate of autolysis if the cell membrane integrity is lost. Cell lysis means death of cell due to external factors such as poison or change in external osmotic pressure etc. and loss of cell integrity but autolysis means self-disruption of cell due to starvation or death. As autolysis is the function of cell's self-integrity but lysis occurs due external effects thus lysis includes both dead and viable cells which cannot be considered as function for concentration of dead cells. Thus lysis should be distinguished and can be represented as [5]:

$$r_l = k_l x_d \quad \text{(Eq. 2.34)}$$

where, r_l = volumetric rate of cell lysis (kg m^{-3} h^{-1})
x_d= concentration of dead cells (kg m^{-3})
k_l= rate constant (h^{-1}).

2.3.3.2.4 Formation of Products

In this section of chapter, the estimation of products formed by the microbial cells extracellularly or intracellularly are studied. To continue the cellular process microbial cell needs energy which is generated by the cell in the form of chemical energy that is Adenosine triphosphate (ATP). ATP can be produced by the cell anaerobically and aerobically using carbon sources as substrate. Aerobic process of ATP generation is done by breakdown of carbohydrates into CO_2 and water in presence of oxygen whereas, anaerobic process, ATP is generated by breaking down of carbon sources into simple sources such as ethanol, lactic acid, CO_2, water, etc. but in absence of oxygen. The resulted products formed are excreted out by the cell in the medium. Other than alcohols and acids microbial cells also secretes many exoenzymes, polysaccharides and antibiotics depending on the specific media and other external conditions. For example, in carbon source enriched media and with few metal ions lipids and glycogens are produced as storage energy.

Thus, some of the expression of product formation kinetics by microbial cells [5] can be expressed as follows:

Growth-associated:

$$r_p = \alpha r_x \quad \text{(Eq. 2.35)}$$

Non-growth associated:

$$r_p = \beta x_v \quad \text{(Eq. 2.36)}$$

Mixed kinetics:

$$r_p = \alpha r_x + \beta x_v \quad \text{(Eq. 2.37)}$$

Equation 2.37 is called Luedeking–Piret expression for product formation and α and β are constants. Combination of Luedeking–Piret expression [28] with Monod's equation gives the following expression:

$$r_p = \left[\frac{\alpha \mu_{max} S}{K_S + S} + \beta \right] x_v \quad \text{(Eq. 2.38)}$$

2.3.3.2.5 Degradation and Inhibition of Product Formation

Degradation or inhibition of product formation can occur by external inhibitors or by the production concentration itself. Such as in case of ethanol production, after a threshold concentration of ethanol, it itself inhibits the production of ethanol [29]. Following is the equation [5] used to estimate product formation rate in case of inhibition by product concentration itself:

$$r_p = \left(1 - \frac{P}{P_m} \right) (\alpha r_x + \beta x_v) \quad \text{(Eq. 2.39)}$$

Where, P_m = maximum achievable product concentration under inhibition conditions.

2.3.3.2.6 Maintenance Energy and Endogenous Respiration

Maintenance energy is the total energy required by the cells to maintain the concentration gradient required between the interior and exterior of the cell and to initiative turnover reactions involved in continuous resynthesize of cell liable components [30]. This energy is used to maintain viability of cell and in cell motility and is not responsible for the production of any product or energy storage components [31]. The kinetic expression for maintenance energy [5] is written as:

$$r_{Sm} = m_S x_v \qquad \text{(Eq. 2.40)}$$

Where r_{Sm} = volumetric rate of consumption of substrate (kg m^{-3} h^{-1}); m_S = rate constant (kg substrate kg cells^{-1} h^{-1}).

Now, when the external source of energy is exhausted, then the cells starts using its internal stored energy which are in the form of lipids and glycolipids leading to loss of biomass of the cells due to starvation. This is called endogenous respiration and the volumetric rate of endogenous respiration can be written as:

$$r_e = k_e x_v \qquad \text{(Eq. 2.41)}$$

Where, r_e = volumetric rate of endogenous respiration (kg cell m^{-3} h^{-1});
k_e = rate constant (kg cell matter kg cells^{-1} h^{-1}).

Even though r_{Sm} represents substrate balance and r_e represents cell mass balance but they don't have any difference in practical. Thus, it can be written as:

$$k_e = m_S Y'_{x/S} \qquad \text{(Eq. 2.42)}$$

Where, $Y'_{x/S}$ = yield of biomass on substrate.

2.3.3.2.7 Stoichiometric Aspects

Though cell composition may vary due to different cell type and the difference in their physiological/environmental conditions, but protein, RNA, DNA, lipids, lipopolysaccharides, peptidoglycan, and glycogen are known to present in all ideal cells [32]. These are also involved in cell metabolism mechanism. These are all stoichiometric aspects.

2.3.3.2.8 Elemental Balance

Cellular biomass consists of some macromolecules such as RNA, DNA, proteins etc. The average element composition can be estimated from the average content of the individual building block of the cell [33]. Some of the estimations of elementary compositions [5] of building blocks are: protein $CH_{1.58}O_{0.31}N_{0.27}S_{0.004}$; DNA $CH_{1.15}O_{0.62}N_{0.39}P_{0.10}$; RNA $CH_{1.23}O_{0.75}N_{0.38}P_{0.11}$; carbohydrates $CH_{1.67}O_{0.83}$; phospholipids $CH_{1.91}O_{0.23}N_{0.02}P_{0.02}$; neutral fat $CH_{1.84}O_{0.12}$, and Biomass $CH_{1.81}O_{0.52}N_{0.21}$.

To explain the relation of elementary balance and the use stoichiometry in the mathematical modeling of biological activities, production of single cell protein (SCP) from methane source can be used as an example. The metabolic products formed can be assumed to be CO_2 and water and ammonia can be considered as nitrogen sources. Experimental oxygen consumption is assumed to be 1.35 mol oxygen per mol methane consumed. Then, the expression for the reaction system can be written as:

$$CH_4 + aO_2 + bNH_3 \rightarrow cCH_{1.81}O_{0.52}N_{0.21} + dCO_2 + eH_2O$$

Where, a, b, c, d, e is stoichiometric coefficient indicating number of mole of the molecules involved to balance the reaction. Thus, the stoichiometric balance equation of the above reaction can be written as:

$$\text{Carbon: } 1 = c + d \quad \text{(Eq. 2.43)}$$

$$\text{Hydrogen: } 4 + 3b = 1.81c + 2e \quad \text{(Eq. 2.44)}$$

$$\text{Nitrogen: } b = 0.21c \quad \text{(Eq. 2.45)}$$

$$\text{Oxygen: } 2a = 0.52c + 2d + e \quad \text{(Eq. 2.46)}$$

Here, number of unknown stoichiometric coefficient = a, b, c, d, e = 5

Number equation formed = carbon, hydrogen, nitrogen, oxygen = 4
Degree of freedom = 1

Here, mole of oxygen (a) = 1.35 mol oxygen used per mol methane used is provided.

Thus, Equation 2.43, 2.44, 2.45, and 2.46 can be solved using the value of 'a' are as below:

a = 1.35 (given)
b = 0.13
c = 0.63
d = 0.37
e = 1.63

Thus, in this way other reaction system can be solved and the stoichiometric balance can be estimated.

2.3.3.2.9 Yield Coefficient and Factors in Rate Expressions

Yield coefficient and factors are related to substrate utilization. Substrate is utilized for cell's own maintenance, energy production as well as metabolite synthesis. Thus, rate of substrate utilization is stoichiometric related to metabolite production [33]. This process can be explained by using an example of ethanol production [5]. The reaction equation of ethanol production from glucose is as follows:

$$C_6H_{12}O_6 \rightarrow 2C_2H_5OH + 2CO_2$$

Thus, if 1 kg of glucose is used for the reaction then, 0.5 kg of ethanol is produced according to the calculation. Thus rate of product formation (r_p) and rate of substrate utilization or uptake (r_{SP}) for product formation can be written as:

$$r_{SP} = \frac{r_P}{0.5}$$

or more generally

$$r_{SP} = \frac{r_P}{Y_{P/S}} \qquad \text{(Eq. 2.47)}$$

Where, $Y_{P/S}$ = coefficient of yield for product on substrate and its unit is kg product/kg substrate converted to product.

Now, in case of an anaerobic cell growth an example of the generalized equation can be written as:

$$CH_2O + 0.2NH_4^+ + e^- \rightarrow CH_{1.8}O_{0.5}N_{0.2} + 0.5H_2O$$

Where, CH_2O is the carbon source and NH_4^+ is the ammonia source. $CH_{1.8}O_{0.5}N_{0.2}$ is approximated as the formula of many cells. Thus, from the calculations we can found 0.82 kg of cells are produced per kg of carbon source and 8.8 kg of cells are produced from per kg of nitrogen source. Thus, the equations can be written as:

$$r_{Sx} = \frac{r_x}{0.82}$$

or

$$r_{Sx} = \frac{r_x}{Y_{x/S}} \quad \text{(Eq. 2.48)}$$

$$r_{Nx} = \frac{r_x}{8.8} \text{ or } r_{Nx} = \frac{r_x}{Y_{x/N}} \quad \text{(Eq. 2.49)}$$

where, r_{Sx} = volumetric rate of carbohydrate utilization for cell growth,
r_{Nx} = volumetric rate of nitrogen utilization for cell growth,
r_x = volumetric rate of cell growth,
$Y_{x/S}$ = yield coefficient for cells on carbohydrate,
$Y_{x/N}$ = yield coefficient for cells on nitrogen

Thus, the balance equation of carbon source as substrate and nitrogen source as nitrogen supply can be written as:

$$r_S = r_{Sx} + r_{Sp} + r_{So} \quad \text{(Eq. 2.50)}$$

$$r_N = r_{Nx} + r_{Np} \quad \text{(Eq. 2.51)}$$

Yield factors and yield coefficient are not same which is generally confused. Yield coefficient the stoichiometric constant that relates the products and the reactants whereas, yield is the ratio between one product with one reactant that enters into other reactions and produce final products [33]. As in fermentation process, the substrate is consumed by cell for its maintenance as well as for product formation, thus, yield is calculated over total amount of substrate consumed. Thus, in complex reaction where yield coefficients are unknown yield factors are preferred. In the above explained ethanol reaction system the yield coefficient of carbon source has been discussed but the use of oxygen for the maintenance of cell is not explicated. Thus the common expression for volumetric rate of substrate utilization [5] from Equation 2.40 and 2.50 can be written as:

$$r_S = r_{Sx} + r_{Sp} + r_{Sm} \quad \text{(Eq. 2.52)}$$

$$r_S = \frac{r_x}{Y'_{x/S}} + \frac{r_P}{Y'_{P/S}} + r_{Sm} \quad \text{(Eq. 2.53)}$$

Where, $Y'_{x/S}$ and $Y'_{P/S}$ are yield factors.

2.4 Conclusion

Modeling the kinetics of the system is best explained with mathematical expressions. Modeling includes lots of factors and variables which needs to be taken care of while optimizing a system. There are some basic mathematical expressions which are explained in this chapter, depending on which a particular system can be modeled. The basics of the possible kinetics, parameters are explained in this chapter on which or combination of which is needed to generate novel model with respect to the system. The use of computer with fermentation system and concept of its use has been discussed which will to under better and design novel model for particular system. The main application and use of modeling is to increase the productivity and to predict the results of the system before hands.

References

1. Olaoye O.S, Kolawole O.S., Modeling of the kinetics of ethanol formation from glucose biomass in batch culture with a non structured model. *IJERA*, 3 (4), 562–565, 2013.
2. Deindoerfer, F.H., Fermentation kinetics and model processes. In *Advances in applied microbiology*, W.W. Umbreit, (ed.), Academic Press, 321–334, 1960.
3. Nielsen, J., Villadsen, J., Modelling of microbial kinetics. *Chem. Eng. Sci.*, 47 (17), 4225–4270, 1992.
4. De Buyser, D.R., Spriet, J.A., Vansteenkiste, G.C., Computer-control of fermentation processes. In *Industrial aspects of biochemistry and genetics*, N.G. Alaeddinoğlu, A.L. Demain, G. Lancini, (eds.), Springer US, Boston, MA, 189–236, 1985.
5. Mavituna, F. Sinclair, C.G., *Modelling the kinetics of biological activity in fermentation systems*. In *Practical fermentation technology*, John Wiley & Sons, Ltd., 167–230, 2008.
6. Morales-Rodriguez, R., *et al.*, Dynamic model-based evaluation of process configurations for integrated operation of hydrolysis and co-fermentation for bioethanol production from lignocellulose. *Bioresour. Technol.*, 102 (2), 1174–1184, 2011.
7. Mitchell, D.A., *et al.*, A review of recent developments in modeling of microbial growth kinetics and intraparticle phenomena in solid-state fermentation. *Biochem. Eng. J.*, 17 (1), 15–26, 2004.
8. Nielsen, J., Nikolajsen, K., Villadsen, J., Computer controlled system for online monitoring of a fermentation process. In *Computer applications in fermentation technology: modelling and control of biotechnological processes*, N.M. Fish, R.I. Fox, N.F. Thornhill, (eds.), Springer Netherlands, Dordrecht, 53–57, 1989.

9. Fox, R., Introduction. In *Computer applications in fermentation technology: modelling and control of biotechnological processes*, N.M. Fish, R.I. Fox, N.F. Thornhill, (eds.), Springer Netherlands, Dordrecht, 1–2, 1989.
10. Beck, M.B., Young, P.C., An introduction to system identification, parameter and state estimation. In *Computer applications in fermentation technology: modelling and control of biotechnological processes*, N.M. Fish, R.I. Fox, N.F. Thornhill, (eds.), Springer Netherlands: Dordrecht, 129–158, 1989.
11. Zyl, J.M., *et al.*, A kinetic model for simultaneous saccharification and fermentation of Avicel with Saccharomyces cerevisiae. *Biotechnol. Bioeng.*, 108, 2011.
12. Unrean, P., Bioprocess modelling for the design and optimization of lignocellulosic biomass fermentation. *Bioresour. Bioprocess.*, 3 (1), 1, 2016.
13. Oliveira, S.C., Oliveira, R.C., Tacin, M.V., Gattás, E.A.L., Kinetic modeling and optimization of a batch ethanol fermentation process. *J. Bioprocess. Biotech.*, 6 (266), 2016.
14. Morales-Rodriguez, R., *et al.*, Dynamic model-based evaluation of process configurations for integrated operation of hydrolysis and co-fermentation for bioethanol production from lignocellulose. *Bioresour. Technol.*, 102, 2011.
15. Heinzle, E., Kaufmann, T., Griot, M., Modelling of kinetics, mass transfer and mixing phenomena in 45, 450 and 4500 L tank fermenters. In *Computer applications in fermentation technology: modelling and control of biotechnological processes*, N.M. Fish, R.I. Fox, N.F. Thornhill, (eds.), Springer Netherlands: Dordrecht, 105–109, 1989.
16. Kumara Behera, B., Varma, A., Material-balance calculation of fermentation processes. In *Microbial biomass process technologies and management*, Springer International Publishing: Cham, 257–298, 2017.
17. Nagy, Z.K., Model based control of a fermentation bioreactor using optimally designed artificial neural networks. *Chem. Eng. J.*, 127, 95–109, 2007.
18. Dana Cervinkova, V.B., Marosevic, D., Kubikova, I., Jaglic, Z., The role of the qacA gene in mediating resistance to quaternary ammonium compounds. *Microb. Drug Resist.*, 19 (3), 160–167, 2013.
19. Lorenzo Pasotti, S.Z., Casanova, M., Micoli, G., De Angelis, M.G.C., Magni, P., Fermentation of lactose to ethanol in cheese whey permeate and concentrated permeate by engineered Escherichia coli. *BMC Biotechnol.*, 17, 2017.
20. Ménoret, S., *et al.*, Homology-directed repair in rodent zygotes using Cas9 and TALEN engineered proteins. *Sci. Rep.*, 5, 14410, 2015.
21. Chaignon, P., *et al.*, *Photochemical reactivity of trifluoromethyl aromatic amines: the example of 3,5-diamino-trifluoromethyl-benzene (3,5-DABTF). Photochem. Photobiol.*, 81 (6), 1539–1543, 2005.
22. Slavov, N., *et al.*, Constant growth rate can be supported by decreasing energy flux and increasing aerobic glycolysis. *Cell Rep.*, 7 (3), 705–714, 2014.
23. Pardo, D., Bovee, J.P., A software tool for fermentation modelling. In *Computer applications in fermentation technology: modelling and control of biotechnological processes*, N.M. Fish, R.I. Fox, N.F. Thornhill, (eds.), Springer Netherlands: Dordrecht, 111–114, 1989.

24. Humphrey, A.E., Fermentation process modeling: an overview. *Ann. N. Y. Acad. Sci.*, 326 (1), 17–33, 1979.
25. Oliveira, S.C., *et al.*, Kinetic modeling of 1-G ethanol fermentations. In *Fermentation processes*, A.F. Jozala, (ed.), InTech: Rijeka, chap. 6, 2017.
26. Hadiyanto, *et al.*, Optimization of ethanol production from whey through fed-batch fermentation using kluyveromyces marxianus. *Energy Proce.*, 47 (Supplement C), 108–112, 2014.
27. Srimachai, T., *et al.*, Optimization and kinetic modeling of ethanol production from oil palm frond juice in batch fermentation. *Energy Proce.*, 79 (Supplement C), 111–118, 2015.
28. Garnier, A., Gaillet, B., Analytical solution of Luedeking-Piret equation for a batch fermentation obeying Monod growth kinetics. *Biotechnol. Bioeng.*,. 112 (12), 2468–274, 2015.
29. Zhang, Q., *et al.*, Substrate and product inhibition on yeast performance in ethanol fermentation. *Energy Fuels*, 29 (2), 1019–1027, 2015.
30. Van Hoek, P., Van Dijken, J.P., Pronk, J.T., Effect of specific growth rate on fermentative capacity of baker's yeast. *Appl. Environ. Microbiol.*, 64 (11), 4226–4233, 1998.
31. Roslev, P., King, G.M., Aerobic and anaerobic starvation metabolism in methanotrophic bacteria. *Appl. Environ. Microbiol.*, 61 (4), 1563–1570, 1995.
32. Tsai, S.P., Lee, Y.H., A criterion for selecting fermentation stoichiometry methods. *Biotechnol. Bioeng.*, 33 (10), 1347–1349, 1989.
33. Farges Bérangère, P.L., Agnès, P., Claude-Gilles, D., *Methodology for bioprocess analysis: mass balances, yields and stoichiometries, stoichiometry and research – The importance of quantity in biomedicine*, D.A. Innocenti, (ed.). InTech, 2012.

3

Sterilization Techniques used in Fermentation Processes

Shivani Sharma, Arindam Kuila and Vinay Sharma*

Department of Bioscience & Biotechnology, Banasthali University, Rajasthan, India

Abstract
Fermentation is a microbial process that must be perfectly sterilised. Apart from the normal sterilization sometimes this process is being operated on a large scale that employs large fermenting vessels that need to be sterilised. Media that are being used also requires sterilization. Major problem in the process of fermentation is contamination, in this chapter we will be reviewing some of the processes of sterilization that includes sterilization of the reactor vessel, media, air, and pipes through which the media passes.

Keywords: Fermentation, sterilization, microbial death

3.1 Introduction

Fermentation comes from a Latin word 'fermentare' that means to boil, that can be seen as a boiling appearance of the most common fermenting microorganism Yeast on the fruits, grains and other materials that undergo fermentation. The carbon dioxide produced during the anaerobic process that degrades the sugars in the fruits and grains forming bubbles that give boiling appearance. Fermentation is done by the inoculation of desirable organism or a group of organisms in the suitable media. But if this fermenting media comes in contact with any foreign microorganism it may lead to a loss in productivity of the actual fermenting microorganism due to a couple of reasons that are listed below.

**Corresponding author*: vinaysharma30@yahoo.co.uk

Arindam Kuila and Vinay Sharma (eds.) Principles and Applications of Fermentation Technology, (45–52) © 2018 Scrivener Publishing LLC

1. The foreign microorganism can outgrow the actual fermenting microorganism
2. The final product can be contaminated
3. The extraction of final product can become difficult due to the formation of some complex compound by the help of foreign microorganisms.

This can be avoided by the following methods.

1. Inoculation in a pure culture
2. Sterilization of the media used
3. Sterilizing the equipments and containers used
4. Aseptic conditions should be maintained during the whole process of fermentation.
5. In order to obtain proper fermentation first the media used is to be sterilised. Sterilization of media can be done by use of radiation, chemical treatment, filtration, ultrasonic treatment or heat. The sterilizing treatment given is referred to as sterilant. Heating is the most commonly used form of sterilization as it kills most of the endospores present in the media. Little amount of heat is given in the form of commercial sterilization that kills most of the microorganisms present in the media. Mostly sterilization with the help of steam is recommended for fermentation and all the other work best for animal or plant cultures.

3.2 Rate of Microbial Death

Death at constant rate occurs when the media is contaminated by some microorganism and heat or any other sterilizing treatment is being given. For example, when heat treatment is given for one minute then 90% of the population dies. To kill the remaining population again the same treatment is given then again 90% of the population dies. To kill the whole population we have to repeat the treatment and for this process if a graph is plotted then we get a constant curve for the death rate of microorganism.

Some factors listed below are responsible for the way a sterilizing treatment works.

Total microbes: if the number of microbes present is large then too much time would be required to kill them.

Environment required: warm environment is best suited for most of the sterilizing actions.

Nature of the media: if the media is rich in proteins and fats then the microbes will not be easily killed; mostly heating treatment helps in these cases.

Duration of treatment: different sterilizing treatments need different time of exposure as chemical sterilizing treatments require more time compared to normal heating methods.

Characteristics of the contaminant present: the type of microorganism present in the contaminant also affects the choice of the treatment that would require to eradicate it.

3.3 How do Sterilants Work?

Most of the sterilizing treatments used are aimed at killing the contaminants present in the media used. There are different ways in which the contaminants can be killed.

Changing the permeability of cell membrane: The plasma membrane present inside the cell membrane of the contaminant is the target of sterilizing agents, as it is responsible for the exchange of nutrients in the cell and removal of wastes. The sterilizing agents used, cause disturbance to the plasma membrane that leads to improper exchange of nutrients and leakage of the cell components leading to death of the contaminant.

Destroying the proteins and nucleic acids present: The microbes that act as contaminants are mostly bacteria. Bacteria are rich source of enzymes and enzymes are proteins that are very fragile to heat treatment as they easily denature. DNA and RNA that carries the genetic information of the microbe can also be easily destroyed by any kind of sterilizing treatment, like heating, radiation or chemical treatment.

3.4 Types of Sterilization

During the early Stone Age, humans used several methods of sterilization like salting and drying. Heating is one of the most favourable methods of sterilization since early days. But time introduced some microbes that are heat resistant, so various other methods are required. Heat, pressure and radiation all play their role in different types of physical sterilization.

3.4.1 Heat

Heat is preferred because it kills the microorganisms by denaturing the enzymes present in them, hence changing the three dimensional structure of proteins which therefore inactivates them. Three factors that decides the action of heat sterilization are as follows.

Thermal death point: minimum temperature required to kill microbes present in a suspension in 10 minute.
Thermal death time: minimum time required to kill microbes present in a suspension at a given temperature.
Decimal reduction time: time, in minutes that is required to kill microbes at a given temperature.

Heat treatment can be given using moist heat or dry heat sterilization. Dry heat sterilization involves killing of the oxidation effects by flaming or incineration. Moist heat sterilization works by coagulating the proteins present in the microbes thereby killing them. Autoclaving is the most common method of moist heat sterilization in which steam is used to kill microbes at high temperature and under appropriate pressure. This helps in sterilizing glassware, culture media, equipments, solutions etc.

3.4.2 Pressure

At high pressure molecular structure of proteins and carbohydrates are altered leading to inactivation of cells.

3.4.3 Radiation

Depending on the intensity, wave-length and duration radiation has various effects on the microbes. Two types of radiations kill microbes, they are Nonionizing and Ionizing radiations.

Nonionizing radiation: They have wavelength greater than 1 nm. Ultraviolet light is a form of nonionizing radiation, it kills the microbes by degrading their DNA by making thymine dimers. These dimmers prevent the replication of DNA during cell reproduction. UV wavelength of 260 nm is very effective in killing the microbes. These radiations are effective in sterilizing glassware, air, culture media etc.

Ionizing radiation: They have wavelength less than 1 nm, like the X rays, gamma rays, electron beams of high energy. These radiations work by ionising the water and forming highly reactive hydroxyl radicals. These radicals kill microbes by causing some damage to DNA or RNA by a series of reactions.

3.4.4 Filtration

It helps in removal of microbes by making any liquid or gas pass through a membrane with pores. Pressure is applied to make the liquid or gas pass through the filter kept on a container in which vacuum is being created.

3.4.5 Steam Sterilization

Sterilization by steam can be done by the use of normal steam created by preventing evaporation of the medium or with the help of steam under pressure used in autoclaves. Koch or Arnold steamer is used that contains a copper cabinet with lagged walls and conical lid that helps in the drainage of steam generated and a perforated tray is placed above water level that helps in proper distribution of steam. Temperature 100 °C for 20 minutes for three days is required and the process is known as intermittent sterilization. This method efficiently kills bacteria but is inefficient in killing anaerobic spores.

Another type of steamer used is autoclave that generates steam under pressure where water boils when its pressure is equal to the atmospheric pressure in the vessel and with the increase in pressure, temperature of water increases.

3.5 Sterilization of the Culture Media

This can be done by heating treatment, that is, autoclaving or by the following two ways.

3.5.1 Batch Sterilization

This can be done in a fermentation vessel or separated cooker. When the fermenters are being cleaned and prepared for fermentation the media used can be sterilized in cookers. One cooker can be used for several fermenters. Concentrated form of media can be used for sterilization and that can be diluted when placed within the fermenter, this helps in saving time. As one cooker is mostly used in this type of sterilization so a complex pipeline network is required that connects all the fermenters delivering the sterilized media. This type of sterilization makes sure that there is minimal loss of nutrients as it is operated at 121 °C that is almost feasible for all nutrients present in the media.

3.5.2 Continuous Sterilization

In this type of sterilization the medium is heated to a particular temperature, it is kept at the same temperature for some time and then cooled. The time at which media is kept at a certain temperature is known as holding time. A continuous heat exchanger is used to heat the media and then insulated serpentine coiled tube is used to keep it at that temperature for the holding period. Sequential heat exchangers are used to cool the media to suitable fermentation temperature. Two types of continuous sterilizers are used, one that uses direct heat by steam injection and the other uses indirect heat. Small increments of media are done continuously at certain time intervals and the heating and cooling times are properly managed. Spiral heat exchangers are used in cooling the heated media with the help of unsterile media that come for sterilization. This unsterile media gets heat from the sterile hot media thus preserving the heat before reaching the sterilizer.

Continuous steam injectors are used to directly inject the heat into the media, but it has certain pros and cons [2]: this is very affordable and manageable procedure with very short heating time but it produces foam on heating that can cause contamination. Sometimes this steam injector is combined with flash cooling, in which a vacuum chamber is made having an expansion valve through which the media is passed, this causes instant cooling of the media. Sometimes a combination of direct and indirect heat exchangers is being used mainly for starch rich media [3]. Spiral heat exchangers are the best choice for continuous sterilization. A plant of interconnected spiral heat exchangers and holding tube is made. Hot water is used to sterilize this plant in a closed circuit before unsterilized media comes in. The fermenter and holding tube are steam sterilized. Heat is conserved in this procedure by cooling the sterilized heated media with the help of cool unsterilized media.

3.6 Sterilization of the Additives

During the course of fermentation some additives are being introduced in order to increase the efficiency of the process. Continuous and batch sterilization methods are used for the sterilization of additives but the process depends on the biochemical nature and the volume to be introduced. Continuous sterilization is preferred when large volume of additive are used [1]. Batch sterilization uses steam injection into the additive liquid used.

3.7 Sterilization of the Fermenter Vessel

The fermenter vessel is to be sterilized before the fermentation begins as the media used is sterilized in separate cookers. The coils or jackets of the fermentation vessel are heated with the help of steam that is introduced into it from various entries and one valve is opened through which the steam goes out of the vessel very slowly. Steam is introduces at 15 psi pressure for 20–25 minutes. Sterile air is removed out of the vessel by sparing and a positive pressure is maintained that prevents formation of vacuum in the vessel.

3.8 Filter Sterilization

Suspended fluids that are present in the media can be sterilized with the help of filter sterilization by the following methods.

3.8.1 Diffusion

Brownian motion is introduced for the separation of very small particles present in the fluid by collision with each other. Small particles deviate from the normal flow of fluid and get caught in the filter fibres. This process is more effective in the filtration of gases.

3.8.2 Inertial Impaction

The suspended particles present in the fluid have some kind of momentum based on which they are separated with the help of this method. A route of less resistance is created in which the fluid flows through the filter and the suspended particles, due to their momentum flow in a straight line and get trapped in the filter.

3.8.3 Electrostatic Attraction

The filtration membrane has charge that attracts the opposite charged suspended particles.

3.8.4 Interception

Interwoven fibres make the filter with openings of different sizes. Particles that are larger than the pore size are excluded and sometime smaller particles are also unable to cross the filter membrane as their passage gets blocked by other particles and they are removed.

3.9 Sterilization of Air

Anaerobic fermentation needs continuous addition of sterile air that can either be sterilized by heat or filtration. Filters with fixed pores are mostly used for the sterilization of air. Hygroscopic materials like PTFE are used in theses filters for this purpose. Other filters used to purify air are as follows.

Depth filters: they consist of glass wool in which the microbes present in the air gets trapped. These filters work by inertia, gravitation, diffusion, electrostatic attraction etc. These glass wool filters can be easily reused after sterilization. But after sometime due to the action of steam theses glass wool filters get damaged, so now glass fibre cartridges are used.

Membrane cartridge filters: these filters are made of nylon, cellulose or polysulfone and are pleated. They can be easily used and operated.

References

1. Anustrup K., Andresen O., Flach E. A., Nielsen T. A., Production of microbial enzymes. *Microb. Technol.*, 1, 282–309, 1979.
2. Banks G. T., Scale up fermentation process. *Top. Enzyme Ferment. Biotechnol.*, 3, 170–266, 1979.
3. Svensson R., Continuous media sterilization in biotechnological fermentation. *Dechema Monogr.*, 113, 225–237, 1988.

4

Advances in Fermentation Technology: Principle and Their Relevant Applications

Monika Choudhary[1], Sunanda Joshi[1], Sameer Suresh Bhagyawant[2] and Nidhi Srivastava[1]*

[1]Department of Bioscience and Biotechnology, Banasthali University, Rajasthan, India
[2]School of Studies in Biotechnology, Jiwaji University, Gwalior, Madhya Pradesh, India

Abstract

Fermentation is a chemical procedure by which molecules such as glucose are broken down anaerobicaly. This process begins with an appropriate microorganism and in particular conditions, cautious modification of nutrient concentration which results in the fabrication of energy occurs in the cells of the body, in plants and some bacteria. Fermentation has emerged as a prospective technology for the manufacture of microbial foodstuffs feed, food, industrial chemicals, and pharmaceutical products. Fermentation is the progression in which numerous microorganisms have altered pyruvic acid into carbon dioxide and ethanol in the absence of oxygen, in order to replenish NAD^+ expedition during glycolysis. Production of several metabolites has been relevant for the food dispensation industry centered on flavors, enzymes, organic acids, and xanthum gum.

Keywords: Fermentation, glycolysis, microorganism, aerobic, anaerobic

4.1 Introduction

In the mid-1970s, term "biotechnology" came into common use, in particularly fermenter control, design, purification, and product recovery to define chemical engineering processes with the help of microorganisms

**Corresponding author*: nidhiscientist@gmail.com

Arindam Kuila and Vinay Sharma (eds.) Principles and Applications of Fermentation Technology, (53–64) © 2018 Scrivener Publishing LLC

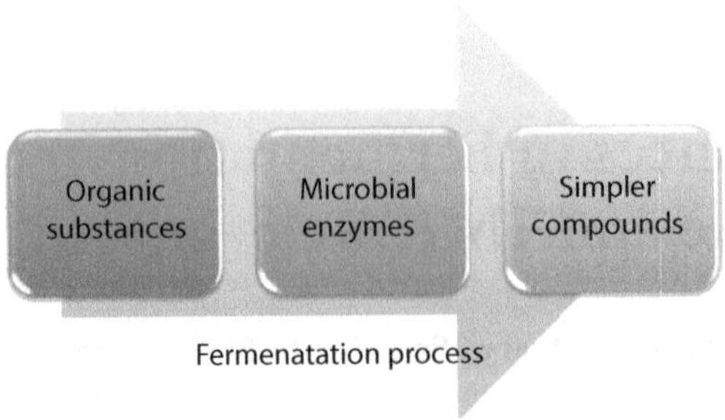

Figure 4.1 Fermentation process.

and their products [1]. Technology has been defined in a very simple form as applied science to the complex form of scientific study of the practical or industrial arts [1]. Fermentation is the area of biotechnology, technology which is very pulsating and fast growing, absorbing an ever-growing processes and products. Fermentation technology has a brighter future and longer history than biological sciences by covering significant areas of food and medicine in the mankind services [2].

Fermentation technology has been widely used in the industry of food, pharmaceuticals, and alcoholic beverages on huge scale. Term fermentation has been derived from the Latin, meaning boil, because bubbling and forming of fermenting beverages appear as boiling. Zymology is identified as the science of fermentation (Figure 4.1).

Modern industries have supplemented with a more favored biological development due to high specificity, easiness of reaction, low cost, and application adaptability of fermentation. The basic fermentation principle with advanced techniques of genetic engineering has been applicable in the biochemical, biomolecules, and biofuels production [3].

Some microbiologists used the term fermentation to depict the mass culture production process with the help of microorganism. The product referred to as biomass production (can either be the cell itself) and referred as a product from a natural or genetically enhanced strain [4].

4.2 Basic Principle of Fermentation

The basic principle of fermentation is also based on the microbial compatibility along with the medium constituents. As soon as microbes uptake the medium nutrients, they started to produce primary metabolites for their growth maxima (log phage). Primary metabolites further acts for the production of their secondary metabolites (stationary phase). So fermentation process is reflected by microbial growth kinetics.

Fermentation is the progression of rising microorganisms in a nutrient media by maintaining the physiochemical situation and thereby converting feed into desired products.

The end products of a metabolic form could be further oxidized for example, yeast cell evolved two molecules of ATP during metabolic process of ethanol production. According to Pasteur "life without air" means in organism an anaerobe redox process. This process is the simple chemical conversion of bioorganic compounds into simpler products with the help of bacteria, yeast or fungi either alone or together. This type of organism's action (bioconversion) can be seen throughout their expansion, development, imitation even senescence, and demise. A smaller number of genera, that is, more or less five of yeasts, six of bacteria, and eight genera of molds participate for the vast greater division of these microbes [5].

High demand products (wine, beer, cider, vinegar, ethanol, cheese, hormones, antibiotics, complete proteins, enzymes, and other useful goods) are released into the media throughout growth/non-growth metabolic period of the microbes and these end products of their life cycles are of use of human being. The basis of the fermentation is that organism grows in the media suspended with all the important necessities under suitable condition [6]. During this process complex organic molecules are converts into smaller compounds or nutrients. For example, the role of three most important enzymes, that is, protease breaks down enormous protein molecule first into polypeptides, peptides, followed by numerous amino acid, similarly amylase coverts carbohydrates, reducing starches and complex sugars into simple sugars, and lipase hydrolyzes complex fat molecules into simpler free fatty acids. There are more than thousand examples of such activities of more enzymes (both internal and external part of our bodies). Some other important by-products are also released during the fermentation process [5].

Fermentation is an extensive pathway, but it is not the only way to get energy in the absence of oxygen (from fuels anaerobically). Aerobic means "with air". This type of fermentation needs oxygen for it to arise so it is called aerobic fermentation. Their life duration is released into the media and the end product formed as a outcome of their metabolism that have a high industrial value which are extracted for utilize by human being (Figure 4.2).

For the fermentation technique, it has been essential to make sure that only desirable microorganism should start to grow on the substrate. This process has ensured the other microorganisms suppression that can cause

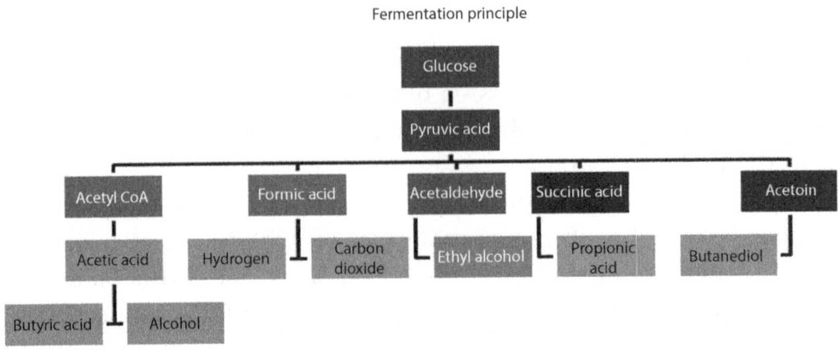

Figure 4.2 Principle of fermentation.

either food poisoning or pathogenic and can destroy the fermentation process, resulting in undesirable and unexpected end product. The best example to determine this difference has been seen in between the pasteurized and unpasteurized milk spoilage. The unpasteurized milk has been destroyed naturally and produces a sour tasting product that has been used in the baking industry to improve the breads texture. However, pasteurized milk has been destroyed and produces an unpleasant product that has been disposed off. The main reason for this difference has been described as pasteurization has changed the microbial environment and if pasteurized milk has been kept unrefrigerated, the undesirable microbes grow and multiply and desirable microbes died. While in unpasteurized milk, lactic acid, non-pathogenic bacteria grow and multiply with greater rate than any other pathogenic bacteria [7].

4.3 Biochemical Process

Glycolysis is a complement process which makes it possible for production of ATP continuously even when oxygen is absent. Fermentation regenerates NAD^+ by NADH oxidization, which can take part in glycolysis to produce more ATP. At the end of glycolysis, pyruvic acid is formed which has the property of reacting eagerly with hydrogen. Pyruvic acid ceases to be a fuel and if the normal path of hydrogen to oxygen is blocked it becomes a hydrogen acceptor. By the use of pyruvic acid as hydrogen acceptor most microorganisms and ethyl alcohol in plants produces, and forms lactic acid in certain bacteria and animals. In alcoholic beverages the active ingredient is ethyl alcohol. Fermentation yields only about 5% of the energy obtained by aerobic respiration. This small amount of energy is adequate to maintain

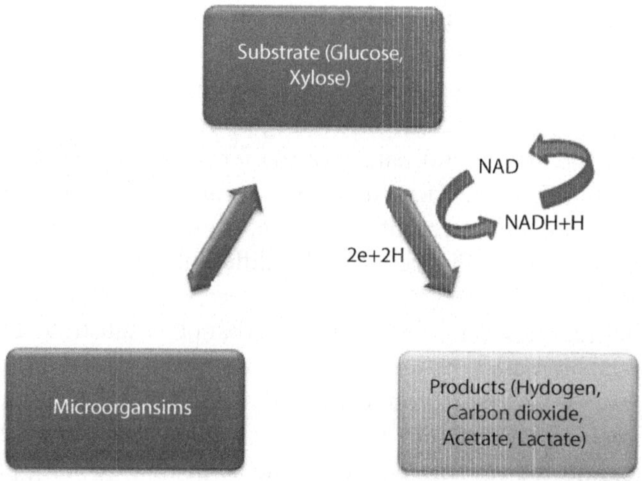

Figure 4.3 Detailed biochemical process of fermentstion.

the life of organisms such as bacteria, yeasts, and other anaerobes (organisms that normally live or can live in the absence of oxygen). Enormous majority of organisms are, however, aerobes, that is, need oxygen for respiration. To maintain the life of these organisms, fermentation energy is too little. In the absence of oxygen they die within minutes. Fermentation can add-on the aerobic energy in them (Figure 4.3).

Result of the enzymes activity is the changes that take place during the fermentation. Biochemical reactions have been precised by enzymes (compounds of proteins created by living cells). Rather than being used in a reaction, enzymes have been recognized as catalysts because they organize the reactions. In nature, they are proteinaceous and fluctuated by moisture content, temperature, pH, concentrations of substrate and inhibitors, and ionic strength. Each enzyme has required optimum conditions at which it can function most resourcefully. Edges of pH and temperature will knock down enzyme activity and denature the protein. They are so sensitive and their enzymatic reactions can be simply controlled by minor adjustments to moisture content, pH, temperature, or extra reaction conditions. Enzymes have several roles in the food industry as the conversion of sugars, the modification of proteins, and the liquefaction and saccharification of starch. In the fermentation of fruits and vegetables microbial enzymes play an important role.

With any fermentation it is essential to make sure that only the desired yeasts, moulds, or bacteria start to multiply and grow on the substrate.

4.4 Fermentation Methodology

Fermentation process is carried out in a container which is known as the fermenter or bioreactor. Depending upon the type of fermentation carried out the design and nature of the fermenter varies. Invariably all the fermenters have services to measure various fermentation parameters such as beyond fermentation time, liquid, temperature, pressure, pH, level, mass, etc. (Figure 4.4). The different types of fermenters are as follows.

External recycles airlift fermenter—methanol as substrate, for producing bacterial biomass.

Abundant fermented products are devoted around the humankind. Each homeland has its own types of fermented food, representing the staple diet and the raw ingredients available in that exacting place. They may not be associated with fermentation, while the products are well known to the individual. Certainly, it is to be expected that the methods of producing abundant of the worlds fermented foods are unidentified and came concerning through possibility. Several more vegetable products and fermented fruit arise from lactic acid fermentation. The more obvious fermented products are the alcoholic beverages, wines and beers (Figure 4.5). However, they are enormously important in meeting the dietary requirements of a large proportion of the world's population.

Organisms which are responsible for fermentation are as follows.

- Prokaryotic: cyanobacteria and bacteria
- Eukaryotic: Algae, yeasts, and fungi

Figure 4.4 Process of fermentation in laboratory.

- Microorganisms are the important part of process in fermentation. For commercial production of lactic acid and citric acid, *Lactobacillus delbrueckii* and *Aspergillus niger* are used.

4.5 Biochemical Mechanism

When there is no oxidative phosphorylation to keep the production of ATP (Adenosine triphosphate) by glycolysis then the development that involved in the fermentation technology is significant in anaerobic conditions.

- Homolactic fermentation is the production of lactic acid from pyruvate.
- The conversion of pyruvate into ethanol and carbon dioxide is alcoholic fermentation.
- Production of lactic acid as well as other acids and alcohols is known as heterolactic fermentation. During fermentation pyruvate is metabolized to various different compounds.
- By fermentation more exotic compounds like butyric acid and acetone can be produced. There are some examples of fermentation products such as lactic acid and ethanol.

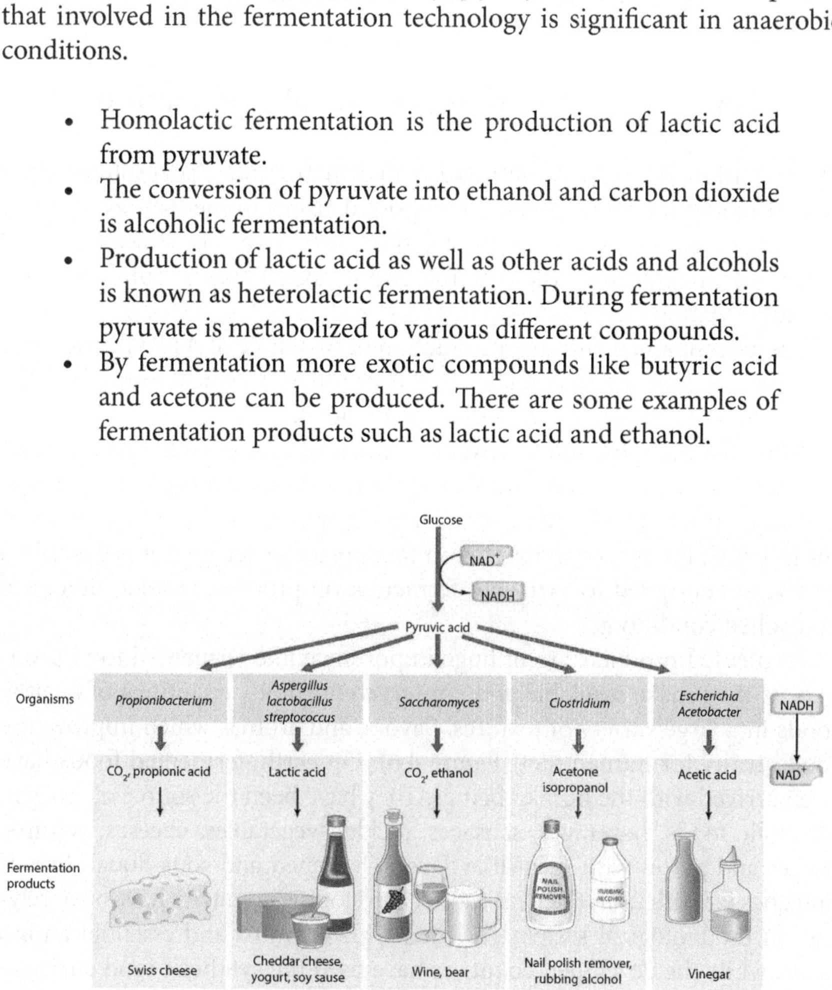

Figure 4.5 Conversion of glucose into different fermented products using micro-organisms.

The assortment of fermentation process is in five major groups of commercially very important fermentations. They are as follows.

- Individuals that produce microbial metabolites.
- Those that produce microbial cells (or biomass) as the product.
- Those that produce microbial enzymes.
- Those that produce recombinant products.
- Those that transform a compound which is supplementary to the fermentation is known as the transformation [8].

4.6 Fermentation and its Industrial Applications

Fermentation is a metabolic process in which energy is resultant from partial oxidation of an organic compound using immediate as electron donors or electron acceptors. In which the substrate is oxidized and ATP is formed directly from the reaction and it takes place by substrate level phosphorylation.

It is accompanied with the production of two mole of ATP Glucose → 2 pyruvate → 2 lactate (resultant 2ATP is produced), typical example is the conversion of glucose to two molecules of lactate.

Note that plentiful and renewable sources of fermentable carbohydrate are plant cellulose, starch, from agricultural wastes; molasses from sugar and whey from cheese manufacturing industries. A fermentor or bioreactor in which process of fermentation takes place. A fermentor is described as a vessel designed to carry out fermentation processes under biological controlled conditions.

Fermented products are of huge importance like fermented food products as they make available and conserve enormous quantities of healthy foods in a large variety of textures, flavors, and aromas which improve the food quality for human diet (Figure 4.6). On earth, fermented foods have been arrived with the human beings. They have been the source of vinegar, alcoholic foods, beverages, sausages, pickled vegetables, cheeses, yogurts, sauces and pastes with meat-like flavors, leavened and sour-dough breads and goes far with us in the near future. Developing countries have been relying on the biological system enriched with vitamins and essential amino acids, while the developed countries have used the synthetic food enriched with vitamins [9].

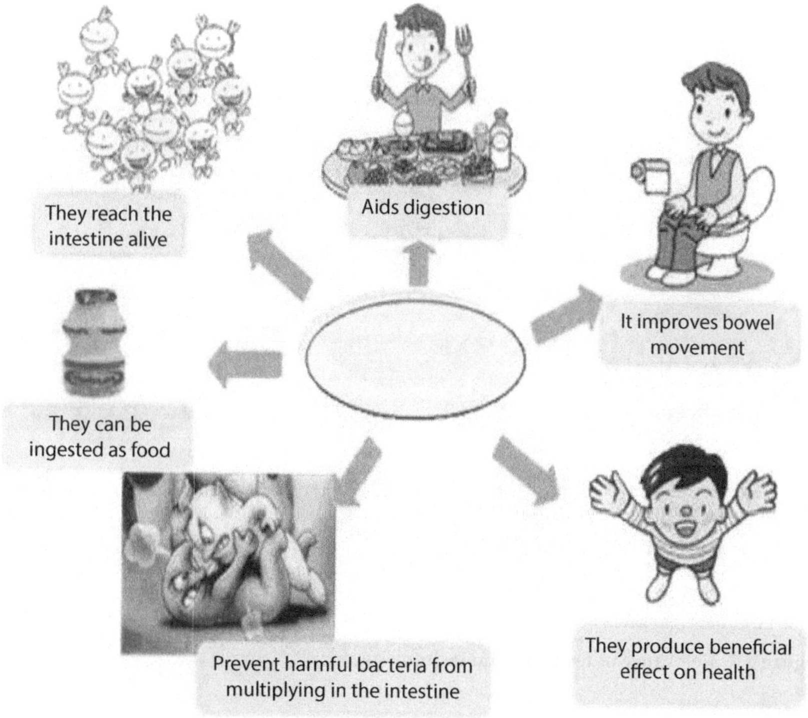

Figure 4.6 Various uses of fermented probiotics.

4.7 Relevance of Fermentation

The applications of fermentation have been rapidly breaching in the food stuff as cheese, wine, beer, and bread to high-value products, food chemicals and pharmaceutical ingredients (Figure 4.7). Due to high cost of petrol and reduction in fossil fuel has been the robust case of cost efficient and easy fermentation processes. According to some research analysts, a burst has there in academic and corporate field, determined by fermentation applications in the field of pharmaceutical industries, energy, utilities, and special chemicals.

Biofuels or commonly energy has gained more significant attention with the help of fermentative techniques by producing cost effective bioethanol [1, 3].

The main relevance of the fermentation techniques have been in the field of fermented food. Iru, Ogi, Fufu, and Gari are such examples of fermented

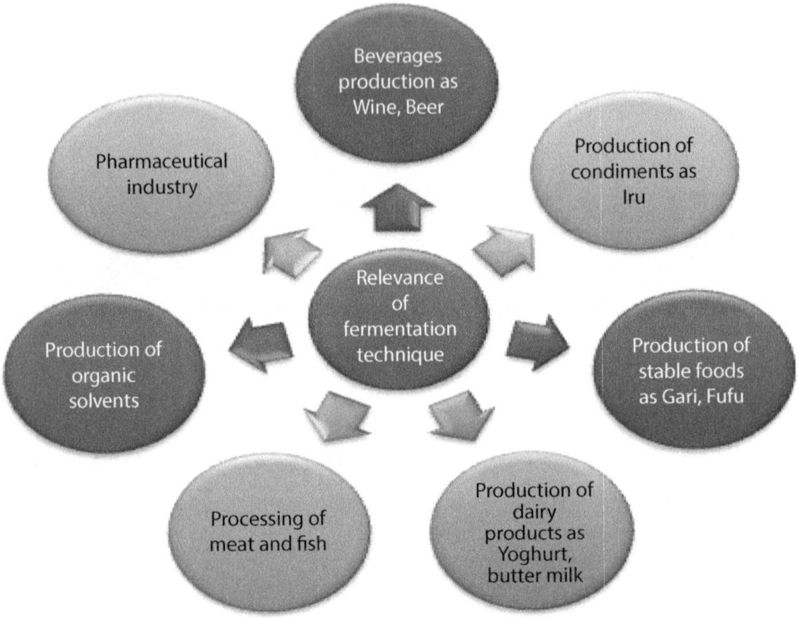

Figure 4.7 Relevance of fermentation technology.

foods. Irú has been a type of processed fermented food made from the locust beans (*Parkia biglobosa*) and has been used as a condiment in cooking. Fufu has been used as a staple food in countries like Africa, Ghana, and Nigeria. It has been made from the equal portions of cassava and green plantain flour thoroughly with water. Gari is the fine granular flour made from the cassava tubers by fermentation procedure and it serves as a major staple food [10].

4.8 Conclusion

Principle of fermentation techniques, their theory of control, processes, and analysis has been studied in this chapter. The fermentation principle has been described as an equational system model with consumption of substrate and their microbial growth under ideal mixing conditions with and without culture broth reuse. This specifically identifies the special characteristics of fermentation techniques, uses, and suitable modes of regulation.

References

1. Crafts-Lighty, *Information sources in biotechnology*, 1983.
2. Goals, Kameswara Rao, C., *Fermentation biotechnology*, 2009.
3. Frost & Sullivan Research Service, *Advances in fermentation technologies: an industry overview technical insights*, 2015.
4. Pumphrey, B., *An introduction to fermentation basics* by New Brunswick Scientific (UK) Ltd.; and Christian Julien New Brunswick Scientific Benelux BV (The Netherlands), May 1996.
5. Shurtleff, W., Aoyagi, A., *A brief history of fermentation*, East and West, 2004.
6. Biology Discussion, http://www.biologydiscussion.com/fermentation/fermentation-technology-meaning-methodology-types-and-procedure/17492, 2017.
7. FAO Corporate Document Respository, http//www.fao.org.docrep/x0560e/x05660e07.htm, 2017.
8. Whitaker, A., *et al.*, *Principle of fermentation technology*, 1995.
9. Steinkraus, K.H., *Classification of fermented foods: worldwide review of household fermentation techniques*, 8, 5–6, 1997.
10. FANNAB Community, http://www.unaab.edu.ng, 2014.

5

Fermentation Technology Prospecting on Bioreactors/Fermenters: Design and Types

Gauri Singhal[1], Vartika Verma[1], Sameer Suresh Bhagyawant[2] and Nidhi Srivastava[1]*

[1]Department of Bioscience and Biotechnology, Banasthali Vidyapith, Rajasthan, India
[2]School of Studies in Biotechnology, Jiwaji University, Gwalior, Madhya Pradesh, India

Abstract
The basic requirement for the fermentation process is the bioreactor or fermenter. A fermenter or bioreactor may refer to an engineered device or system to support a biologically active environment. They have been the necessary and basic thing to enable the scientific assistance in the biotechnological research to produce the desirable products. In this chapter, the comprehensive design of fermenter has been elaborated from inside and outside with different fermenter type to differentiate the fermentation processes.

Keywords: Fermentation, bioreactor, fermenter design

5.1 Introduction

Among oldest techniques, fermentation technology has been the popular technique used in the biotechnology field. In fermentation technology, fermentation process has been the heart of it [1] and used for preparation of vine, bread, idli, cheese, sausages, and other food products traditionally. In other words, fermentation is the biological process caused by microorganisms as a result of which heat and carbon dioxide is produced. Fermentation has been carried out in small or large size fermenters or bioreactors depending upon the quantity of product (Figure 5.1).

**Corresponding author*: nidhiscientist@gmail.com

Arindam Kuila and Vinay Sharma (eds.) Principles and Applications of Fermentation Technology, (65–84) © 2018 Scrivener Publishing LLC

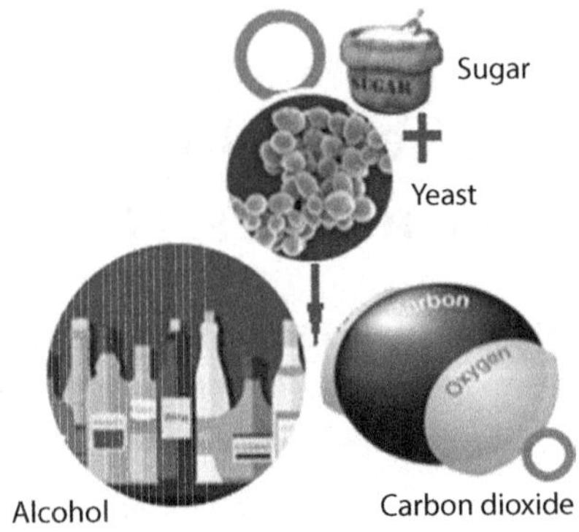

Figure 5.1 Fermentation process.

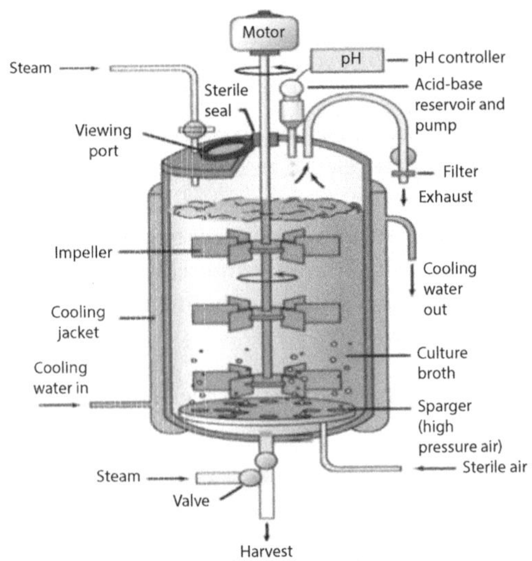

Figure 5.2 Fermenter [2].

Fermenters and bioreactors are two similar terms with different purposes. Fermenters are type of closed-vessel bioreactors with a controlled and sterile environment for the growth of the microorganisms in a liquid medium for the production of various compounds.

Fermenters are a type of closed system to carry out biological processes under controlled environmental conditions. In another words, fermenters are considered as a vessel to carry out chemical practices involving microorganisms and/or substances produced from those microorganisms (Figure 5.2). Fermenters usually control the metabolic products from the cells by transfer of oxygen and nutrients to cells and maintain a biomechanical as well as biochemical environment [3–5]. Fermenters can also be used for the recombinant products (such as vaccines and hormones); biotransformation of the products (L-sorbitol, steroid biotransformation), and production of enzymes (lipase, amylase, and cellulase). The reactions inside these fermenters may be aerobic or anaerobic. These fermenters are made up of stainless steel and cylindrical in shape, ranging from liters to cubic meters in size. Varied bioreactor designs have been developed to cater a wide array of substrate products and biocatalysts [6].

Microorganisms developed in these fermenters can be submerged (suspended or immobilized) or attached to the surface of a solid medium according to their nature. A wide variety of microorganisms can be used in suspension fermenters since all microorganisms do not need special attachment surfaces and because of that these fermenters can be handled at much greater level than immobilized fermenters. Immobilization is known as a process to entrap or attachment of wide variety of cell or particle with some immobilizable matter. In contrast, the microorganism will be removed from the fermenter in a continuously operated process with the effluent [7].

Bioreactors differ from conventional chemical reactors to the extent that they support and control biological entities. The bioreactor conditions should be favorable for the living microorganisms to exhibit their activity under defined conditions [8]. The term bioreactor is often used synonymously with fermenter, which is a type of bioreactor using a living cell as the biocatalyst. The sizes of the bioreactor can vary from the microbial cell (few mm^3) to pilot level fermenters (0.3–10 m^3), plant scale fermenters (2–500 m^3), shaking flask (100–1000 ml), and fermenters of laboratory scale (1–50 l) for applications in production of large volume industrial products [9].The conditions of bioreactors including flow rates, gases level (air, oxygen, nitrogen, and carbon dioxide), pH, temperature, foam production, agitation speed or circulation rate etc. needs to be closely monitored [10].

5.2 Bioreactor and Fermenter

The fermentation process has been known for thousands of years but its first scientific study was conducted by a French scientist Louis Pasteur in 1850s through lactic acid formation. The development of fermenters and bioreactors has been led by the advances in fermentation techniques. Although they are similar in terms of working principle. they also have differences.

Table 5.1 Differences between fermenter and bioreactor [11].

Differences	Fermenter	Bioreactor
Size	Small, upto 2 liters	Large, range from few liters to cubic meters
Nature	Fermenters are not sensitive in nature as bacterial cells are robust and have strong cell walls	Bioreactors are sensitive in nature as mammalian cells are fragile and have shear sensitive cell membrane
Production rate	Fast as microbes are fast growing.	Slow as mammalian cells double in 24 hours
Oxygen demand	High	Low
Contamination	No viral threat in microbial cells, so no need of inactivation or removal	Threat is present, need inactivation or removal
Sterilization	Full sterilization	Empty sterilization

The fermentation process has been mostly used for brewing and production of alcoholic beverages earlier but with our increasing knowledge about fungi and bacteria, more productive fermenters have been developed. In terms of design and construction, bioreactors are one step ahead. Fermenters have been used for the growth of bacterial and fungal cells with maintenance in controlled manner while bioreactors have been used for the growth of mammalian and insect cells and their maintenance. It has been clear, that fermenters and bioreactors are naturally different [11]. Table 5.1 shows some differences between the two.

5.3 Types of Fermenter and Bioreactor

There are basically three different kind of fermenters available.

5.3.1 Laboratory Scale Fermenters

It has been the smallest type of fermenter and it has limited capacity, that is, 1 to 15 l. Mainly these types of fermenters are used for the research and development purposes. Laboratory scale fermenters are used for

Figure 5.3 Laboratory scale fermenter [12].

determination of optimum condition of growth and biosynthesis of microorganism (Figure 5.3).

5.3.2 Pilot Scale Fermenters

These fermenters are of intermediate size and used for large scale studies in fermentation process. They have minimum capacity of 94 liters and maximum capacity of 7570 liters (Figure 5.4).

5.3.3 Industrial Scale Fermenter

These fermenters are used for large scale production in industries for fermentation of important products.Their capacity ranges from 37854.12–378541.18 l (Figure 5.5).

5.4 Design and Operation

A good bioreactor design should address improved productivity, validation of desired parameters towards obtaining consistent and higher quality products in a cost effective manner [15]. Designing of bioreactors have been comparatively complex engineering task and have studies in biochemical

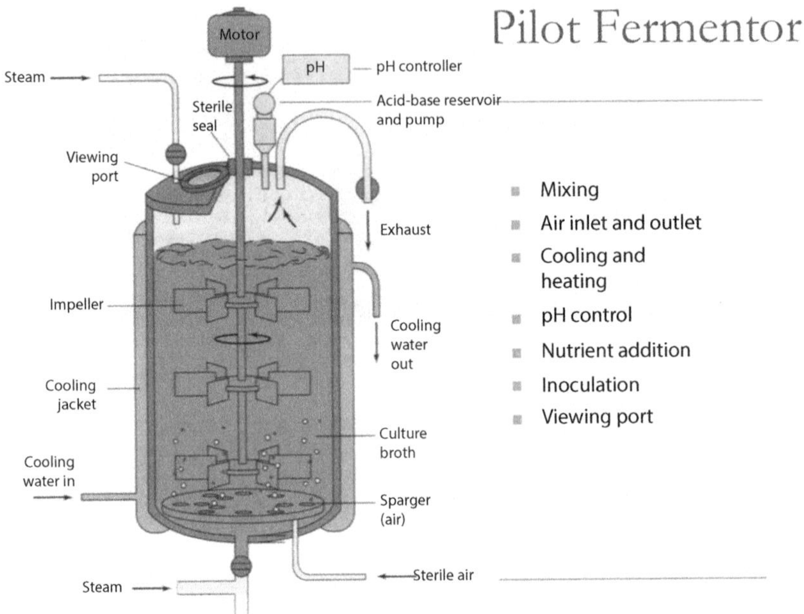

Figure 5.4 Pilot scale fermenter [13].

engineering. In bioreactors, microorganisms have performed their function under optimum conditions with insignificant impurities. However, the growth and production of microorganisms have been typically affected by the environmental conditions including nutrient concentrations, temperature, concentration of dissolved gases (especially oxygen for aerobic fermentations), and pH inside the bioreactor. Cooling jackets, coils or both are used to maintain the temperature for fermentation medium and external heat exchangers may also be used in case of exothermic fermentations. The bioreactor design and operation mode depends on the optimal conditions required for chosen product, microorganism production, product quality, and its production scale. An effective bioreactor must influence and control the biological reaction and prevent the contamination [15].

In a fed-batch fermenter, nutrients have been continuously added or may be charged at the beginning of fermentation. Depending on fermentation, pH of the medium has been adjusted and maintained with small amounts of acid or base. Reactant gases especially oxygen must be added in the aerobic fermenters, hence oxygen is comparatively insoluble in water, air or purified oxygen must be added continuously. The rising bubbles in the fermenter helps in mixing of fermentation medium and strip out waste gases as carbon dioxide. In aerobic fermentation, the rate limiting step is

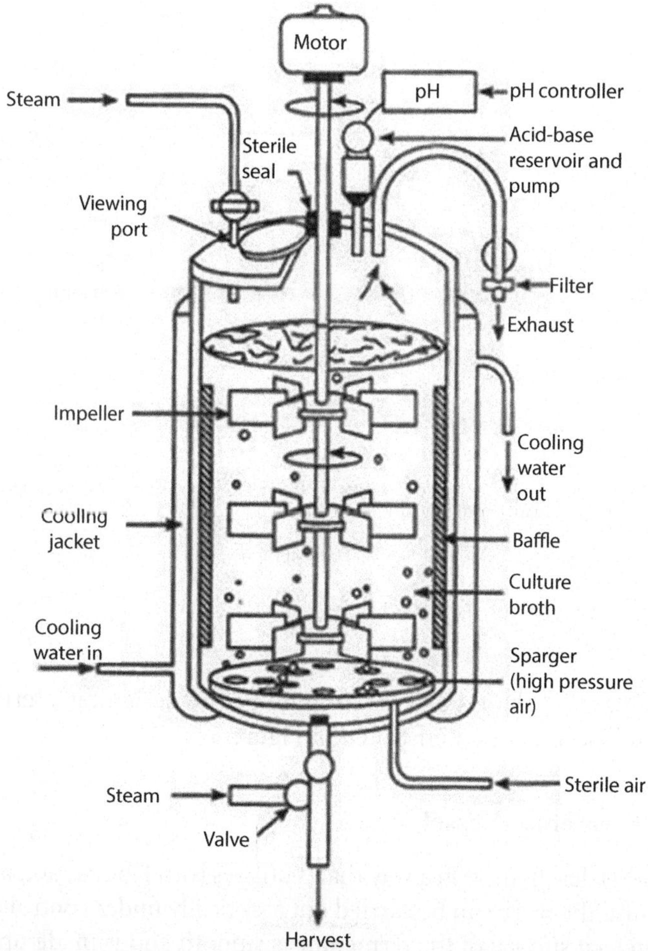

Figure 5.5 Industrial scale fermenter [14].

the transfer of optimal oxygen level. Sometimes, agitation helps in oxygen transfer to mix the nutrients and keep the fermentation homogeneous. The bioreactor attributes such as sterilization, process control devices, simple construction and measuring, scale-up, regulating techniques, flexibility and compatibility with upstream and downstream processes, antifoaming measures, etc. are essential factors to fulfill the design requirements [3].

The design of a fermentor should be in such a way that it should provide proper sterilized conditions that can be built within it, proper aeration and agitation, proper harvesting of the products and proper drainage. There are

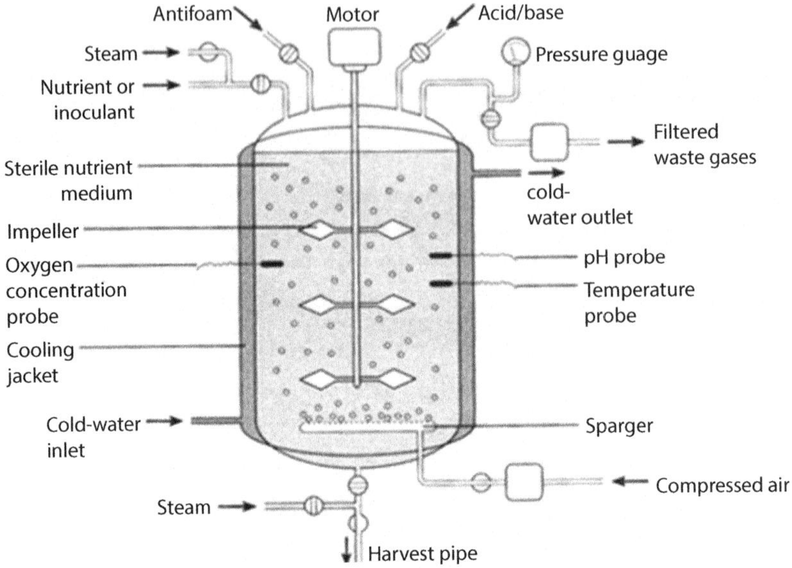

Figure 5.6 Fermentor with its parts [17].

other factors that should be kept in mind before designing a fermentor as described below and demonstrated in Figure 5.6.

5.4.1 Fermenter Vessel

The vessel is design in such a way that it utilizes least labor action and maintenance and the work can be carried out aseptically under controlled conditions. Internal surface of the fermentor is smooth and is made up of cheap materials with best results. There are two types of fermentor vessels used, for small scale glass is preferred and for industrial purpose stainless steel is used.

Glass is nontoxic and corrosion proof. It is easy to examine the interior reaction in the vessel. Sterilization is done with the process of autoclave. These are very small fermentor with the diameter of around 60 cm.

Stainless steel is mostly used for large scale fermentations. These vessels have the potential to resist pressure and corrosion. The sterilization is achieved *in-situ* [16].

5.4.2 Heating and Cooling Apparatus

Heat in the fermentor vessel is produced due to microbial activity and agitation. Temperature in vessel is maintained by either adding or removing heat from the system. Thermostatically controlled bath or internal coils are generally used to provide heat while silicone jackets are used to remove excess heat. It has double silicone mat with heating wires sandwiched between the mats. If the size is exceeded, resulting in covering the surface by the jacket, heat removal is tedious then in the internal coils, cold water has to be circulated to maintain the exact temperature [16].

5.4.3 Sealing Assembly

Sealing assembly is used for the sealing of stirrer shaft to offer proper agitation and it can function for a longer period aseptically. There are three types of sealing assembly in the fermenter.

Packed gland seal: In this, shaft has been sealed with several packing rings of asbestos, pushed by gland against the shaft. To prevent insufficient heat penetration, packing rings have been regularly checked and replaced.

Mechanical seal: This type of seal consists of two portions, stationary portion in the bearing and rotating portion on the shaft. Two parts are pushed together with the help of springs. During operation, stem condensate are used to lubricate and cool the seals.

Magnetic drives: These drives involve two types of magnets, that is, driving and driven magnet. The driving magnet will be seized on the external part of head plate in bearing and associated to the drive shaft. Another, the driven magnet will be located at the end of the impeller shaft and seized in bearings on the head plate's inner surface [16].

5.4.4 Baffles

Baffles prevent vortex to expand aeration capacity and consist of metal strips attached radially to the wall. Baffles have minimized the microbial growth on the sides of the fermenter [16].

5.4.5 Impeller

Impellers are used to offer uniform microbial cells suspension in homogeneous nutrient medium by agitation. Impellers mix the bulk fluid, solid particles, and gas phases in the suspension culture. Variable impellers are used in the fermenters and are classified as follows.

Disc turbines: They contain disc with a series of rectangular vanes. They provide air from the sparger to hit underside of the disc and displace the air towards the vanes to break large air bubbles into smaller ones.

Variable pitch open turbine: They also consist of vanned disc attached with the marine propeller blades on the agitator shaft. Air bubbles in this turbine do not hit any surface before dispersion [16].

5.4.6 Sparger

Sparger provides proper aeration in the vessel so that sufficient oxygen is supplied to the microorganisms for metabolic process. Three types of sparger are used.

Porous sparger: These are made up of ceramic or sintered glass and used in non-agitated vessel on the laboratory scale.

Nozzle sparger: It has open or partially open single pipe. This type of sparger is generally used because they do not get blocked and provide lower pressure.

Combined sparger–agitator: They introduce air by hollow agitator shaft and release it from the holes of the drilled disc to connect to the base of the main shaft. When the agitator is operated at a range of rpm, they deliver good aeration in a baffled vessel [16].

5.4.7 Feed Ports

Feed ports are tubes made up of silicone. They are used to add nutrients and acid/alkali in the fermenter. *In-situ* sterilization is performed before removal or addition of the products [16].

5.4.8 Foam Control

This is one of the important parts of the fermenter as the level of foam in the vessel has to be reduced to avoid contamination. Foam is controlled by two units, foam sensing and control unit. In the fermenter, a probe has been inserted through the top and set at a distinct level above the broth surface. When the foam level rises and touches the probe tip, a current will be passed through the circuit. This current will activat the pump and antifoam will immediately be released to combat the situation [16].

5.4.9 Valves

Valves are used in the fermentor to control the movement of liquid in the vessel. There are around five types of valves are used, that is, globe valve, butterfly valve, ball valve, and diaphragm valve. Globe valves are suitable for general purposes but they do not regulate flow. Butterfly valves are not suitable for aseptic conditions and are used for large diameter pipes which operate under low pressure. Ball valves are suitable for aseptic condition. They handle mycelial broths and are operated under high temperature. Diaphragm valves help in flow adjustments [16].

5.4.10 Safety Valves

Safety valve is built-in in air and pipe layout to operate under pressure. With the help of these valves the pressure is maintained within safe limits [16].

5.5 Classification of Bioreactor

In fermenters and bioreactors, various biochemical processes take place to produce different biological products with the help of microbes such as bacteria, fungus, mammalian cell, and plant cells systems in the form of initial product. They provide promising environment for the metabolite production.

Mainly fermenter and bioreactor are divided into four main groups on the basis of oxygen and stirring requirement, operation mechanism, microbial growth, and process requirement as shown in Figure 5.7. The nonstirred non aerated bioreactors are used to manufacture traditional products such as wine, cheese, and beer. While stirred aerated reactors are based on modern technology.

5.6 Types of Fermenter/Bioreactor

The fermentor types used are the continuous stirred tank, airlift, fluidized bed, packed bed, photobioreactor, membrane fermenter, and bubble column fermenter.

5.6.1 Stirred Tank Fermentor

In this fermenter, difference in concentration of the components of the medium and microorganisms of the fermentor is not with respect to time. A fixed state can be achieved either by chemostatic or turbidostatic principles. An appropriate and constant value can be accomplished by modifying flow rate of the fermentor so that the microorganisms, substrates, and product(s) amount remains at their optimum levels. One which uses yeasts and bacteria's, is the most successful continuous system which helps to obtain desired products. The most widely used continuous process is based

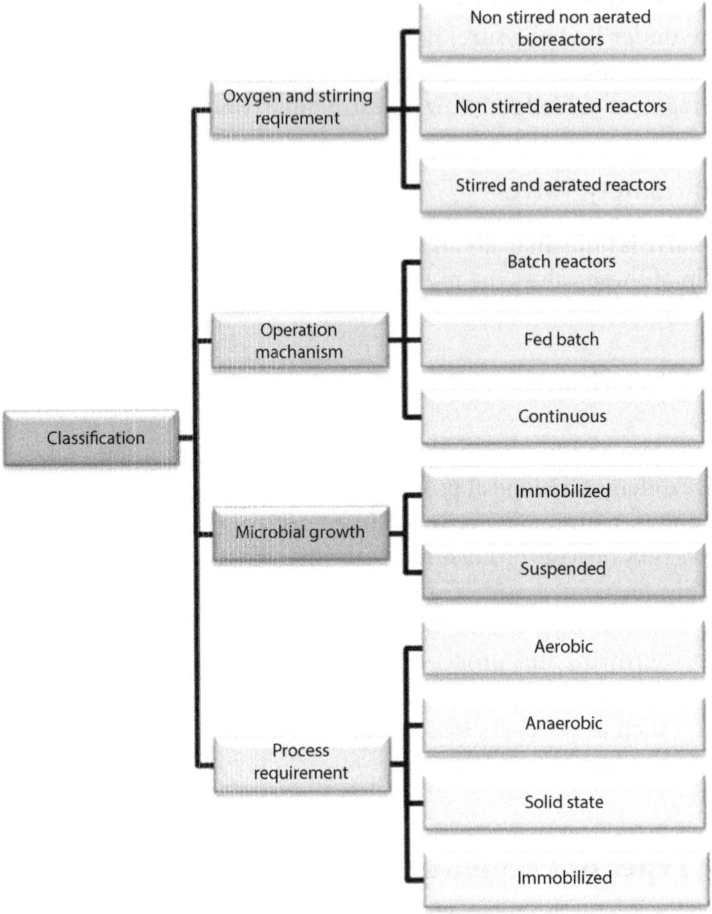

Figure 5.7 Classification of fermenter and bioreactor.

on Continuous Stirred Tank Fermentor (CSTF) is the activated sludge process used in waste water treatment industry (Figure 5.8).

Stirred tank bioreactor has various benefits as continuous operation can be performed in the fermentor, temperature control can be easily done, construction cost is cheap, easily operated so labor cost is reduced, and cleaning can be easily done [1].

5.6.2 Airlift Fermentor

Airlift reactor is generally for gas–liquid or gas–liquid–solid contacting devices. They have different fluid circulation, which is a definite cyclic

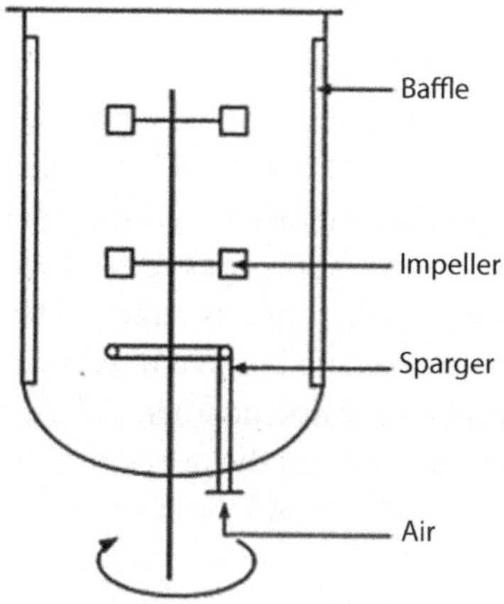

Figure 5.8 Stirred tank fermentor [18].

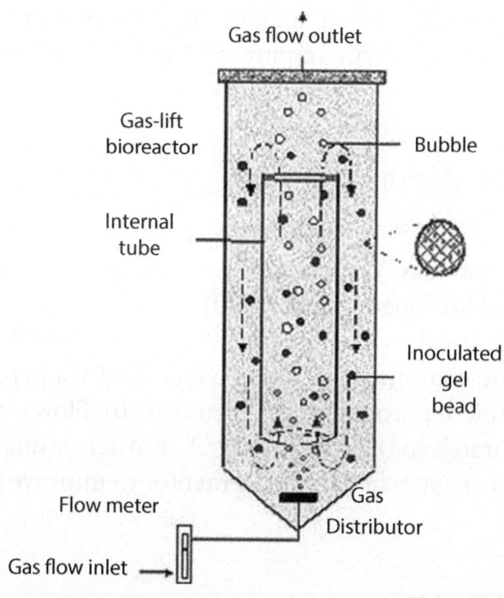

Figure 5.9 Airlift Fermentor [18].

pattern via built channels. Stream of air or other gases provides agitation to the content in side channels. The gas stream help swap over of material between the gas phase and the medium, oxygen is usually transferred to the liquid. Products formed after reactions are excreted when the gas phase is inserted (Figure 5.9).

Airlift reactors consists of two main types of reactors, that is, external-loop vessels that provides circulation through separate and distinct channels and internal-loop vessels, in which baffles are placed in a single vessel which provides circulation.

Variations in the fluid dynamics can be achieved by modifying the pattern of both types of vessels. External- and internal-loop vessel include four distinct parts with diverse flow characteristics such as riser, downcomer, base, and gas separator. Base of riser allows gas to be injected and gas and liquid flow in upward direction. Downcomer is attached to the riser from top and bottom. Direction of flow of gas and liquid is downwards in this. Recirculation of the gas and liquid between the downcomer and the riser is due to the mean density between them which creates the pressure gradient required for recirculation. Base effects only gas holdup, liquid velocity, and solid phase flow. Basic design of base of riser and downcomer is very simple. While gas separator located on top of the reactor which connects the riser to the downcomer, enabling proper liquid recirculation and gas disengagement [19].

Airlift fermenter has different benefits as it is very easy to design as it doesn't have any moving part or agitator, sterilization can be easily done as it do not have agitator, energy requirement is very less, and it is very low cost [1].

5.6.3 Bubble Column Fermentor

Bubble column reactors are used in many chemical, petrochemical, and biochemical industries. These reactors are simple in construction, easy maintenance, and low operating cost [20].

They are cylindrical in shape with ratio of 4:6 (height:diameter ratio) and at base of the column air or gas is introduced via perforated pipes or plates, or metal micro porous sparger (Figure 5.10). Flow rate of air or gas is maintained accuratelyso that the proper O_2 transfer or mixing is achieved. Perforated plates are attached in the fermentor to improve performance of the reactor [21].

5.6.4 Packed Bed Reactors

Packed bed reactor is also called as fixed bed reactor which are used in many chemical processing applications like absorption, distillation, stripping, separation processes, and catalytic reactions. It consists of partition like tube or channel which has catalyst particles or pellets on to which liquid flows through the catalyst (Figure 5.11). Chemical composition of the substance gets altered when the liquid reacts with the catalyst [22].

This reactor has many advantages as its conversion rate is high for the catalyst, easy to manage and build, more efficient contact between reactant

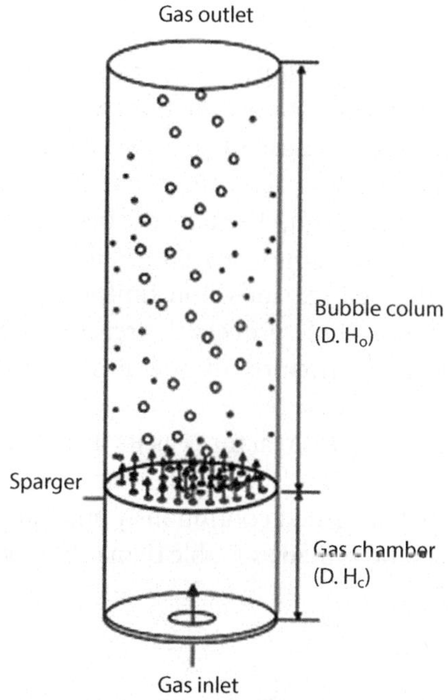

Figure 5.10 A bubble column fermentor [18].

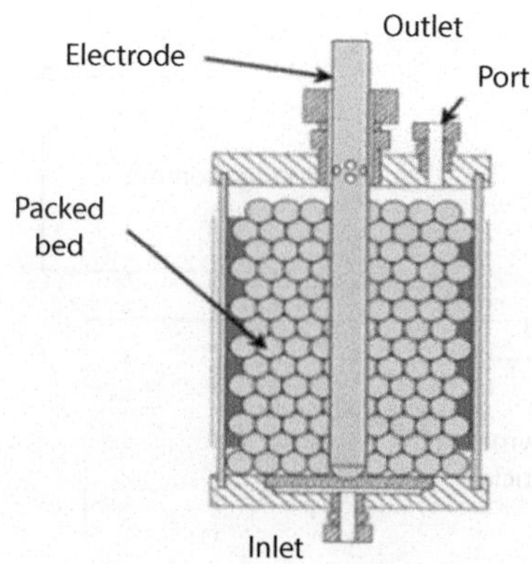

Figure 5.11 Packed bed reactor [22].

and catalyst is made compared to other types of reactors, product formation is more due to increased contact of reactant/catalyst, low construction, operation, and maintenance cost, and it work effectively even on high temperatures and pressures [23].

5.6.5 Fluidized Bed Bioreactor

They are very large reactor ranging in size from 10–300 microns. Design of this reactor must be proper so that the fluid flow rate is sufficient to suspend the catalyst particles (Figure 5.12).

Catalyst is laid on the bottom of the reactor and the reactants are pumped into reactor via distributor pump to make the bed fluidized. If the reactant is liquid then bed expands uniformly and make homogeneous fluidization and if its gas, the bed expands non-uniformly to make aggregative fluidization. During this whole process the reaction between the reactant and catalysts leds to the formation of new products which are retrieved continuously during the course of time.

The main advantages of this reactor is as its distribution of the temperature is even, regeneration and replacement of the catalyst can be easily done, the operations can be performed continuously and automatically, and contact between gas and solid is made suitable than other catalytic reactors [25].

5.6.6 Photobioreactor

Although some models of photobioreactors are planned, only few of them can utilize biomass from algae. The main applications of photobioreactors are in photosynthetic processes, involving vegetable biomass growth

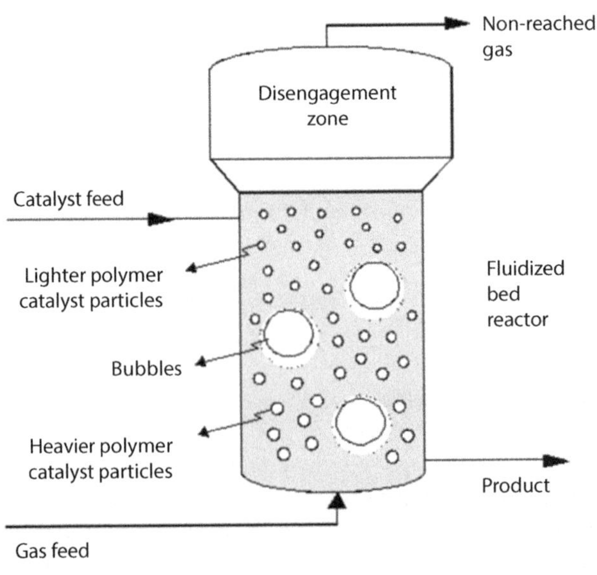

Figure 5.12 Fluidized reactor [24].

or microalgae growth under restricted conditions. The photobioreactors array from laboratory to industrial scale models and more over they are classified into closed photobioreactor, open ponds, flat-plate, horizontal/serpentine tubular airlift, and inclined tubular photobioreactors [26]. The introduction of more complicated cultivating methods of microalgae with higher production value and capable of providing sterile conditions, which is accessible by different types of closed photobioreactors, applied outdoors. In general, laboratory-scale photobioreactors are artificially illuminated using fluorescent or other light lamp distributors. Some of these reactors include open ponds, flat-plate, tubular, bubble column, airlift column, helical tubular, conical, torus, stirred-tank, seaweed type photobioreactors. According to Ugwu *et al.* [26], the only disadvantage which limits their practical application in algal mass cultures is mass transfer that is required for proper processing of mass algal cultures. The algal biomass is mainly used in water treatment, in aquaculture, production of fine chemicals and useful supplements in humans and animals, for biosorption of heavy metals and CO_2 fixation.

5.6.7 Membrane Bioreactor

Membrane bioreactors (MBR) are been used since 90s. It basically combines traditional treatment system with filtration via membranes resulting in removal of organic and suspended solid matters that also removes high level of nutrients as shown in Figure 5.13.

Membranes in the MBR system are submerged in an aerated biological reactor. The pore size of the membrane ranges from 0.035 microns to 0.4 microns [28]. After mid-90s, with the development of submerged MBR system the use has widely extended [29] and is rapidly growing both in

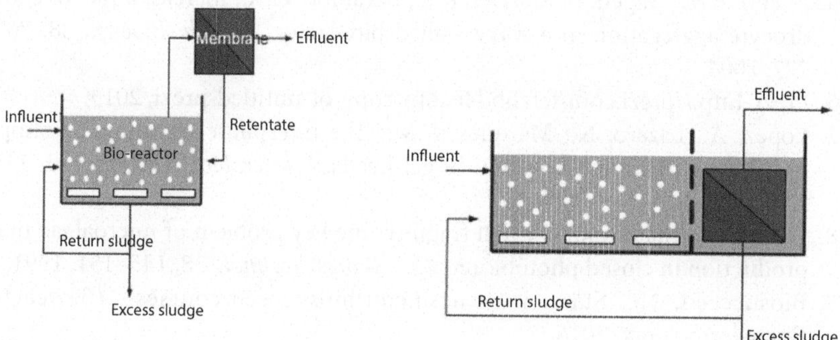

Figure 5.13 Membrane bioreactor method diagram [27].

research and commercial applications. Several variations of MBR systems have evolved and presently, an MBR system is widely used in treatment of waste water from several sources. Therfore in coming years, MBR systems will increase its capacity size and will broaden its application area [29–31]. However, membrane fouling is a chief obstacle to the extensive application of MBRs. Moreover large-scale use of MBRs in waste water treatment will involve a notable worthy decrease in price of the membranes [29].

5.7 Conclusion

Bioreactors have been used for decades to produce a range of therapeutic biomolecules and other high-value products. They provide the opportunity to monitor and control environmental conditions continuously throughout the culture/reaction period along with the added benefits of maintaining a closed system. They are critical and integral part of the development of many new processes.

References

1. Bio-Resource, http://technologyinscience.blogspot.in/2012/08/different-types-of-fermentors.html#.Wi0bMtKWbIU, 2012.
2. Slide Share, Aeration and agitation in fermentation, https://www.slideshare.net/151212345/aeration-agitation-in-fermentation, 2014.
3. Biotech Articles, https://www.biotecharticles.com/Biotech-Research-Article/Design-and-Operational-Key-Factors-of-Bioreactor-1558.html, 2012.
4. Haj, A.J.E., Wood, M.A., Thomas, P., Yang, Y., Controlling cell biomechanics in orthopaedic tissue engineering and repair. *Pathologie. Biologie.*, 53, 581–589, 2005.
5. Bueno, E.M., Bilgen, B., Carrier, R.L., Barabino, G.A., Increased rate of chondrocyte aggregation in a wavy-walled bioreactor. *Biotech. Bioeng.*, 88, 767–777, 2004.
6. Prezi, http://prezi.com/fefahbd4eq6m/copy-of-untitled-prezi, 2013.
7. Lopez, A., Lazaro, N., Marques, A.M., The interphase technique: a simple method of cell immobilization in gel-beads. *J. Microbiol. Methods*, 30, 231–234, 1997.
8. Gudin, C., Chaumont, D., Cell fragility: the key problem of microalgae mass production in closed photobioreactors. *Biores. Technol.*, 38, 145–151, 1991.
9. Biosucceed, NC State, www.ncsu.edu/biosucceed/courses/.../Bioreactor Engineering.pptx, 2016.
10. Chen, H.C., Hu, Y.C., Bioreactors for tissue engineering. *Biotech. Letters*, 28, 1415–1423, 2006.

11. Difference Between, http://www.differencebetween.com/difference-between-bioreactor-and-vs-fermentor/, 2011.
12. Bio- Age Equipments and Services, http://www.bioageindia.org/bio-fermenters.html, 2015.
13. Player, http://slideplayer.com/slide/7804510/, 2016.
14. Biology Discussion, http://www.biologydiscussion.com/industrial-microbiology-2/fermentor-bioreactor-history-design-and-its-construction/55756, 2017.
15. Singh, J., Kaushik, N., Biswas, S., Bioreactors – technology & design analysis. *Sci. tech. J.*, 1, 28–36, 2014.
16. Bodduman, Information creates wealth, https://www.boddunan.com/articles/education/42-general/1040-construction-of-a-fermenter.html, 2016.
17. Andrew Biology, http://andrewbiology.blogspot.in/2012/02/58-fermenter.html, 2012.
18. Fermenters, https://sites.google.com/site/fermentersin/types, 2016.
19. Bioreactor Sciences, https://www.bioreactorsciences.com/uploads/1/8/5/9/18594674/bioreactors_airlift_reactors_merchuk_gluz.pdf, 2015.
20. Scribed, https://www.scribd.com/doc/26979880/Bubble-Column-Reactors, 2015.
21. Biology Discussion, http://www.biologydiscussion.com/biotechnology/bio-process-technology/bioreactors-types-6-types-of-bioreactors-used-in-bio-process-technology/, 2015.
22. Cosmol, https://www.comsol.com/blogs/packed-bed-reactor/, 2015.
23. Visual encyclopedia of chemical engineering, http://encyclopedia.che.engin.umich.edu/Pages/Reactors/PBR/PBR.html, 2011.
24. Fernandes, F.A.N., Lona, L.M.F., Fluidized bed reactor for polyethylene production. The influence of polyethylene prepolymerization. *Braz. J. Chem. Eng.*, 17, 1–4, 2000.
25. Visual encyclopedia of chemical engineering, http://encyclopedia.che.engin.umich.edu/Pages/Reactors/FBR/FBR.html, 2013.
26. Ugwu, C.U., Aoyagi, H., Uchiyama, H., Photobioreactors for mass cultivation of algae. *Biores. Technol.*, 99, 4021–4028, 2008.
27. Emis, energie-en milieu-informatiessysteem, https://emis.vito.be/en/techniekfiche/membrane-bioreactor, 2010.
28. SSWM, Sustainable sanitation and water management, https://www.sswm.info/category/implementationtools/wastewatertreatment/hardware/semi-centralised-wastewater-treatments/m, 2015.
29. Meng, F., Chae, S.R., Drews, A., Kraume, M., HangSik, H.S.S., Yang, F., Recentadvances in membrane bioreactors (MBRs): membrane fouling and membrane material. *Water Res.*, 43, 1489–1512, 2009.
30. Visvanathan, C., Ben, R.A., Parameswaran, K., Membrane separation bioreactors forwastewater treatment. *Crit. Rev. Env. Sci. Technol.*, 30, 1–48, 2000.
31. Yang, W., Cicek, N., Ilg, J., State-of-the-art of membrane bioreactors: worldwide research andcommercial applications in North America. *J. Mem. Sci.*, 270, 201–211, 2006.

Part II
APPLICATIONS OF FERMENTATION TECHNOLOGY

6

Lactic Acid and Ethanol: Promising Bio-Based Chemicals from Fermentation

Andrea Komesu[1,2*], Johnatt Oliveira[3], Luiza Helena da Silva Martins[4], Maria Regina Wolf Maciel[2] and Rubens Maciel Filho[2]

[1]*Departamento de Ciências do Mar (DCMar), Federal University of São Paulo (UNIFESP), Santos, SP, Brazil*
[2]*School of Chemical Engineering, University of Campinas (UNICAMP), Campinas, SP, Brazil*
[3]*Instituto de Ciências da Saúde, Faculdade de Nutrição, Federal University of Pará (UFPA), Belém, PA, Brazil*
[4]*Centro de Ciências Naturais e Tecnologia, State University of Pará (UEPA), Belém, PA, Brazil*

Abstract

Production of bio-based chemicals by fermentation is one important technological platform because of several advantages, such as low substrate costs, production temperature, and energy consumption. Current global bio-based chemical and polymer production (excluding biofuels) is estimated to be around 50 million tons. In 2010, the US Department of Energy issued a report which listed the lactic acid and ethanol as potential building blocks chemicals for the future. This work will summarize information about the lactic acid and ethanol properties and applications, and raw materials used for its biotechnological production. Fermentation process and challenges in lactic acid and ethanol production will also be pointed out. In addition, integrated ethanol and lactic acid production will also be discussed.

Keywords: Lactic acid, ethanol, bio-based chemicals, fermentation, raw materials, biotechnology production

**Corresponding author*: andrea_komesu@hotmail.com

Arindam Kuila and Vinay Sharma (eds.) Principles and Applications of Fermentation Technology, (87–116)

6.1 Introduction

An increasing interest for discovering new environment-friendly sources of chemicals has been observed due to the current concerns related to the cost and environmental impact of using traditional petrochemical processes [1]. One important technological platform is the production of bio-based chemicals by fermentation that potentially offers several advantages: low substrate costs, production temperature, and energy consumption [2]. Based on current growth, the market for bio-based chemicals is projected to reach $19.7 billion in 2016 [3]. In 2010, the US Department of Energy issued a report which listed the chemicals building blocks which is considered as potential building blocks for the future [4]. Lactic acid (LA) and ethanol are among promising bio-based chemicals.

LA or 2-hydroxypropionic acid (CAS 50-21-5), the simplest hydroxyl carboxylic acid, is a bulk chemical with two optically active enantiomers (L(+) or D(−) LA) [5]. In recent years, the demand for L-LA has been increasing considerably owing to its use as a monomer in the preparation of polylactic acid (PLA) [6, 7], biodegradable and biocompatible polymer which is used in food packaging, disposable tableware, shrink wrap, 3-D printers, and elsewhere [3]. The environmental performance of PLA-based polymers is better than that of petrochemical polymers in terms of global warming, dependency on fossil energy, and human toxicity [8].

LA can be produced by fermentative or chemical synthesis. Most commercial production of LA is by microbial fermentation of carbohydrates. However, a relatively small amount is produced by chemical synthesis, using acetaldehyde as a starting material [3].

Raw materials used as substrates in microbial fermentation for LA production are varied: starch (wheat, maize, cassava, potato, rice, rye, and barley), lignocellulose, whey, sugar beet or sugarcane molasses. The use of inexpensive, non-food substrates, including lignocellulosic biomass, food-waste materials, or algal substrates, is highly recommended to aid development of a cost-effective LA production process [9].

Alcohols are oxygenated fuels with one or more oxygen in the molecule, decreasing its combustion heat. Practically any organic molecules in alcohol`s family can be used as fuel to engines, the most common being methanol (CH_3OH), bioethanol (C_2H_5OH), propanol (C_3H_7OH), and butanol (C_4H_9OH). However, only bioethanol and methanol are economically and technically suitable to intern combustion engines [10].

Bioethanol or ethyl alcohol has high octane index (108) [11], allowing high compression rate and lower emission of greenhouse gases [12].

Thereby, bioethanol is used as a modern biofuel, both in gasoline addictive or gasoline replacement functions. It can also be used as ETBE (ethyl tert-butyl ether) currently added in synthetically produced octanes as enhancers and in gasoline bioethanol mixtures in order to reduce scape gases emissions [13, 14].

Bioethanol is, by far, the most widely used biofuel in transportation all around the world and can be produced from different feedstock types. These feedstocks are classified in three agricultural categories: simple sugars, starch, and lignocellulose. The feedstock price is highly volatile, which directly affect bioethanol production costs. One of the major problems in bioethanol production is feedstock availability. Lignocellulosic biomass is the most promising feedstock considering its wide availability and low cost, and it does not compete with food production, however, large scale commercial production of ethanol from lignocellulosic materials has not yet been implemented [14].

Bearing the importance of LA and ethanol in mind, this chapter will summarize information about the LA and ethanol properties and applications, and raw materials used for its biotechnological production. Fermentation process and challenges in LA and ethanol production will also be pointed out. Integrated ethanol and LA production, and future potentials will also be discussed.

6.2 Generalities about LA and Ethanol

LA, also known as milk acid, is the most widely occurring carboxylic acid in nature. It was first isolated in 1780 by a Swedish chemist, Carl Wilhelm Scheele, but it was first produced commercially by Charles E. Avery at Littleton, MA, USA in 1881 [15].

LA is a simple chiral molecule that exists as two enantiomers, L- and D-LA (Figure 6.1), which differ in their effect on polarized light [16]. LA can be produced either by chemical synthesis or by fermentation. The chemical synthesis pathway produces an optically inactive racemic mixture of the L and D isomers, while the fermentation pathway generally yields optically pure L or D isomer, depending on the microorganism used [5]. L-(+)-LA is also naturally produced from pyruvate in the normal metabolism of microorganisms, animals, and humans [3].

Optically pure LA (L- or D-isomer) is more valuable than racemic DL-LA [9, 17], and each form has specific applications such as production of poly-L-LA and poly-D-LA.

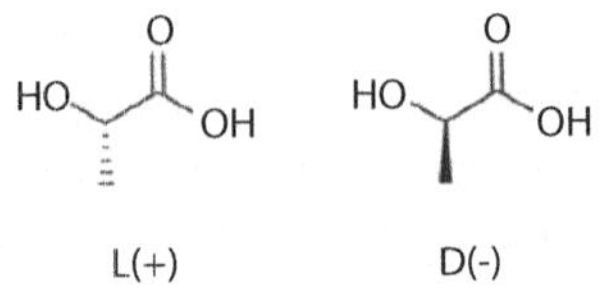

Figure 6.1 LA enantiomers.

Table 6.1 Properties of LA [15].

CAS number	D/L: 50-21-5 L: 79-33-4 D: 10326-41-7
Molar mass (g/mol)	90.08
Melting point (°C)	L: 53 D: 53 D/L: 16.8
Boing point (°C)	122 (12 mmHg)
Specific gravity (g/mL)	1.2

Chemical and physical properties of LA are summarized in Table 6.1. LA is a colorless or yellowish syrupy liquid, odorless, and hygroscopic. It is miscible with water, alcohol, glycerol, and furfural; insoluble in chloroform, petroleum ether, carbon disulfide. Cannot be distilled at atmospheric pressure without decomposition; when concentrated above 50% it is partially converted to lactic anhydride [18]. LA commercially occurs in aqueous solutions of 20–90 wt% [19].

LA has a wide range of applications in chemicals, pharmaceuticals, and food, and it is a precursor to several products [20]. The many uses of LA are shown in Figure 6.2.

The major parts of LA (39%) are used for the synthesis of the PLA. Another large application of LA (35%) is in the food industry. They are found in beverages, candy, meat, sauces, and others, for their mildly acidic taste [19]. LA is also used as an emulsifying agent in bakery goods and as inhibitors of bacterial spoilage. Other uses are found in pharmaceutical industry and cosmetics (13%) as well as in chemical industry (13%). LA is a molecule which can be converted to different products (Figure 6.2) showing its flexibility as a feedstock. Moreover, different industries use LA on a small scale for specific tasks. LA is for instance used as a pH adjuster, terminating agent, in adhesive formulation and lithographic printing [19].

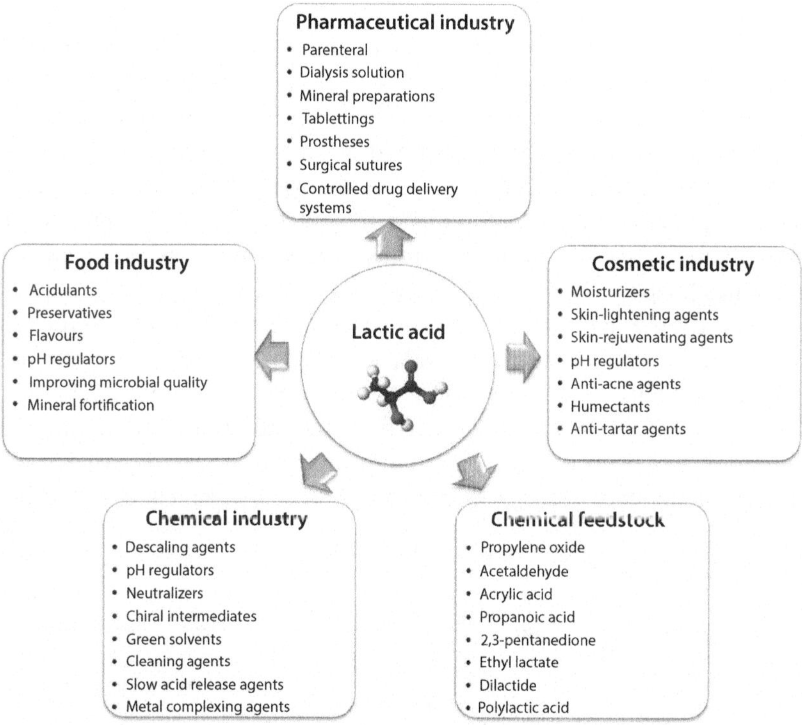

Figure 6.2 Commercial uses of LA [21].

Global LA demand was estimated to be 714.2 kilo tons in 2013, and it is expected to grow annually by 15.5% to reach 1,960.1 kilo tons by 2020 [9]. The three largest consumer market in the world are the United States (31% of total LA consumption in 2013), followed by China and Western Europe. China surpassed Western Europe due to export demand as well as consumption in the food and beverage industry [20]. PURAC, Cargill, and Henan Jindan Lactic Acid Technology Co., Ltd. as the world's top three LA manufacturers boasted a combined capacity of 505,000 tons in 2013. Cargill mainly supplies LA products to its subsidiary—NatureWorks for production of PLA [22].

The most commonly used renewable fuel in the transportation sector is ethanol and has a long history as alternative fuels [23]. Ethanol production worldwide has strongly increased since the oil crises in 1970 [24]. Global ethanol production increased from 13.12 billions of gallons in 2007 to 25.68 billions of gallons in 2015 with a slight decreased in 2012 and 2013 [25].

Ethanol is a liquid substance that is volatile, colorless and has a slight odor. It apparently burns with smokeless blue flames that are not always visible in normal light. It can be produced either by chemical or microbiological processes: fermentation of ethanol, indirect hydration (esterification-hydrolysis) process, and direct hydration of ethylene [26, 27].

It presents widely in nature and has many applications in the industrial and pharmaceutical sectors as a solvent of substances intended for human contact consumption, including scents, flavorings, colorings, and medicines [28].

Currently, the main industrial route used for ethanol production worldwide is the microbiological process, also referred as alcoholic or ethanolic fermentation with yeast, for example Saccharomyces cerevisiae [29].

A variety of feedstocks from the first, second, and third (Figure 6.3) generation has been used in bioethanol production. The first-generation bioethanol involves feedstocks rich in sucrose (sugar cane, sugar beet,

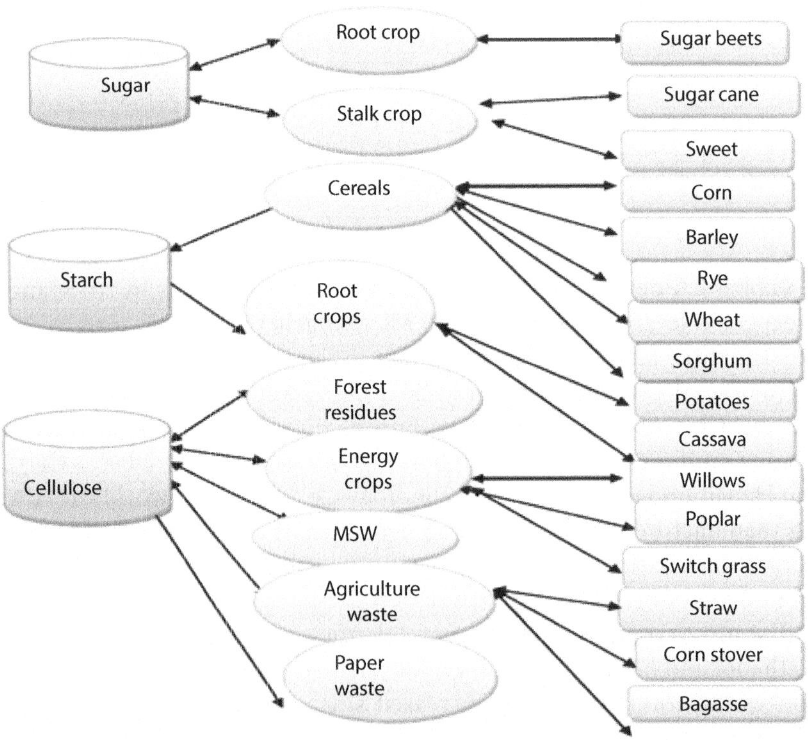

Figure 6.3 Feedstock's for ethanol production [30].

sweet sorghum, and fruits) and starch (corn, wheat, rice, potato, cassava, sweet potato, and barley). Second-generation bioethanol comes from lignocellulosic biomass such as wood, straw, and grasses. Third-generation bioethanol has been derived from algal biomass including microalgae and macroalgae [31].

Lignocellulosic biomass is more preferred than starch or sugar-based crops for production of ethanol, since it does not compete with food and takes care of agricultural and plant residues in an environmentally sustainable process [32, 33]. Lignocellulosic biomass ethanol production demands good knowledge of the material structure used in the biotechnological transformation [34].

6.3 Fermentation Methods to LA and Ethanol Production

Over 90% of the current commercial production of LA is performed via fermentation [19, 35]. The fermentation process is characterized by the biological degradation of the substrate (glucose) by a population of microorganisms (biomass) into metabolites, such as ethanol, citric acid, and LA [20, 36].

Several microorganisms have been isolated and used in the production of LA from the genera *Lactobacillus*, *Streptococcus*, and *Pediococcus* [37], which are LA bacteria (LAB). Although the majority of LA processes are carried out with LAB, LA is also produced by filamentous fungi, particularly *Rhizopus sp.* [38].

LAB can be classified into two groups according to fermentation end-product: homo- and hetero-fermentative strains. Homo-fermentative LAB converts glucose almost exclusively to LA, while hetero-fermentative LAB catabolize glucose into ethanol, CO_2, and LA [20]. Homo-fermentative strains that produce optically pure L (+) or D (-) LA are industrially attractive compared to hetero-fermentative strains due to the higher LA yield and easier downstream processing [38]. The temperature range for optimal growth of mesophilic LAB is from 28 to 45 °C and that of thermophilic LAB is from 45 to 62 °C [37]. In addition, LAB is strongly inhibited at pH 5 and ceases at pH values below 4.5 [37]. So, fermentation temperature and pH affects LA concentration, yield and productivity. In most studies, LA productivity have been conducted at temperatures ranging from 30 to 43 °C [39] while the optimal pH for LA production varies between 5.0 and 7.0 [35].

Other than temperature and pH, some parameters may affect the fermentation efficiency such as fermentation operation, nitrogen and vitamins sources, and by-products formation.

The fermentation is mainly a batch process and takes around 2–4 days to complete, providing a lactate yield of up to 90% with dextrose. However, fed-batch, repeated batch, and continuous systems are also reported [19].

Nitrogen sources and vitamins are important primarily because of the limited ability of LAB to synthesize B vitamins [20]. The main sources of nitrogen are yeast extract and peptone. Yeast extract is rich in vitamin B and is known to enhance LA production. An economic analysis of LA production at the industrial scale shows that yeast extract accounts for 38% of the medium cost [38]. Inexpensive nitrogen sources supplements have been studied to replace yeast extract and peptone, such as flour of pigeon pea, red lentil gram, black gram, bengal gram, green gram, soya bean, baker's yeast, and corn steep liquor [40, 41].

Production of other organic acids (e.g., acetic acid and formic acid), carbon dioxide, ethanol, during LA fermentation depends on the purity and quality of the inoculum, metabolic route used, and process conditions, which should prevent external contamination [20]. For efficient industrial production of LA, by-product formation should be avoided, or kept to a minimum [35].

Nowadays, the USA and Brazil are the world's largest ethanol producers. Together, both countries account for more than 94 billion liters of ethanol produced per year, accounting for around 85% of worldwide with huge differences in the fermentation processes [42].

There are three processes that are commonly used in ethanol production which are separate hydrolysis and fermentation (SHF), simultaneous saccharification and fermentation (SSF) and simultaneous saccharification and co-fermentation (SSCF) [43].

Microorganisms such as yeasts play an essential role in ethanol production by fermenting a wide range of sugars to ethanol. They are used in industrial plants due to valuable properties in ethanol yield (>90.0% theoretical yield), ethanol tolerance (>40.0 g/L), ethanol productivity (>1.0 g/L/h), growth in simple, inexpensive media and undiluted fermentation broth with resistance to inhibitors and retard contaminants from growth condition [44].

Temperature, pH, oxygen, initial sugar concentrations, organic acids, dissolved solids, and immobilization of the yeast are greatly essential parameters that influence the specific rate of yeast growth and ethanol production. Medium conditions direct the viability of yeasts, specific rate of fermentation, and nutrient uptake [45].

6.4 Potential Raw Materials for Biotechnology Production

6.4.1 Potential Raw Materials for LA Production

Raw material cost is one of the major factors in the economic production of LA. Since substrate cost cannot be reduced by process scale-up, extensive studies currently are underway to search for novel substrates for LA production [7].

Starchy materials, such as wheat, corn, maize, cassava, potato, rice, rye, barley, and others [20, 46–48] are typical substrates for LA production. Among them, cassava starch is the most commonly used substrate [38]. Although starchy materials can avoid glucose repression [49], they are unfavourable because of the high price, their alternative applicability in the food industry [50], and cannot be directly utilized by most microbes for metabolism [38].

Lignocellulose biomass is one of the most abundant renewable feedstock and has recently attracted substantial interest for biofuel and biochemical production [51] due to its advantages of inexpensiveness, abundance, and renewable resource [52]. In addition, lignocellulose does not compete with other food sources [51]. Several studies have recently reported LA production using lignocellulose biomass including corn stover [51–55], sugarcane bagasse [56, 57], rice straw [58], wood [58], paper sludge [59], sweet sorghum [60], and others.

Lignocellulose is a complex structure which consists primarily of cellulose (~35–45%), hemicellulose (~20–25%), and lignin (~15–20%). Pretreatment and hydrolysis of this complex lignocellulosic structure is necessary for high LA production by fermentation. Pretreatments are employed to remove lignin, separate cellulose and hemicellulose, increase the accessible surface area, partially depolymerize cellulose, and increase the porosity of the materials to aid in the subsequent access of the hydrolytic enzymes [39].

Traditional lignocellulosic biomass pretreatments include physical pretreatments (size reduction), physicochemical pretreatments (liquid hot water, steam explosion, and ammonia fiber explosion), chemical pretreatments (acid, alkaline, alkaline/oxidative, wet oxidation, and ozonolysis), and biological pretreatments [51]. Such pretreatment processes could further lead to production of inhibitors (sugar degradation products). These inhibitors are weak acids, furans and phenolic compounds which are considered as potential fermentation inhibitors for LA producers [55]. Acetic acid is formed by the deacetylation of hemicellulose. Furfural and

5-hydroxymethylfurfural (HMF) (furans) are from pentose and hexose dehydration, respectively. Subsequent degradation of furfural and HMF generates formic acid and levulinic acid, respectively. Various phenolic compounds are formed from lignin breakdown [61].

Besides the production of inhibitors, if pretreatment is not sufficiently efficient, the resultant residue is not easily saccharified by hydrolytic enzymes [39]. In addition, pretreatment can influence the downstream costs.

After pretreatment, enzymatic hydrolysis are used to depolymerizing lignocellulose to fermentative sugars, such as glucose and xylose, by means of hydrolytic enzymes [39]. A multi-enzyme mixture increases the efficiency of the hydrolysis process, but this represents more cost.

Although the lignocellulosic biomass has many drawbacks to be overcome, numerous investigations with the objective of making the LA production process more economical have been developed. Pretreatment of biomass using enzymatic routes instead chemical routes is advantageous because is conducted under milder conditions and does not generate degradation products [38]. Another pretreatment is the wet explosion, which is a combination of wet oxidation with oxygen and steam explosion. It a method developed to pretreat any lignocellulosic biomass material demanding no addition of harmful chemicals and resulting in minimal production of sugar degradation products [55]. Other approach is to remove or neutralize the inhibitors produced by pretreatment. Many studies have been devoted to the development of different detoxification processes [62] or strains resistant to these inhibitors [58]. Solid acid, which is safe powder acid composed mainly by sulfamic acid, sodium chloride, and metal oxide, can replace conventional acids in acid pickling [54].

Food-waste is also a potential material for LA production, because it has high carbohydrate content and effective method of environmental waste management [9]. Several attempts have been made to produce LA from food waste [63–67]. Mixtures of acids are often produced from food-waste, for example, acetic and butyric acid, which have to be separated before utilization of LA [63].

Glycerol is a very versatile raw material for producing various chemicals, polymers, and fuels [68]. It is a by-product in a biodiesel production process that discharges 1 kg crude glycerol for every 10 kg of biodiesel produced [69]. So, it is available in abundant quantities and is inedible [69]. The conversion of glycerol to LA can be categorized into hydrothermal and heterogeneous catalysis methods [20]. Recently, LA production by glycerol have been studied by many researches [68–71].

Recently, the manufacturing of cheese has been reported to produce large volumes of whey as a byproduct [72] and its disposal is currently a major pollution problem for the dairy industry [7]. Whey is a rich source of lactose, nitrogenous substances including vitamins and other essential nutrients for the growth of certain bacteria. Availability of the lactose and other nutrients make whey a potent raw material for the production of LA [73] as reported by published works [73, 74].

Another approach for production of LA is from algal biomass, which is rich in carbohydrates and proteins. Algal biomass does not contain lignin, which simplifies its conversion into fermentable sugars [7]. In addition, microalgae grow almost anywhere, have an extremely short harvesting cycle of approximately 1 to 10 days, and have high fermentable sugar contents [7, 20]. Prior to LA fermentation, the microalgae biomass needs to be processed in order to extract lipid and release sugar using methods such as chloroform-methanol, chloroform-methanol-water, and hexane [75].

Sugarcane molasses is an agro-industrial by-product generated from sugar industry, which can be converted to LA by the use of microorganisms [76]. It contains 40–60% of sucrose available to fermentation and other nutrients. Moreover, sugar manufacturing process of cane molasses will inevitably generate some hazardous substances such as 5-hydroxymethylfurfural and bring in excessive metallic ions, all of which are toxic to cells [77].

The potential raw materials used for LA production are summarized in Table 6.2.

6.4.2 Potential Raw Materials for Bioethanol Production

Bioethanol is considered one of the most promising renewable fuels in the world as it can replace fossil fuels. It is most commonly produced by microbial fermentations (most often yeast) catalyzed using plant biomass as feedstock. Starch raw materials (i.e., corn, wheat, and sorghum) are still the most common raw materials for the production of ethanol fuel in temperate regions of the world (Europe, North America, and Central Asia). However, their use as fuel production resources can affect prices and compete with the food products manufactured from them. The use of non-edible parts of plants (straw and stalks) or lignocellulosic biomass as the raw material at the distillery is now considered the most promising opportunity for ethanol production that does not affect food prices [78].

A wide variety of biomasses can be used to produce bioethanol because they have a high sugar content or materials that can be converted to sugars such as starch or cellulose. The most commonly used today are based on

Table 6.2 LA production using different substrates and microorganism.

Substrate	Microorganism	Fermentation mode	LA			References
			Concentration (g/L)	Yield (g/g)	Productivity (g/L/h)	
Corn stover hydrolysate	*Bacillus coagulans* AD	Continuous	–	0.95	3.69	[55]
Corn stover	*Lactobacillus pentosus FL0421*	Fed-batch (SSF)	92.30	0.66	1.92	[51]
Wood hydrolysate	*Lactobacillus paracasei 7BL*	Fed-batch	99	0.96	2.25–3.23	[58]
Corn stover hydrolysate	*Bacillus coagulans NBRC 12714*	Continuous	92	0.91	13.8	[52]
Food waste	*Streptococcus sp.*	Batch (SSF)	–	0.81	2.16	[64]
Food waste	*indigenous microbiota*	Batch	–	0.46	0.2781	[65]
Glycerol	*Enterococcus faecalis*	Fed-batch	55.3	0.991	0.772	[69]
Hydrolysed cheese whey	*L. bulgaricus*	Batch	31.7	0.645	0.660	[73]

SSF: simultaneous saccharification and fermentation

sucrose, among these raw materials times sugarcane, sugar beet, sweet sorghum, and also starch raw materials such as corn, cassava, wheat, rye, etc. being corn, sugar beet, and sugar cane among the most productive crops in the world, either in terms of ethanol production or in terms of productivity per area. Corn is the most ubiquitous crop with the third largest extension in the world [79, 80].

One of the most well-established raw materials for the production of bioethanol is sugarcane, which fits into biorefinery concepts involving multi-production facilities, which include the production of fuel (ethanol), food (sugar), energy (electricity), and other useful by-products. By-products include sugarcane bagasse used as fuel after pre-treatment and hydrolysis, the bagasse can also be used as feed for cattle and as fertilizer. First-generation (1G) sugarcane biorefineries are common in Brazil also in other countries such as India and Thailand. In these facilities, ethanol can be produced by fermentation of sucrose-rich sugarcane juice in "autonomous" distilleries, which produce only ethanol from sugarcane juice or by fermentation of molasses in attached plants, which produce sugar and ethanol from sugarcane [81].

Another raw material approach for the production of bioethanol is corn, a grain plant of the C4 group of the Poaceae family (gram family) and the genus Zea, originating in Mexico, but widespread in all continents. Corn crops currently occupy about 147 million hectares worldwide and are widely used as human or animal food because of their nutritional properties. Based on the world's largest corn ethanol producer in the US, the harvesting process is performed mechanically, where the corn ear is separated from the culm so that the grains are extracted while the tang with the straw is left in the fields to improve the soil fertility. The grains are transported to the corn mills by truck or rail and can be stored in silos before processing. There are two processes of ethanol production from maize: dry milling and wet milling. The initial treatment of the grain is that it will defer processes, however, both involve prior hydrolysis to break down the starch chains (α-glucose polymer) to obtain glucose syrup (d-glucose isomer), which can be converted into bioethanol by yeast according to Eq. (6.1) below [80].

$$(C_6H_{10}O_5)\ n + nH_2O \rightarrow n\ C_6H_{12}O_6 \qquad \text{(Eq. 6.1)}$$

In the past, most of the raw materials used to produce cellulosic by-products were from wood species. Non-timber resources were consolidating somewhat further in mid-2013 due to the crucial need to replace timber, as this led to economic and environmental problems, which led to a shift from timber sources to non-timber sources as a source pulp fiber in some part of the world. These materials are the lignocellulosic, already discussed above: as bagasse, wheat straw, etc. Lignocellulosic biomass is not only promising for LA production but also for the production of second-generation ethanol [82].

Second-generation ethanol is obtained through the fermentation of lignocellulosic materials, the most classic example being sugarcane bagasse, which has a composition of 50% cellulose, 25% hemicelluloses, and 25% lignin, where such compounds can be hydrolyzed after pretreatment of the recalcitrant biomass. In addition to sugar production, inhibitory compounds such as furfural and hydroxymethylfurfural (HMF), the degradation of pentoses and hexoses, respectively, and acetic acid from the hydrolysis of the acetyl radical present in the hemicelluloses are formed in the pre-treatment phase. Thus, the use of yeasts such as Saccharomyces cerevisiae may show reduced yields with respect to the fermentation of ethanol due to the presence of inhibitors such as acetic acid, so the pre-treatment step should not be too severe [83].

The production of ethanol from lignocellulosic feedstocks can generate multiple streams since only a part of the material is converted to sugars and subsequently fermented into ethanol. This requires a "polygeneration" approach, in which by-products must also be of high value (e.g., lignin and hemicellulose sugars). To achieve the large scale required for profitability, it is proposed that the best way is to integrate the new processes with existing industries, preferably those that already operate biomass plants for materials or biomass for fuels. One of the largest agencies in the industry in this regard is the pulp and paper industry. The production of second-generation ethanol (or other products) through sugars of lignocellulosic materials includes a relatively expensive pretreatment of the feedstock to separate the lignin from the cellulose according to [84].

Currently, lignocellulosic ethanol costs approximately US $1.00/lge on the pilot scale, assuming a base price of US $3.60/GJ for biomass supply. The cost is expected to decline by half over the next ten years with the development of the method, co-production of other by-products, low-cost raw materials, and even plant expansion [82, 85].

Coffee cut-stems (CCS) is a promising raw material for the production of fuel ethanol in tropical countries due to its high availability and high biomass yield per hectare. Triana *et al.* [86] evaluated the pre-treatment of this agricultural residue with diluted sulfuric acid and liquid hot water (LHW) and an integration and simulation, also evaluating the economic part of the second-generation ethanol production process. The high concentration of reducing sugars and ethanol yields were obtained with LHW pretreatment with high-energy costs. Acid pretreatment is still one of the most applied technologies for lignocellulosic materials because of its efficiency and lower energy consumption. For the determination of the environmental impact index of the process pollution with and without cogeneration system, the waste reduction algorithm (WAR) was used. Thus, the high lignin and low water contents of coffee stalks meant a high potential for cogeneration of energy. Coffee cut-stems are an interesting material for the production of bioethanol.

Soy is a plant that has shown potential for a biorefinery. Besides the use of conventional products from its seeds, stem, and foliage, soy can also be used as raw material for the production of different products, mainly for the production of ethanol. The soybean plant contains substantial amounts of biopolymers such as carbohydrates, proteins, lipids, and lignin. So, it can be considered as a unique feedstock that could be used in a complete biorefinery for a variety of biochemicals, biomaterials, and valuable biofuels. This approach may increase the benefit margin of currently available soybean processing plants [82].

One of the most promising food wastes that can be processed into ethanol is bread residue. These bread residues contain a significant amount of starch which can be easily hydrolyzed into monomeric sugars with the aid of amylases, so the amount of starch and simple sugars in the bread loaves are 500–750 and 3–50 g kg^{-1}, respectively. In addition, bread contains interesting nutrients such as protein that, after hydrolysis releases peptides and amino acids that may be essential for the growth of the fermentative microorganism. Waste bread is also highly accessible raw material for the processing of ethanol. The estimated waste for bakery products can range from 7 to 10% of their total output. The main factor for the formation of bread residues is that part of the product produced remains in the stores because it is not sold and then returned. Due to the significant level of storage and large amount of assortment available from bakery products that are overproduced to meet the demands of consumers. There are few possibilities of reprocessing bread residues in bakeries. Some waste can be processed into bread crumbs, such as a replacement of flour in the preparation of needles or as feed. However, due to microbial deterioration, its use for human and animal nutrition can be hazardous to the health of consumers. These problems are the reason why the loaves are most often left in landfills or used as fuel for combustion. Residual bread is a high yield material for ethanol fermentation [78].

The typical pretreatment method for the enzymatic hydrolysis of starches to fermentable sugars is based on a two step method. In the first step, the starch is liquefied by thermostabilized α-amylase (EC 3.2.1.1) to reduce the viscosity of the gelatinized starch solution and to produce short chain dextrins by breaking the α-1,4-glycosidic linkages in the chains of amylose and amylopectin. During the second step of the hydrolysis of the starch (saccharification), the dextrins are saccharified by glucoamylase (amyloglucosidase, EC 3.2.1.3) to obtain monomeric sugars (glucose). Many support enzymes, such as proteases, cellulases, pullulnases, and others, are used to increase the amount of fermentable sugars, decrease the viscosity of the mass, and produce free amino nitrogen which is used as a nutrient for yeast [78].

Most of first and second-generation energy crops can affect the food chain as well as having to use more arable land for cultivation. These issues have led to the failure of commercial bioethanol production. For example, due to the increased demand for palm for food and biodiesel production, more land for the cultivation of oil palm plantations in Malaysia has led to rapid deforestation and destruction of the habitat of many animals and birds [87, 88]. Hence, researchers are now focusing on creating new potential alternative food sources for the large-scale production of biofuels without disrupting the environment.

Algae are considered as the only alternative to food crops for the production of renewable fuel because they contain lipids and energy-rich carbohydrates. In addition, some cell walls of the microalgae are composed of cellulose, mannans, xylans, and sulfated glycans. These polysaccharides can be chemically or enzymatically broken into simple sugars and then converted into ethanol. Algae are the fastest growing plants in the world and are generally divided into macroalgae and microalgae based on morphology. Macroalgae or "sea algae" are larger, multicellular oceanic plants that grow up to 60 m in length. Microalgae are microscopic, mostly existing as small cells of about 2–200 μm and inhabitants of fresh, marine wastewater systems. The algae are able to efficiently convert solar energy into biomass by photosynthetic process, using sunlight, CO_2, and water nutrients [88].

Dinoflagellates, green algae (*chlorophytes*), golden algae (*chryosophyceae*), and diatoms (*bacillariophyceae*) are several types of microalgae that have protein, carbohydrate, and lipid contents that can vary in different species. Most microalgae can store highly concentrated lipids that can exceed 70% by weight of the dry biomass. And the carbohydrate content in some species can correspond in up to 50% of dry weight, as example we have: *Scenedesmus*, *Chlorella*, and *Chlamydomona*. Some factors such as light, temperature, nutrient content, pH, O_2, and CO_2, salinity and toxic chemical parameters may influence the composition of microalgae. The

Table 6.3 Ethanol production from various feedstocks.

Source	Ethanol yield (gal/acre)	Ethanol yield (L/ha)
Corn stover	112–150	1050–1400
Wheat	277	2590
Cassava	354	3310
Sweet sorghum	326–435	3050–4070
Corn	370–430	3460–4020
Sugar beet	536–714	5010–6680
Sugarcane	662–802	6190–7500
Switch grass	1150	10,760
Microalgae	5000–15,000.	46,760–1,402,900

Adapted from [90]

most common components in the cell wall of microalgae are cellulose, protein, lignin, pectin, hemicelluloses, and other carbohydrates that can be converted into monomers through acid or enzymatic hydrolysis to produce third-generation bioethanol [89].

Table 6.3 shows the ethanol production from different feedstocks.

6.5 Challenges in LA and Ethanol Production

Several limitation and drawbacks of LA and Bioethanol production have been reported. This includes the fermentative LA and Bioethanol production and the downstream processes. To fermentative LA and Bioethanol production, the challenges are as follows.

- Utilization of cheap raw materials for LA and Bioethanol production that non-compete with food resources;
- Direct LA and Bioethanol production from cellulose or xylan by LA producers (wild strain or genetically engineered);
- Pretreatment steps and enzymatic hydrolysis for saccharification optimized for the production of second generation ethanol and operating at high solids loading;

- Modification of pre-treatment strategies, isolation of new inhibitor-tolerant strains, and modification of existing LA and Bioethanol production strains for inhibitor-tolerance/detoxification for LA and Bioethanol production from lignocellulosic biomass [9];
- Isolation and development of powerful strains with high substrate tolerance;
- Reduce the generation of by-product, such as ethanol, acetic acid, and carbon dioxide, during LA fermentation;
- In the case of Bioethanol, reduce the generation of by-product, such as LA, acetic acid, carbon dioxide, and glycerol during ethanol fermentation;
- Development of strains able to utilize xylose or mixed sugars in lignocellulosic biomass;
- LA fermentation at acidic pH (at or below the pKa of LA, 3.78) without the use of neutralizing agents [9];
- Production of optically pure LA.

Abdel-Rahman and Sonomoto [9] discussed in their review the many efforts that have been done by researches to overcome these LA fermentation challenges.

Another technology barrier in cost-effective production of high-purity LA is its downstream processing (DSP) [91]. To DSP, and it can be applied as well also in the ethanol processing [92] the challenges are as follows.

- Production of LA and Bioethanol with high levels of purification;
- Production of LA and Bioethanol with high levels of efficiency;
- Production of LA and Bioethanol with high levels of yields;
- More efficient and cheaper pre-treatment steps for biomass deconstruction for better enzymatic hydrolysis efficiency for saccharification and subsequent bioethanol fermentation [92];
- Reduced number of separation units;
- DSP with simple operation;
- No generation of waste (gypsum);
- Lower energy consumption;
- Great flexibility in scale of production;
- Reduction of LA thermal decomposition;
- No use of hazardous solvents.

The literature reported many LA separation and purification technologies, such as liquid–liquid extraction, membrane process, molecular distillation, reactive distillation, and others, but there are some drawbacks that limit the application of these technologies at industrial level [91]. However, it is necessary to develop more efficient and viable separation technologies to bring out the potential of LA [91].

6.6 Integrated Ethanol and LA Production

A biorefinery is a facility that integrates biomass conversion processes and equipment to produce fuels, power, materials and/or chemicals from biomass [93]. The product streams include ethanol, bio-diesel, hydrocarbon-like oils, agricultural chemicals, food ingredients, and bio-commodities that are identical or functionally similar to those produced by the petrochemical industry [94]. By producing multiple products, a biorefinery can take advantage of the differences in biomass components and intermediates and maximise the value derived from the biomass feedstock, and optimise the cost effectiveness of its products [93].

The integration of different biorefinery concepts into existing bioethanol and sugar plants represents economic benefit. Co-production of LA and ethanol in biorefineries was investigated in some works [5, 93, 95–97).

The European project BIOREF-INTEG investigated the feasibility and economic impact of production of LA from sugars obtained from the main process of ethanol production. After liquefaction, part of the stream which contains mainly C6-sugars was sent to the LA process and the rest of the stream was directed to the sugars fermentation unit to produce ethanol [93]. The block diagram of this process is shown in Figure 6.4. The techno–economic assessment showed that integration of LA production into an existing grain-to-ethanol leads to much lower production costs of main product bioethanol in comparison with the conventional grain-to-ethanol process via fermentation of sugars, which makes this concept very interesting from the economic point of view [93].

Daful *et al.* [5] studied the LA production by annexing a biorefinery to an existing sugar industry using lignocellulosic residues, sugarcane bagasse, and leaves. This studied was carried out using the Aspen Plus® process simulator. Economic analysis and environmental impact assessment was conducted in SimaPro®. Six process scenarios for LA production from both the hemicellulose liquid fraction and cellu-lignin solid fractions of leaves after steam explosion pretreatment were studied. It is found that production of LA from cellulose fraction instead of the hemicellulose fraction leads to

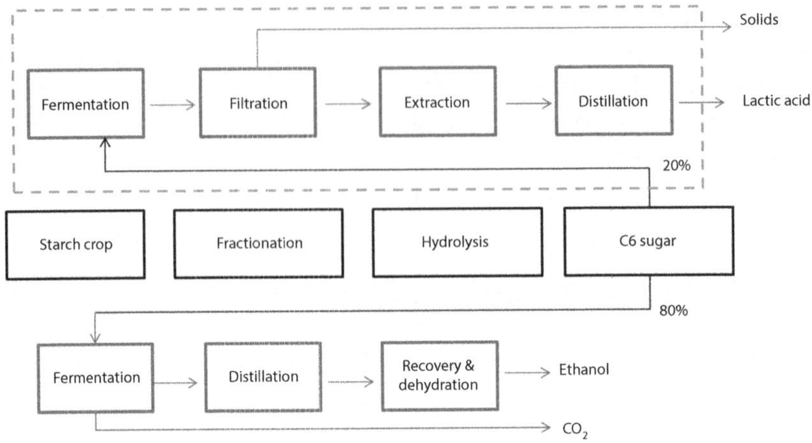

Figure 6.4 Integrated ethanol and LA production [93].

7–10% increment in total capital investment, 58–86% increment in operating cost, and 12–18% increment in revenue [5].

Daful *et al.* [93] evaluated the environmental performance of LA produced from lignocellulosic biomass and petrochemical sources using a life cycle approach. The results showed that biobased LA production has significantly reduced the impact on the environment, giving 80–99% environmental savings compared to fossil-derived LA.

Parajuli *et al.* [96] compared three biorefinery systems: a standalone system producing bioethanol from winter wheat-straw, a standalone system producing biobased LA from alfalfa, and an integrated biorefinery system combining the two standalone systems and producing both bioethanol and LA. The studies highlights the benefits of the system integration for bioethanol and biobased LA productions were in terms of higher net savings of greenhouse gas emissions, non-renewable energy use, and eutrophication potential compared to the standalone systems.

Farzad *et al.* [97] have been investigated six potential biochemical/thermochemical pathways for product diversification in the sugarcane industry by means of biorefineries (bioethanol, bioethanol and LA, bioethanol and furfural, butanol, methanol and Fischer–Tropsch synthesis). The study showed that production of ethanol, LA and electricity pathway, and methanol and electricity pathway had the highest profitability.

Among the materials covered, sugarcane is currently one of the most efficient crops for the production of first-generation bioethanol (sugarcane juice) and second (sugar-cane bagasse and straw) generation. Lignocellulosic (bagasse and straw) materials were formerly sold or used for electricity generation in Brazil, but as discussed in previous topics, the lignocellulosic material is excellent for the production of second generation bioethanol; thus, some commercial units of lignocellulosic bioethanol production are starting to operate. The efficient processes of conversion of lignocellulosic materials to bioethanol, however, remain to be developed, with the pretreatment stage being one of the major bottlenecks in this process. Second-generation ethanol production may be more competitive when considering its integration into a first-generation distillery [98, 99].

As sugarcane is currently used as a fuel for steam and electricity production, second-generation ethanol production from this material is able to compete with its current use as fuel, so that the configuration of the conventional production process can be adapted to include the production of second generation bioethanol, ensuring self-sufficiency in energy and steam production [99].

Therefore, the unitary operations of the ethanol production process of the first generation of sugarcane should be redesigned to increase the feasibility of integrating a second generation ethanol production from a simplified scheme of the integrated ethanol production process of the first and second generation of sugarcane is shown in Figure 6.5 adapted from Dias *et al.* [99].

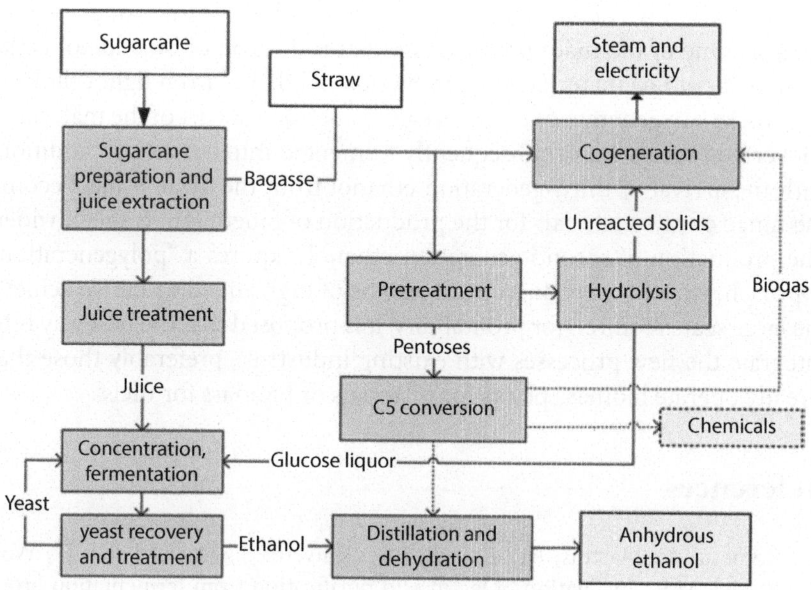

Figure 6.5 Scheme of the integrated first (1G) and second (2G) ethanol production process from sugarcane and sugarcane bagasse [99].

A review of literature suggests that very limited research has been conducted on the combined LA and ethanol production. But, in the near future, the biorefinery concept of complete utilisation of sugarcane biomass will become a pivotal element for a sustainable sugarcane industry [94]. The industry that produces liquid fuels, electricity, and commodity chemicals from a renewable source will contribute with environmental and economic aspects.

6.7 Concluding Remarks

LA and ethanol are promising bio-based chemicals with an exponential grow over the last years because of their widespread use and its applications. LA has many applications, such as chemistry, food, cosmetics, and pharmaceutical products and more recently in the medical area. Production of LA by fermentation is an important technological biomass-based platform, which has attracted much attention from the researches. However, LA production is still limited by the downstream process that requires many steps and makes the process expensive. To address this problem, efficient and viable separation of technologies is needed. Bioethanol is the most widely used biofuel in transport around the world and can be produced from different types of feedstock. These raw materials are classified into three agricultural categories: simple sugars, starch, and lignocellulose. The price of the raw material is highly volatile, which directly affects the costs of producing bioethanol. One of the main problems in the production of bioethanol is the availability of raw material. The production of ethanol from lignocellulosic feedstocks can generate multiple streams, since only a part of the material is converted to sugars and subsequently fermented into ethanol; in addition, with the arrival of third-generation ethanol from the algae it has become the range of raw materials for the production of bioethanol is much wider. The production of second generation ethanol requires a "polygeneration" approach, where by-products must also be of high added value. To achieve the large scale required for profitability, it is proposed that the best way is to integrate the new processes with existing industries, preferably those that already operate biomass plants for materials or biomass for fuels.

References

1. Komesu, A., Martins, P.F., Lunelli, B.H., Oliveira, J., Maciel Filho, R., Wolf Maciel, M.R., Evaluation of lactic acid purification from fermentation broth by hybrid short path evaporation using factorial experimental design. *Sep. Purif. Technol.*, 136, 233–240, 2014.

2. Datta, R., Henry, M., Lactic acid: recent advances in products, processes and technologies-a review. *J. Chem. Technol. Biotechnol.*, 81, 1119–1129, 2006.
3. Biddy, M.J., Scarlata, C., Kinchin, C., Chemicals from Biomass: A market assessment of bioproducts with near-term potential; NREL/TP-5100-65509; NREL: Oak Ridge, TN, 2016.
4. Jong, E., Higson, A., Walsh, P., Wellisch, M., Bio-based chemicals value added products from biorefineries; IEA Bioenergy-Task 42 Biorefinery, 2012.
5. Daful, A.G., Görgens, J.F., Techno-economic analysis and environmental impact assessment of lignocellulosic lactic acid production. *Chem. Eng. Sci.*, 162, 53–65, 2017.
6. Ouyang, J., Ma, R., Zheng, Z., Cai, C., Zhang, M., Jiang, T., Open fermentative productions of L-lactic acid by Bacillus sp. strain NL01 using lignocellulosic hydrolyzates as low-cost raw material. *Bioresource Technol.*, 135, 475–480, 2013.
7. Abdel-Rahman, M.A., Tashiro, Y., Sonomoto, K., Recent advances in lactic acid production by microbial fermentation processes. *Biotechnol. Adv.*, 31, 877–902, 2013.
8. Papong, S., Malakul, P., Trungkavashirakun, P., Wenunun, P., Chom-in, T., Nithitanakul, M., Sarobol, E., Comparative assessment of the environmental profile of PLA and PET drinking water bottles from a life cycle perspective. *J. Clean. Prod.*, 65, 539–550, 2014.
9. Abdel-Rahman, M.A., Sonomoto, K., Opportunities to overcome the current limitations and challenges for efficient microbial production of optically pure lactic acid. *J. Biotechnol.*, 236, 176–192, 2016.
10. Demirbas, A., Producing and using bioethanol as an automotive fuel. *Energy Source B*, 2, 391–401, 2007.
11. Balat, M., Bioethanol as a vehicular fuel: a critical review. *Energy Source A*, 31, 1242–1255, 2009.
12. Çelik, M.B., Experimental determination of suitable ethanol–gasoline blend rate at high compression ratio for gasoline engine. *Appl. Therm. Eng.*, 28, 396–404, 2008.
13. Pejin, D., Mojović, L.C., Vučurović, V., Pejin, J., Denčić, S., Rakin, M., Fermentation of wheat and triticale hydrolysates: a comparative study. *Fuel*, 88, 1625–1628, 2009.
14. Balat, M., Production of bioethanol from lignocellulosic materials via the biochemical pathway: a review. *Energy Convers. Manage.*, 52(2), 858–875, 2011.
15. Ren, J., *Biodegradable poly (lactic acid) synthesis, modification, processing and applications, 1st ed.* Springer: New York, 2010.
16. Lasprilla, A.J.R., Martinez, G.A.R., Lunelli, B.H., Jardini, A.L., Maciel Filho, R., Poly-lactic acid synthesis for application in biomedical devices - a review. *Biotecnol. Adv.*, 30, 321–328, 2012.
17. Grabar, T.B., Zhou, S., Shanmugam, K.T., Yomano, L.P., Ingram, L.O., Methylglyoxal bypass identified as source of chiral contamination in L(+) and D(–)-lactate fermentations by recombinant Escherichia coli. *Biotechnol. Lett.*, 28, 1527–1535, 2006.

18. Larrañaga, M.D., Lewis, R.J., Lewis, R.A., *Hawley's Condensed Chemical Dictionary*, 16th ed. John Wiley & Sons: New Jersey, 2016.
19. Dusselier, M., Wouwe, P.V., Dewaele, A., Makshina, E., Sels, B.F., Lactic acid as a platform chemical in the biobased economy: the role of chemocatalysis. *Energy Environ. Sci.*, 6, 1415–1442, 2013.
20. Komesu, A., Oliveira, J.A.R., Martins, L.H.S., Wolf Maciel, M.R., Maciel Filho, R., Lactic acid production to purification: a review. *Bioresources*, 12 (2), 4364–4383, 2017.
21. Wee, Y.-J., Kim, J.-N., Ryu, H.-W., Biotechnological production of lactic acid and its recent applications. *Food Technol. Biotechnol.*, 44 (2) 163–172, 2006.
22. PR Newswire., Lactic acid market and derivatives 2016 forecasts (global, China) in new research report, 2016. http://www.prnewswire.com/news-releases/lactic-acid-market-and-derivatives-2016-forecasts-global-china-in-new-research-report-279286152.html (accessed 25 February 2017).
23. Azhar, S.H.M., Abdulla, R., Jambo, S. A., Marbawi, H., Gansau, J. A., Mohd Faik, A. A., Rodrigues K. F. Yeasts in sustainable bioethanol production: A review. *Biochemistry and Biophysics Reports*, 10, 52–61, 2017.
24. Tesfaw, A., Assefa, F. Current Trends in Bioethanol Production by Saccharomyces cerevisiae: Substrate, Inhibitor Reduction, Growth Variables, Coculture, and Immobilization. *International Scholarly Research Notices*, 2014, 2014, 532852.
25. Renewable Fuels Association US, Global Ethanol Production, 2015. http://www.afdc.energy.gov/data/10331 (accessed 2 March 2017).
26. Hidzir, N.S., Som, A.S., Abdullah, Z., Ethanol production via direct hydration of ethylene: a review. In: International conference on global sustainability and chemical engineering (ICGSE), 2014. Available from: https://www.researchgate.net/publication/277957681_Ethanol_Production_via_Direct_Hydration_of_Ethylene_A_review.
27. Radecka, D., Mukherjee, V., Mateo, R.Q., Stojiljkovic, M., Foulquie-Moreno, M.R., Thevelein, J.M., Looking beyond Saccharomyces: the potential of non-conventional yeast species for desirable traits in bioethanol fermentation. *FEMS Yeast Res.*, 15, 2015. http://dx.doi.org/10.1093/femsyr/fov053
28. Spivey, J.J., Dooley, K.M., Catalysis. *R. Soc. Chem.*, 68, 2010.
29. Madaleno, L.L., Minari, G.D., Annunzio, F.R., Carvalho, M.R., Bossa, Jr., G.R., Sales, D.C., Frigieri, M.C., Use of antimicrobials for contamination control during ethanolic fermentation. *Científica.*, 44, 226–234, 2016.
30. Parkash, A., Modeling of ethanol production from molasses: a review. *Ind. Chem.*, 3, 108, 2015.
31. Nigam, P.S., Singh, A., Production of liquid biofuels from renewable resources. *Prog. Energy Combust. Sci.*, 37, 52–68, 2011.
32. Gutierrez-Rivera, B., Waliszewski-Kubiak, K., Carvajal-Zarrabal, O., Aguilar-Uscanga, M.G., Conversion efficiency of glucose/xylose mixtures for ethanol production using Saccharomyces cerevisiae ITV01 and Pichia stipitis NRRL Y-7124. *J. Chem. Technol. Biotechnol.*, 87, 263–270, 2012.

33. Ishola, M.M., Taherzadeh, M.J., Effect of fungal and phosphoric acid pretreatment on ethanol production from oil palm empty fruit bunches (OPEFB). *Bioresource Technol.*, 165, 9–12, 2014.
34. Oliveira, J.A.R., Komesu, A., Martins, L.H.S., Maciel Filho, R., Evaluation of microstructure of açaí seeds biomass untreated and treated with H2SO4 and NaOH by SEM, RDX and FTIR. *Chem. Eng. Trans.*, 50, 379–384, 2016.
35. Hofvendahl, K., Hahn–Hägerdal, B., Factors affecting the fermentative lactic acid production from renewable resources. *Enzyme Microb. Technol.*, 26, 87–107, 2000.
36. Silveira, M.S., Utilização do Suco de Caju Clarificado para Produção de Ácido Lático pelo Lactobacillus casei B-442. MSc Dissertation, Universidade Federal do Ceará, Ceará, Brazil, 2009.
37. Vaidya, A.N., Pandey, R.A., Mudliar, S., Suresh Kumar, M., Chakrabarti, T., Devotta, S., Production and recovery of lactic acid for polylactide—an overview. *Crit. Rev. Env. Sci. Technol.*, 35, 429–467, 2005.
38. Juturu, V., Wu, J.C., Microbial production of lactic acid: the latest development. *Crit. Rev. Biotechnol.*, 36(6), 967–977, 2016.
39. Abdel-Rahman, M.A., Tashiro, Y., Sonomoto, K., Lactic acid production from lignocellulose-derived sugars using lactic acid bacteria: overview and limits. *J. Biotechnol.*, 156(4), 286–301, 2011.
40. Altaf, M., Naveena, B.J., Reddy, G., Screening of inexpensive nitrogen sources for production of L(+) lactic acid from starch by amylolytic Lactobacillus amylophilus GV6 in single step fermentation. *Food Technol. Biotechnol.*, 43, 235–239, 2005.
41. Altaf, M., Naveena, B.J., Reddy, G., Use of inexpensive nitrogen sources and starch for L(+) lactic acid production in anaerobic submerged fermentation. *Bioresource Technol.*, 98, 498–503, 2007.
42. Bertrand, E., Vandenberghe, L.P.S., Soccol, C.R., Sigoillot, J.C., Faulds, C., First generation bioethanol. In: Soccol, C.R., Brar, S.K., Faulds, C., Ramos, L.P., eds., *Green fuels technology*. Switzerland: Springer International Publishing, 175–212, 2016.
43. Canilha, L., Chandel, A.K., dos Santos Milessi, T.S., Antunes, F.A.F., Freitas, W.L.C., Felipe, M.G.A., Silva, S.S., Bioconversion of sugarcane biomass into ethanol: an overview about composition, pretreatment methods, detoxification of hydrolysates, enzymatic saccharification, and ethanol fermentation. *J. Biomed. Biotechnol.*, 1–15, 2012.
44. Dien, B.S., Cotta, M.A., Jeffries, T.W., Bacteria engineered for fuel ethanol production: current status. *Appl. Microbiol. Biotechnol.*, 63, 258–266, 2003.
45. Lin, Y., Zhang, W., Li, C., Sakakibara, K., Tanaka, S., Kong, H., Factors affecting ethanol fermentation using Saccharomyces cerevisiae BY4742. Biomass Bioenergy, 47, 395–401, 2012.
46. Tirpanalan, Ö., Reisinger, M., Smerilli, M., Huber, F., Neureiter, M., Kneifel, W., Novalin, S., Wheat bran biorefinery – an insight into the process chain for the production of lactic acid. *Bioresource Technol.*, 180, 242–249, 2015.

47. Gonzalez, K., Tebbani, S., Lopes, F., Thorigné, A., Givry, S., Dumur, D., Pareau, D., Regulation of lactic acid concentration in its bioproduction from wheat flour. *Control Eng. Pract.*, 54, 202–213, 2016.
48. Watanabe, M., Techapun, C., Kuntiya, A., Leksawasdi, N., Seesuriyachan, P., Chaiyaso, T., Takenaka, S., Maeda, I., Koyama, M., Nakamura, K. Extracellular protease derived from lactic acid bacteria stimulates the fermentative lactic acid production from the by-products of rice as a biomass refinery function. *Journal of Bioscience and Bioengineering*, 123(2), 245–251, 2017.
49. Nakano, S., Ugwu, C.U., Tokiwa, Y., Efficient production od D-(-)-lactic acid from broken rice by Lactobacillus delbrueckii using Ca(OH)2 as a neutralizing agent. *Bioresource Technol.*, 104, 791–794, 2012.
50. Reddy, L.V., Kim, Y.-M., Yun, J.-S., Ryu, H.-W., Wee, Y.-J., L-Lactic acid production by combined utilization of agricultural bioresources as renewable and economical substrates through batch and repeated-batch fermentation of Enterococcus faecalis RKY1. *Bioresource Technol.*, 209, 187–194, 2016.
51. Hu, J., Lin, Y., Zhang, Z., Xiang, T., Mei, Y., Zhao, S., Liang, Y., Peng, N., High-titer lactic acid production by Lactobacillus pentosus FL0421 from corn stover using fed-batch simultaneous saccharification and fermentation. *Bioresource Technol.*, 214, 74–80, 2016.
52. Ma, K., Hu, G., Pan, L., Wang, Z., Zhou, Y., Wang, Y., Ruan, Z., He, M., Highly efficient production of optically pure L-lactic acid from corn stover hydrolysate by thermophilic Bacillus coagulans. *Bioresource Technol.*, 219, 114–122, 2016.
53. Qiu, Z., Gao, Q., Bao, J., Constructing xylose-assimilating pathways in Pediococcus acidilactici for high titer D-lactic acid fermentation from corn stover feedstock. *Bioresource Technol.*, 2017. https://doi.org/10.1016/j.biortech.2017.05.128
54. Wang, X., Wang, G., Yu, X., Chen, H., Sun, Y., Chen, G., Pretreatment of corn stover by solid acid for D-lactic acid fermentation. *Bioresource Technol.*, 239, 490–495, 2017.
55. Ahring, B.K., Traverso, J.J., Murali, N., Srinivas, K., Continuous fermentation of clarified corn stover hydrolysate for the production of lactic acid at high yield and productivity. *Biochem. Eng. J.*, 109, 162–169, 2016.
56. Utrilla, J., Vargas-Tah, A., Trujillo-Martínez, B., Gosset, G., Martinez, A., Production of D-lactate from sugarcane bagasse and corn stover hydrolysates using metabolic engineered Escherichia coli strains. *Bioresource Technol.*, 220, 208–214, 2016.
57. Oonkhanond, B., Jonglertjunya, W., Srimarut, N., Bunpachart, P., Tantinukul, S., Nasongkla, N., Sakdaronnarong, C., Lactic acid production from sugarcane bagasse by an integrated system of lignocellulose fractionation, saccharification and fermentation, and ex-situ nanofiltration. *J. Environ. Chem. Eng.*, 2017. http://dx.doi.org/10.1016/j.jece.2017.05.004
58. Kuo, Y.-C., Yuan, S.-F., Wang, C.-A., Huang, Y.-J., Guo, G.-L., Wang, W.-S., Production of optically pure L-lactic acid from lignocellulosic

hydrolysate by using a newly isolated and D-lactate dehydrogenase gene-deficient Lactobacillus paracasei strain. *Bioresource Technol.*, 198, 651–657, 2015.

59. Marques, S., Matos, C.T., Gírio, F.M., Roseiro, J.C., Santos, J.A.L., Lactic acid production from recycled paper sludge: process intensification by running fed-batch into a membrane-recycle bioreactor. *Biochem. Eng. J.*, 120, 63–72, 2017.
60. Ou, M.S., Awasthil, D., Nieves, I., Wang, L., Erickson, J., Vermerris, W., Ingram, L.O., Shanmugam, K.T., Sweet sorghum juice and bagasse as feedstocks for the production of optically pure lactic acid by native and engineered Bacillus coagulans strains. *Bioenerg. Res.*, 9,123–131, 2016.
61. Zhang, L.Z., Li, X., Yong, Q., Yang, S.-T., Ouyang, J., Yu, S., Impacts of lignocellulose-derived inhibitors on L-lactic acid fermentation by Rhizopus oryzae. *Bioresource Technol.*, 203, 173–180, 2016.
62. Cavka, A., Jönsson, L.J., Detoxification of lignocellulosic hydrolysates using sodium borohydride. *Bioresource Technol.*, 136, 368–376, 2013.
63. Bonk, F., Bastidas-Oyanedel, J.-R., Yousef, A.F., Schmidt, J.E., Exploring the selective lactic acid production from food waste in uncontrolled pH mixed culture fermentations using different reactor configurations. *Bioresource Technol.*, 238, 416–424, 2017.
64. Pleissner, D., Demichelis, F., Mariano, S., Fiore, S., Gutierrez, I.M.N., Schneider, R., Venus, J., Direct production of lactic acid based on simultaneous saccharification and fermentation of mixed restaurant food waste. *J. Cleaner Prod.*, 143, 615–623, 2017.
65. Tang, J., Wang, X., Hu, Y., Zhang, Y., Li, Y. Lactic acid fermentation from food waste with indigenous microbiota: effects of pH, temperature and high OLR. *Waste Manage.*, 52, 278–285, 2016.
66. Tang, J., Wang, X.C., Hu, Y., Ngo, H.H., Li, Y., Dynamic membrane-assisted fermentation of food wastes for enhancing lactic acid production. *Bioresource Technol.*, 234, 40–47, 2017.
67. Kim, M.-S., Na, J.-G., Lee, M.-K., Ryu, H., Chang, Y.-K., Triolo, J.M., Yun, Y.-M., Kim, D.-H., More value from food waste: lactic acid and biogas recovery. *Water Res.*, 96, 208–216, 2016.
68. Moreira, A.B.F., Bruno, A.M., Souza, M.M.V.M., Manfro, R.L., Continuous production of lactic acid from glycerol in alkaline medium using supported copper catalysts. *Fuel Process. Technol.*, 144, 170–180, 2016.
69. Murakami, N., Oba, M., Iwamoto, M., Tashiro, Y., Noguchi, T., Onkohara, K., Abdel-Rahman, M.A., Zendo, T., Shimoda, M., Sakai, K., Sonomoto, K., L-Lactic acid production from glycerol coupled with acetic acid metabolism by Enterococcus faecalis without carbon loss. *J. Biosci. Bioeng.*, 121 (1), 89–95, 2016.
70. Arcanjo, M.R.A., Silva Jr., I.J., Rodríguez-Castellón, E., Infantes-Molina, A., Vieira, R.S., Conversion of glycerol into lactic acid using Pd or Pt supported on carbon as catalyst. *Catal. Today*, 279, 317–326, 2017.
71. Palacio, R., Torres, S., Lopez, D., Hernandez, D., Selective glycerol conversion to lactic acid on Co_3O_4/CeO_2 catalysts. *Catal. Today*, 2017. https://doi.org/10.1016/j.cattod.2017.05.053

72. Li, Y., Shahbazi, A., Lactic acid recovery from cheese whey fermentation broth using combined ultrafiltration and nanofiltration membranes. *Appl. Biochem. Biotech.*, 129–132, 985–996, 2006.
73. Juodeikiene, G., Zadeike, D., Bartkiene, E., Klupsaite, D., Application of acid tolerant Pedioccocus strains for increasing the sustainability of lactic acid production from cheese whey. *LWT – Food Sci. Technol.*, 72, 399–406, 2016.
74. Bernardo, M.P., Coelho, L.F., Sass, D.C., Contiero, J. l-(+)-Lactic acid production by Lactobacillus rhamnosus B103 from dairy industry waste. *Braz. J. Microbiol.*, 47, 640–646, 2016.
75. Talukder, M.M.R., Das, P., Wu, J.C., Microalgae (Nannochloropsis salina) biomass to lactic acid and lipid. *Biochem. Eng. J.*, 68, 109–113, 2012.
76. Dumbrepatil, A., Adsul, M., Chaudhari, S., Khire, J., Gokhale, D., Utilization of molasses sugar for lactic acid production by Lactobacillus delbrueckii subsp. delbrueckii mutant Uc-3 in batch fermentation. *Appl. Environ. Microbiol.*, 74, 333–335, 2008.
77. Xu, K., Xu, P., Efficient production of L-lactic acid using co-feeding strategy based on cane molasses/glucose carbon sources. *Bioresource Technol.*, 153, 23–29, 2014.
78. Pietrzak, W., Kawa-Rygielska, J., Ethanol fermentation of waste bread using granular starch hydrolyzing enzyme: effect of raw material pretreatment. *Fuel, 134*, 250–256, 2014.
79. Leff, B., Ramankutty, N., Foley, J.A., Geographic distribution of major crops across the world. *Global Biogeochem. Cycles*, 18(1), 2004.
80. Manochio, C., Andrade, B.R., Rodriguez, R.P., Moraes, B.S., Ethanol from biomass: a comparative overview. *Renewable Sustainable Energy Rev.*, 80, 743–755, 2017.
81. Mendes, F.M., Dias, M.O.S., Ferraz, A., Milagres, A.M.F., Santos, J.C., Bonomi, A., Techno-economic impacts of varied compositional profiles of sugarcane experimental hybrids on a biorefinery producing sugar, ethanol and electricity. *Chem. Eng. Res. Des.*, 125, 72–78, 2017.
82. Abdulkhani, A., Alizadeh, P., Hedjazi, S., Hamzeh, Y., Potential of soya as a raw material for a whole crop biorefinery. *Renewable Sustainable Energy Rev., 75*, 1269–1280, 2017.
83. Bazoti, S.F., Golunski, S., Siqueira, D.P., Scapini, T., Barrilli, É.T., Mayer, D.A., Barros, K.O., Rosa, C.A., Stambuk, V.U., Alves Jr., S.L., Valério, A., Oliveira D., Treichel, H., Second-generation ethanol from non-detoxified sugarcane hydrolysate by a rotting wood isolated yeast strain. *Bioresour. Technol.*, 244, 582–587, 2017.
84. Von Schenck, A., Berglin, N., Uusitalo, J. Ethanol from Nordic wood raw material by simplified alkaline soda cooking pre-treatment. *Appl. Energy*, 102, 229–240, 2013.
85. Consonni, S., Katofsky, R.E., Larson, E.D.,A gasification-based biorefinery for the pulp and paper industry. *Chem. Eng. Res. Des.*, 87(9), 1293–1317, 2009.
86. Triana, C.F., Quintero, J.A., Agudelo, R.A., Cardona, C.A., Higuita, J.C., Analysis of coffee cut-stems (CCS) as raw material for fuel ethanol production. *Energy*, 36(7), 4182–4190, 2011.

87. Ahmad, A.L., Yasin, N.M., Derek, C.J.C., Lim, J.K., Microalgae as a sustainable energy source for biodiesel production: a review. *Renewable Sustainable Energy Rev.*, 15(1), 584–593, 2011.
88. Sirajunnisa, A.R., Surendhiran, D., Algae–A quintessential and positive resource of bioethanol production: a comprehensive review. *Renewable Sustainable Energy Rev.*, 66, 248–267, 2016.
89. Jambo, S.A., Abdulla, R., Azhar, S.H.M., Marbawi, H., Gansau, J.A., Ravindra, P., A review on third generation bioethanol feedstock. *Renewable Sustainable Energy Rev.*, 65, 756–769, 2016.
90. Rodionova, M.V., Poudyal, R.S., Tiwari, I., Voloshin, R.A., Zharmukhamedov, S.K., Nam, H.G., Allakhverdiev, S.I., Biofuel production: challenges and opportunities. *Int. J. Hydrogen Energy*, 42(12), 8450–8461, 2017.
91. Komesu, A., Wolf Maciel, M.R., Oliveira, J., Martins, L.H.S., Maciel Filho, R., Purification of lactic acid produced by fermentation: focus on non-traditional distillation processes. *Sep. Purif. Rev.*, 2016. http://dx.doi.org/10.1080/15422119.2016.1260034
92. Martins, L.S H., Rabelo, S.C., Costa, A.C., Effects of the pretreatment method on high solids enzymatic hydrolysis and ethanol fermentation of the cellulosic fraction of sugarcane bagasse. *Bioresource Technol.*, 191, 312–321, 2015.
93. Bioref-Integ., Development of advanced biorefinery schemes to be integrated into existing industrial (fuel producing) complexes; Project 212831, FP7-ENERGY, Netherlands, 2010.
94. Edye, L.A., Doherty, W.O.S., Blinco, J.A., Bullock, G.E., The sugarcane biorefinery: energy crops and processes for the production of liquid fuels and renewable commodity chemicals. *Proc. Aust. Soc. Sugar Cane Technol.*, 27, 9–22, 2005.
95. Daful, A.G., Haigh, K., Vaskan, P., Görgens, J.F., Environmental impact assessment of lignocellulosic lactic acid production: integrated with existing sugar mills. *Food Bioprod. Process.*, 99, 58–70, 2016.
96. Parajuli, R., Knudsen, M.T., Birkved, M., Djomo, S.N., Corona, A., Dalgaard, T., Environmental impacts of producing bioethanol and biobased lactic acid from standalone and integrated biorefineries using a consequential and an attributional life cycle assessment approach. *Sci. Total Environ.*, 598, 497–512, 2017.
97. Farzad, S., Mandegari, M.A., Guo, M., Haigh, K.F., Shah, N., Görgens. Multi-product biorefineries from lignocelluloses: a pathway to revitalization of the sugar industry? *Biotechnol. Biofuels*, 10, 1–24, 2017.
98. Palacios-Bereche, R., Ensinas A.V., Nebra S.A., Energy consumption in etanol production by enzymatic hydrolysis – the integration with the conventional process using pinch analysis. *Chem. Eng. Trans.*, 24, 1189–1194, 2011. https://doi.org/10.3303/CET1124199
99. Dias, M.O.S., Cavalett, O., Filhob, R.M., Bonomi, A., Integrated first and second generation ethanol production from sugarcane. *Chem. Eng. Trans.*, 2014.

7

Application of Fermentation Strategies for Improved Laccase Production

Priyanka Ghosh[1], Arpan Das[2] and Uma Ghosh[1*]

[1]Food Technology & Biochemical Engineering, Jadavpur University, Kolkata, India
[2]Department of Microbiology, Maulana Azad College, Kolkata, India

Abstract

Laccases are a group of blue multi copper containing enzymes, which have received much attention of researchers in the last decades due to their ability to oxidize both phenolic and non-phenolic lignin related compounds as well as highly recalcitrant environmental pollutants. For this reason these biocatalysts are also known as "Green Catalysts". Recently laccases are increasingly being used in the field of textile, pulp and paper, food industry, biosensors, as bioremediation agent to clean up waste of industries, herbicides, pesticides, and certain explosives in soil, etc. Owing to sustain these biotechnological applications widespread, studies on laccase producing organisms have been intensified and different fermentation strategies have been developed for the improvement of laccase production to meet the industrial demands. Present chapter delineate the recent advances that have taken place in efforts in over expression of laccase in heterologous systems, and various fermentation techniques that have been developed to efficiently produce laccase at the industrial scale.

Keywords: Lignin, polyphenol oxidase, submerged fermentation, solid state fermentation, bioremediation

7.1 Introduction

Lignocelluloses biomass is the single renewable resource on earth, reproduced at 60 billion tons as organically bound carbon per year, which has

**Corresponding author*: ughoshftbe@yahoo.co.in

Arindam Kuila and Vinay Sharma (eds.) Principles and Applications of Fermentation Technology, (117–140) © 2018 Scrivener Publishing LLC

the potential to create a sustainable energy future. Globally 50–60 million tons of lignin is produced from different pulp and paper industry [1]. Removal of this high amount of lignin from lignocellulosic biomass is essential for the production of cellulosic pulp as a raw material for paper manufacture, and also to enhance the efficiency of cellulose and hemicellulose hydrolysis, as a feedstock for chemical synthesis, including biofuel [2]. However, the commercial use of lignin is limited to only 2% of its availability [3] with the rest usually burned to provide steam and process heat for the pulp and paper mills. Therefore, new methods for lignin deconstruction and utilization for value-added products, other than just simply burning it as a solid fuel, are needed. Bio-degradation of lignocellulose is a multi-enzyme process involving both hydrolytic and oxidative transformations, due to complex cross-linked three-dimensional network structure of lignified plant material. Lignin is most abundant naturally occurring aromatic complex oxyphenyl propanoid polymer found in all vascular plants including herbaceous species, and following cellulose, the second most abundant organic polymer on earth. Majority of lignin are generally present in the middle lamella and primary cell wall. Crosslinks between cellulose–hemicellulose matrix through lignin–carbohydrate network structures provide stiffness and glue the cells together thereby protecting the cell wall against microbial degradation. Lignin is synthesized by one-electron oxidation of the precursors; p-coumaryl alcohol, coniferyl alcohol, and sinapyl alcohol, generating phenoxy radicals which then undergo nonenzymatic polymerization. Lignin precursors are linked together through carbon–carbon and carbon–oxygen bonds with a varying degree of methoxylation [4, 5]. In nature, efficient and selective lignin biodegradation is mediated mainly by laccase enzymes produced from different fungi, bacteria, plants as well as insects. Laccases are multicopper containing extracellular and metalloenzymes, characterized by their ability to catalyze one-electron oxidation of four reducing-substrate molecules concomitant with four-electron reduction of molecular oxygen to water. Currently the broad substrate specificity as well as catalytic properties of laccases are being exploited and they have become industrially important enzymes because of their diverse applications: in pulp delignification and bleaching, as a stabilizer in wine production, detergents, adhesives, fibre functionalization, detoxification of wastewaters and organic pollutants, denim bleaching, textile dye decolourization, baking, biosensors, and in biofuel cells etc. [6, 7]. The cost of laccases production is one of the main factors determining the economy of the process. Due to the usefulness of laccase, much effort has been spent in optimization of fermentation process parameters in both submerged and solid state fermentation (SSF) processes trying to reduce production cost and boost yield of laccase from wild-type strains of filamentous fungi.

7.1.1 What is Laccase?

Laccases (benzenediol: oxygen oxidoreductase, EC 1.10.3.2) are glycosylated polyphenol oxidase containing four copper ions per molecule and catalyze single-electron oxidation of a wide range of organic and inorganic substrates with a concomitant four-electron reduction of oxygen to water [7]. Laccases are basically monomeric, dimeric or tetrameric glycoproteins which catalyze monoelectronic oxidation of substrate molecules to corresponding reactive radicals with the assistance of four copper atoms (belonging to three types: 1, 2 or 3) that form catalytic core of the enzyme, accompanied with the reduction of one molecule of oxygen to two molecules of water and the concomitant oxidation of a variety of aromatic (like ortho- and para-diphenols, aromatic amines, methoxy-substituted phenols, etc.) and non-aromatic substrates via a mechanism involving radicals that can undergo further laccase-catalyzed reactions and/or non-catalytic reactions such as polymerization, hydration or hydrogen abstraction. Laccases contain four copper atoms termed Cu T1 (binding site for reducing substrate) and trinuclear copper cluster T2/T3 (electron transfer from type I Cu to the type II Cu and type III Cu trinuclear cluster/reduction of oxygen to water at the trinuclear cluster) [1].These four copper ions are classified into three categories: Type 1 (T1), Type 2 (T2), and Type 3 (T3). At oxidizing state, the Type 1 Cu gives blue colour to the protein at an absorbance of 610 nm. The Type 2 copper and Type 3 copper form a trinuclear centre which is involved in the enzyme catalytic mechanism. The O_2 molecule binds to the trinuclear cluster for asymmetric activation, and it is postulated that the O_2 binding compartment appears to restrict the access of oxidizing agents. During steady state, laccase catalysis indicates that O_2 reduction takes place. Laccase operates as a battery and stores electrons from individual oxidation reactions to reduce molecular oxygen. Hence, the oxidation of four reducing substrate molecules is necessary for the complete reduction of molecular oxygen to water.

Substrate oxidization by laccases can be further increased through a mediator-involved reaction mechanism. The low-molecular weight redox mediators are in some cases, very unstable and reactive cationic radicals, which can oxidize more complex substrates before returning to their original state. The electrons taken by laccases are finally transferred back to oxygen to form water (Figure 7.1).

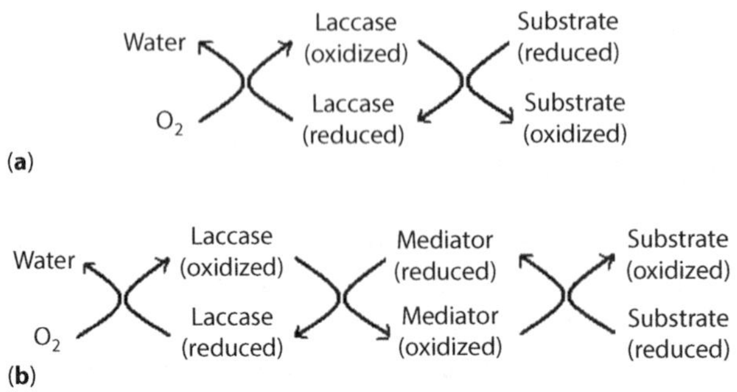

Figure 7.1 Proposed mechanism of substrate oxidation in the absence (a) or in the presence (b) of redox mediators.

Laccase was first detected in exudates of Japanese lacquer tree *Rhus vernicifera* [8]. Among fungi, Ascomycetes, Basidiomycetes, and Deuteromycetes can produce laccase and white-rot fungi have been found to be the most efficient laccase producers. The better degradative efficiency of fungi is due to their hyphal organization, which imparts them penetration capacity. A number of brown-rot fungi like *Postia placenta*, *Antrodia vaillantii*, *Fomitopsis pinicola*, and *Coniophora puteana* have also been known to produce laccase [9, 10]. Beside white-rot fungi, several actinomycetes and bacterial species can also degrade lignin. Examples of some laccase producers are shown in Table 7.1.

Intracellular laccase in bacteria like *Azospirillum lipoferum*, *Bordetella compestris*, *Bacillus subtilis*, *Escherichia coli*, *Caulobacter crescentus*, *Mycobacterium tuberculosum*, *Pseudomonas aeruginosa*, *Yersinia pestis*, and *Stenotrophomonas maltophilia* strain was found to be potent laccase producer [11, 12]. The plants in which the laccase enzyme has been detected include lacquer, mango, mung bean, peach, pine, prune, and sycamore, cabbage, turnip, beet, apple, asparagus, potato, pear, sycamore maple, poplar, tobacco, peach, etc. The laccase enzyme has also been detected in different insects, for example, Bombyx, Diploptera, Calliphora, Lucilia, Drosophila, Manduca, Papilio, Orycetes, Phormia, Musca, etc. [16].

7.2 Major Factors Influencing Fermentation Processes for Laccase Production

7.2.1 Influence of Carbon Source

Laccase production depends on the nature of the carbon source, which in some cases may come from different agro-industrial lignocellulosic

Table 7.1 Example of organisms capable of producing laccase enzymes.

Organisms		Examples	References
Microorganisms	Bacteria	*Azospirillum lipoferum, Bordetella campestris, Caulobacter crescentus, Escherichia coli, Mycobacterium tuberculosum, Pseudomonas syringae, Pseudomonas aeruginosa, Yersinia pestis, Stenotrophomonas maltophilia, Azospirillum lipoferum, Marinomonas mediterranea, Streptomyces griseus, Streptomyces lavendulae, Bacillus subtilis*	[11–14]
	Fungi	*Anthracophyllum discolor, Pycnoporus sanguineus, Monocillium indicum, Phanerochaete chrysosporium, Theiophora terrestris, Lenzites, betulina, Phlebia radiate, Pleurotus ostreatus, Trametes versicolour, Trametes pubescens, T. hirsute, T. gallica, Trichoderma atroviride, T. harzianum, T. longibrachiatum, Pleurotus ostreatus, Pycnoporus sanguineus,* etc.	[14, 15]
Plants		Lacquer, mango, mung bean, peach, pine, prune, sycamore, cabbages, turnips, beets, apples, asparagus, potatoes, pears, sycamore maple, poplar, tobacco, peach, loblolly pine	[16, 17]
Insects		*Bombyx, Calliphora, Diploptera, Drosophila, Lucilia, Manduca, Musca, Oryctes, Papilio, Phormia, Rhodnius, Sarcophaga, Schistocerca, Tenebrio*	[18, 19]

Table 7.2 Example of different carbon sources used by microorganisms during laccase production.

Microorganisms	Carbon source used	References
Streptomyces chartreusis	Dextrose	[20]
Ganoderma sp.	Starch	[21]
Pleurotus sp.	Mannitol	[22]
Lentinus sp.	Glycerol	[23]
Pleurotus sajor-caju PS-2001	Sucrose	[24]
Pleurotus ostreatus DSM 1833 and *Phoma* sp. UHH 5-1-03	Banana peels	[25]
Coriolus versicolor MTCC 138	Glucose+Starch	[26]

residues. So, selection of an appropriate carbon source in the medium is important in growth and metabolism of fungi, hence it plays an important role in enzyme yield. Examples of microbial utilization of different carbon sources during laccase production are shown in Table 7.2.

Hatvani *et al.*, [27] showed that fructose induced 100-fold increase in laccase production of *Basidiomycete* sp. I-62. Johnsy and Kaviyarasan [28] also reported fructose as the best co-substrate for laccase production by *Lentinus kauffmanii*. *Aspergillus fumigatus* was found to be an excellent producer of laccase in fermentation of banana peels [29]. Glucose and cellobiose were efficiently and rapidly utilized by *Trametes pubescens* with high laccase activity [30]. Glucose has been reported as an effective co-substrate for laccase production using *Ganoderma lucidum* by Ding *et al.*, [31]. Patel and Gupte [32] also reported glucose as suitable carbon source for maximum laccase production from *Tricholoma giganteum* AGHP under SSF. However, the excessive concentrations of glucose are also inhibitory to laccase production in various fungal strains [33]. An excess of sucrose also reduced the production of laccase by blocking its induction and only allowed constitutive production of enzyme. Use of polymeric substrates like cellulose and starch was able to alleviate this problem [21].

7.2.2 Influence of Nitrogen Source

Variety of organic nitrogen sources (yeast extract, peptone, tryptone, etc.) and inorganic nitrogen sources (urea, ammonium sulphate, ammonium chloride, ammonium nitrate, potassium nitrate, etc.) are utilized by

Table 7.3 Example of different nitrogen sources used by microorganisms during laccase production.

Microorganisms	Nitrogen source used	Reference
Ganoderma lucidum	yeast extract	[31]
Pleurotus sajorcaju	Ammonium tartrate	[34]
Pleurotus ostreatus HP-1	L-aspargine and NH_4NO_3	[35]
Dictyoarthrinium Synnematicum Somrith	Sodium nitrate	[36]
Trametes trogii	Glutamic acid	[37]
Pycnoporus sanguineus	Asparagine	[38]
Coriolopsis gallica	Peptone	[39]

microorganisms for laccase production (examples are shown in Table 7.3). The effect of nitrogen source on laccase production by different organisms appears to be greatly controversial [40]. In some fungi, the C and N depletion helps laccase production, but this is not a general statement. Also, the initial C:N ratio influences laccase production in different ways by various basidiomycetes [41]. Monteiro and De Carvalho reported high laccase activity with semi-continuous production in shake-flasks using a low carbon to nitrogen ratio (7.8 g/g). *Cerrena unicolors* also produced laccase in the low nitrogen medium [43]. Buswell *et al.*, found that laccases were produced at high nitrogen concentrations. Dong *et al.* [45] have reported improved laccase production using tryptone and peptone. Revankar and Lele [26] reported yeast extract as a suitable nitrogen source for laccase production.

7.2.3 Influence of Temperature

The temperature is an important factor in the development of a biological process is such that it could determine effects such important as protein denaturization, enzymatic inhibition, promotion, or inhibition on the production of a particular metabolite, cells death, etc. [46]. Therefore, temperature in its optimum condition is necessary for enzyme production through both submerged and solid state fermentation process. The optimal temperature of laccase differs greatly from one strain to another. Many researchers have reported an optimum temperature between 25 and 30 °C for laccase production using various white-rot fungi [47–49]. Patel and Gupte [32] found 30 °C as an optimum temperature during laccase production by *T. giganteum* AGHP through SSF. Similar result was found by Nandal *et al.*, [50] during statistical optimization of laccase production through Taguchi design by *Coriolopsis caperata* RCK2011 under SSF. Higher temperature for laccase production also has been reported by Nasreen *et al.* [51]. The optimization studies showed that the laccase yield by *Coriolus versicolor* was maximum at 37 °C during SSF on rice bran. Similarly Vantamuri and Kaliwal [52] reported 40 °C as an optimum temperature for laccase production by *Marasmius* sp. BBKAV79 during submerged fermentation (SmF). Generally higher temperatures lead to adverse effect on the microbial metabolism, thereby leading to the denaturation of the key enzymes. However, lower temperature of 10 and 20 °C did not support the growth of fungi, thus leading to lower enzyme production. Xin and Geng [53] have also reported that the lower temperature retards the metabolic rate of *Trametes versicolor* resulting decrease in laccase production.

7.2.4 Influence of pH

The pH of the culture medium is an important physical regulatory factor affecting the fermentation process because both cell growth and the production of enzymes can be affected by it. The optimum value of fermentation pH varies according to the substrate because different substrate causes different reaction for laccases. Sun *et al.*, [54] found the optimum pH for laccase production by *Coriolus hirsutus* was ~4.5 during its growth in molasses distillery wastewater. Similar results were reported by Youshuang *et al.*, [55] where laccase production by *Trametes versicolor* was statistically optimized at pH 4.5 during SmF. Besides acidic pH ranges, alkaline optimum pH for laccase production also has been reported by Ding *et al.*, [56]. They found that an initial alkaline pH was beneficial for laccase production by *Pleurotus ferulae* and 6,832.86 U/L was obtained using an initial pH of 9.0. Most studies show that pH between 4.5 and 6.0 is suitable for laccase production. Further increasing the pH of the culture media may be adverse for enzyme activation and responsible for the decrease of laccase production. This finding might be attributed to the accumulation of microbial metabolic products in the growing culture, which inactivates laccase or inhibits its biosynthesis, or due to the action of proteolytic enzymes [55].

7.2.5 Influence of Inducer

Laccase production has been seen to be highly dependent on fungus cultivation. Laccases are generally produced in low concentrations by laccase-producing microbes, but enhanced concentrations were obtained with the addition of various inducer supplements to media [57]. Supplementation of an appropriate inducer can greatly enhance the laccase production. Different compounds such as phenolic and non-phenolic substrates can act as inducers. The most reported effective inducers used for laccase production are copper, guaiacol, ethanol, Gallic acid, 2,5-xylidine, ferulic acid, Catechol, Veratryl alcohol, Pyrogallol, anisidine, Vanilic acid, tannic acid, etc. (Table 7.4). Manavalan *et al.* [63] and Mann *et al.* [64] reported that the addition of copper is significant to induce laccase production in *Cerrena consors* and *Ganoderma lucidum*, respectively. Galhaup and Haltrich [30] showed that extracellular laccase formation could be greatly stimulated by the addition of Cu^{2+} to the growth medium. Many other authors have found copper as a suitable inducer for laccase production. The addition of xenobiotic compounds such as 2,5xylidine, lignin, and veratryl alcohol increased and induced laccase activity [59, 62]. A similar observation has been reported by several researchers [32, 65, 66]. It was demonstrated that cultures of *Fomes annosus, Pholiota mutabilis, Pleurotus ostreatus*, and *Trametes versicolor* were stimulated for laccase production by addition of low concentration of 2,5-xylidine [67]. At higher concentrations the 2,5-xylidine had a reducing effect due to toxicity. The promoter regions of the genes encoding for laccase contains several sites of recognition that are specific for xenobiotics and heavy metals; they bind to the recognition sites and induce laccase production [68].

Table 7.4 Example of different inducers used by microorganisms during laccase production.

Microorganisms	Inducer used	Reference
Streptomyces psammoticus	Pyrogallol	[57]
Fusarium incarnatum LD-3	ortho-di-anisidine	[48]
Pycnoporus cinnabarinus	ethanol vapours	[58]
Trametes pubescens MB89	2, 5-xylidine	[59]
Trametes versicolor ATCC 200801	ABTS	[60]
Agaricus blazei	Ethanol and guaiacol,	[61]
Botryosphaeria sp.	3,4-dimethoxybenzyl (veratryl) alcohol	[62]

7.3 Type of Cultivation

Laccases are the enzymes which are secreted out in the medium by several fungi [25] during the secondary metabolism. Laccases have been produced vividly in both submerged and solid state modes of fermentation.

7.3.1 Submerged Fermentation

SmF involves the cultivation of microorganisms in liquid medium with appropriate nutrients and high oxygen concentrations when operated in aerobic conditions. Production of laccase through this method requires relatively short time and the physiological regulation of laccase production using SmF is also comparatively simpler than in SSF. Viscosity of broth is the major problem associated with the fungal SmFs. Fungal SmF is also challenging as the mycelial growth hinders impeller action, which in turn limits oxygen and mass transfer in the fermenter. Different strategies have been employed to deal with viscosity, oxygen and mass transfer limitations. Cell immobilization has been proven as a good alternative technique in this regard. A continuous and successful laccase production for a period of 4 months using immobilized *Neurospora crassa* on membrane has been reported by Luke and Burton [69]. Sedarati *et al.* [70] reported that immobilization of *Trametes hirsute* on stainless steel showed the highest laccase activity in fixed bed bioreactors. Schliephake *et al.* [65] used nylon sponge cube immobilized *Pycnoporus cinnabarinus* to produce laccase in batch culture by in packed bed bioreactor. Park *et al.* [71] found that *Funalia trogii* immobilized in Na-alginate beads was efficient in decolourizing Acid Black 52. Similarly microbial immobilization on plastic n*et al*so has been reported for laccase production [72]. Utilization of different agro-industrial lignocellulosic residues in SmF instead of synthetic carbon source was also carried out for laccase production by fungi, which showed promising results to reduce the cost of production of the enzyme and to allow large-scale industrial applications [73]. Examples of successful laccase production through SmF are shown in Table 7.5.

7.3.2 Solid-State Fermentation

SSF is defined as fermentation process occurring in absence or near absence of free liquid, employing an inert substrate (synthetic materials) or a natural substrate (organic materials) as a solid support [76]. SSF processes have shown to be particularly suitable for the production of enzymes by filamentous fungi, since they reproduce the natural living conditions of

Table 7.5 Example of microbial laccase productions through SmF.

Microorganisms	Optimized fermentation condition	Enzyme yield	References
Pycnoporu ssanguineus	C source: glucose N_2 source: yeast extract and peptone Inducer: $CuSO_4$	1.6 U/mL	[74]
Neolentinus kauffmanii	C source: Xylan N_2 source: Peptone Inducer: copper Temperature: 25 °C pH: 6.0	70.51 U/mL	[75]
Pleurotus ferulae JM30X	Metal ions: CuSO 4 inhibitors: DTTand NaN3 Inducer: ABTS Temperature: 50 °C–70 °C pH: 3	6,832.86 U/L	[56]
Streptomyces chartreusis	N_2source: yeast extract Metal: Cupric sulfate Inducer: Pyrogallol	330.00 U/gm	[20]
Ganoderma lucidum	C source: Pomelo peelN_2 source: yeast extract Temperature: 60 °C pH: 3	11842.13 U/L	[31]
Aspergillus flavus PUF5	C source: Ribbedgourd peel N_2 source: yeast extract Temperature: 25 °C pH: 4 Metal source: NaCl	15.96 U/ml	[73]

such fungi due to which they may be more capable of producing certain enzymes with high productivity in comparison to SmF. Porosity and particle size of the substrate affect the surface area accessible to the organism. Small substrate particles provide a large surface area for microbial attachment while larger particles provide better aeration but a limited surface for microbial attachment. Recently, there has been an increasing trend towards the utilisation of different agrowastes (Table 7.6) as raw materials to produce value-added products by SSF technique as they mimic the

Table 7.6 Example of microbial laccase productions through SSF using different agro wastes as substrate.

Microorganisms	Substrate	Enzyme yield	References
Pleurotus ostreatus 1804	Pulse husk waste	2200 U/g	[76]
Trichoderma longibracheatum	Rice bran	45.24 U/gds	[77]
Trametes hirsuta	Banana skin	4010 U/l	[78]
Coriolus versicolor	Rice bran	0.98 U/ml	[51]
Pleurotus ostreatus PVCRSP-7	Black gram husk (BGH)	3186 U/gds	[79]
Aspergillus fumigatus VkJ2.4.5	Banana peel	6281.4 ± 63.60 U/1	[29]
Streptomyces psammoticus	Rice straw	55.4 U/g	[80]
Streptomyces chartreusis	Rice bran	72 U/g	[20]

conditions under which the fungi grow naturally [29]. Furthermore, the presence of lignin and cellulose/hemicellulose in these substrates act as inducers of the ligninolytic activities. Moreover, most of them are rich in sugars, promoting better fungal growth and thus making the process more economical. Furthermore, SSF processes have several potential economic and environmental advantages: different agro industrial by-products and residues may be used as solid substrates, lowering the production costs, and avoiding pollution issues related to waste disposal; the effluent generation and the demands on energy and sterile water are lower; the enzymes are obtained at higher concentrations, reducing downstream processing [51]. As shown in Table 7.6, many agricultural wastes such as pulse husk waste, grape seeds, grape stalks, barley bran, cotton stalk, coffee husk, banana peel, corncobs, and wheat bran are also used as substrate for laccase production. Zilly *et al.* [81] successfully utilized passion fruit waste for laccase production by white-rot fungi. Sun *et al.* [82] reported that the white-rot fungus *Trametes* sp. AH28-2 grown on agro-byproducts produced laccase with high ability to decolourize textile reactive dyes. Couto [83] also found that laccase produced by *T. hirsuta* grown on paper cuttings successfully decolourized synthetic dyes at alkaline pH. Despite the numerous advantages of SSF offers over SmF, the major disadvantage with SSF is lack of any established bioreactor designs. This is principally due to several problems encountered in the robust control mechanisms for different parameters such as pH, temperature, aeration and oxygen transfer, moisture, and agitation. Recently different bioreactor configurations have been studied for laccase production such as immersion configuration, expanded bed, tray, inert (nylon), and non-inert support (barley bran), but lot of progress is still to be made for a better controlled SSF operation.

7.4 Biotechnological Application of Laccases

Laccases have enormous applications due to their oxidizing ability towards a broad range of phenolic and non-phenolic compounds. With their broad substrate specificity, the enzymes can be used for degradation of various industrial wastes such as textiles, paper, pulp, petrochemical, oil-refining, pharmaceutical, pesticides, resin products, coking plants, and leather [84–86]. Laccases are also used in medical diagnostics, biolinkers, biofuels, synthetic detergents, antimicrobial materials, and for clearing herbicides, pesticides, and some explosives in soil [87, 88]. Currently, researchers are working on enzymatic synthesis of organic compounds, laccase based bio-oxidation, biotransformation, and biosensor development [88]. Examples of different industrial applications of laccase are shown in Table 7.7.

7.4.1 Food Industry

In the food industry, wine stabilization is the main application of laccase. Phenolic compounds are responsible for the color and taste of the wine while the organic acids are responsible for the aroma. Polyphenols have undesirable effects on wine production and on its organoleptic characteristics. Due to oxidation reactions it can help in color intensification, turbidity, aroma, and flavor changes in the wine. Hence, their removal from the wine is really necessary. Laccase from *Trametes versicolor* have shown phenol removal efficiency to stabilize wine/must [101]. The degradation of phenols by laccase was very fast for catechins, and slowly for stilbenes (cis- and trans- resveratrol) and derivatives of cinnamic (ferulic and caffeic) and benzoic (syringic, vanillic, and gallic) acids [102]. Laccase has been successfully applied in bakery to increase strength, stability, and reduced stickiness and thereby improved the dough consistency, increase in volume, improvisation in crumb structure, and softness in the baked product. Besides these, laccase has been used in fruit juice clarification, stabilization

Table 7.7 Biotechnological applications of microbial laccases.

Sources of laccase enzyme	Applications	References
Paraconiothyrium variabile	Biosynthesis of gold nanoparticles	[89]
Trametes versicolor	Paper biosensor for the detection of phenolic compounds	[86]
Yarrowia lipolytica	Removal of phenolic compounds	[90]
Trametes sp.	Development of bioactive hydrogel dressing	[91]
Trametes versicolor	Wine stabilization	[92]
Coriolopsis gallica	Beer factory waste water treatment	[84]
Trametes versicolor	Development of microbial fuel cells (MFC) cathode	[93]
Trametes versicolor (ATCC 32745)	Development of biosensors	[94]
Lentinula edodes	Polyaromatic hydrocarbon biodegradation	[95]
Pleurotus eryngii	Lignin and organopollutant degradation	[96]
Aspergillus oryzae	Biosensor and gold nanoparticle	[88]
Pleurotus florida NCIM 1243	Nanofiber preparation	[97]
Myceliophthora thermophila	Dough conditioner	[85]
Trametes pubescens	Bioremediation of a mixture of pentachlorophenol (PCP),2-chlorophenol (2-CP), 2,4-dichlorophenol (2,4-DCP)and 2,4,6-trichlorophenol (2,4,6-TCP)	[98]
Fusarium incarnatum UC-14	Bioremediation of bisphenol A	[48]
Aspergillus flavus	Decolorization of Malachite green dye	[87]
Streptomyces cyaneus	Decolorize and detoxify azo dyes	[99]
Coriolus versicolor	Degradation of textile dyes	[100]

of beer through inhibition of haze formation, production of thermo-irreversible gels through gelation of sugar beet pectin etc. [92, 103].

7.4.2 Textile Industries

Textile industry effluents are of great threat to the environment due to the presence of different residual dyes and xenobiotic compounds. Decolourization of dye wastewaters using the laccase of white-rot fungi has found to have an enormous potential for wastewater treatment [100]. Fungal laccase was used for degradation of triarylmethane, indigo, azo, and anthraquinone dyes, which resulted in very effective reduction of these dyes [99, 104]. Vijaykumar [105] isolated a new fungus from coal sample which decolourized five different azo and triphenylmethane dyes like acid blue 193, acid black 210, crystal violet, reactive black, and reactive black BL/LPR. Laccase from *Cerrena unicolor* is also capable of decolorization of different industrial dyes like acid blue 62, acid blue 40, reactive blue 81, direct Black 22, and acid red 27 [57]. Kirby *et al.* [106] reported that laccase from *Phlebia tremellosa* decolorized eight synthetic textile dyes added to culture under stationary conditions. Beside degradation of dyes, it has profound application in the removal of indigos during denim finishing in the denim industry, rove scouring of garments, biobleaching of cottons, etc. [107, 108].

7.4.3 Paper Industry

In the manufacture of paper, separation of lignin from cellulose fibers is required. A mild strategy of delignification using laccase has replaced conventional and polluting chlorine, sulphite and oxygen based methods without affecting the integrity of cellulose. Laccases from white-rot fungi has been applied as biopulping agents to wood chips which can improve strength of the paper and save energy consumption. Laccases also has been used for reduction of resins, binding fiber, particle and paper-board, deinking, and decolorization of printed paper [14].

7.4.4 Bioremediation

Different aromatic xenobiotics and pollutants in the effluents generated by pulp and paper, petrochemical, coal conversion, alcohol distilleries, dyeing, and textile industries are responsible for toxicity to mammals and fishes. Treatment of these wastes by different physico-chemical processes including ultrafiltration, ion exchange, and lime precipitation, are expensive, so alternative biotreatment processes are now being considered. White-rot fungi that produce lignin-degrading enzymes are reported to be the most efficient in detoxification and decolorization of such effluents [96]. The organism usually obtains its carbon, nitrogen or energy from the pollutant and reduces it to undetectable, nontoxic or acceptable levels. Laccase mediated degradation of a wide variety of aromatic xenobiotics, including polyaromatic hydrocarbons (PAHs), polychlorinated biphenyls, pentachlorophenol, and various groups of pesticides has been reported [90]. Laccases were able to mediate the coupling of reduced 2,4,6-trinitrotoulene metabolites to an organic soil matrix, which resulted in detoxification of munition residue [109]. Moreover, PAHs, which arise from natural oil deposits and utilization of fossil fuels, were also found to be degraded by laccases [86].

7.4.5 Pharmaceutical Industry

Laccases have been used for synthesis of several products with pharmaceutical importance because of their high oxidation potential. Hence laccases are used for preparation of some important drugs like anticancer drugs, medical diagnosis, synthesis of antioxidants, hormone derivatives, and antiviral derivatives and they are also added in cosmetics to reduce their toxicity [94]. Laccase is used in oxidization of iodide to produce iodine, preparation of less irritant hair dyes replacing hydrogen peroxide as an oxidizing agent in the dye formulation [84]. Recently, proteins for skin lightening have also been developed. Laccase from *Aspergillus oryzae*, in the immobilized form, is used to synthesize and characterize gold nanoparticles on PEI coatings to prepare a biosensor [88]. The electrode was tested for detection of phenolic compounds in streams, and the biosensor demonstrated high sensitivity, good repeatability and reproducibility, and long-term stability.

7.5 Conclusion

Laccases are promising enzymes that are widespread in nature and can replace different conventional chemical processes involved in various industrial, agricultural, and medicinal applications due to their high reduction potential that makes them potential candidate to catalyze the oxidation of a broad range of phenolic and non-phenolic compounds. Emerging research areas include the laccase-based enzymatic synthesis, bio-oxidation, bioremediation and detoxification, biosensor development, etc., which have been successfully carried out in either laboratory or pilot plant scale. However, the major drawbacks in large-scale applications are lack of sufficient enzyme stocks and their high price. To overcome this drawback, overexpression of laccases in heterologous systems has been actively pursued to enhance their titers and to improve their catalytic activity. Proper optimization of various vital fermentation process parameters, use of appropriate inducers and the usage of inexpensive sources like agro-wastes could bring additional benefits of higher production with expenditure of minimum resources. Both submerged and solid state cultivation techniques have been employed by the researchers and most of them found SSF as better than the other for industrial production of laccase. In spite of that, SSF have some limitations of control mechanisms for different parameters such as pH, temperature, aeration and oxygen transfer, moisture, and agitation, so future efforts should be involved in improving the SSF bioreactor designs to make SSF more potent and competitive. Therefore, it is not surprising that this enzyme has been studied intensively and yet remains a topic of the intense research today.

References

1. Christopher, L.P., Bin, Y., Yun, J., Lignin biodegradation with laccase-mediator systems. *Front. Energy Res.*, 2, 12, 2014.
2. Siqueira, G., Várnai, A., Ferraz, A., Milagres, A.M., Enhancement of cellulose hydrolysis in sugarcane bagasse by the selective removal of lignin with sodium chlorite. *Appl. Energy*, 102, 399, 2012.
3. Gosselink, R.J.A., Jong, E., Guran, B., Abacherli, A., Co-ordination network for lignin—standardisation, production and applications adapted to market requirements (Eurolignin). *Ind. Crops Prod.*, 20, 121, 2004.
4. Adler, E., Lignin chemistry: past, present and future. *Wood Sci. Technol.*, 11, 169, 1977.
5. Karhunen, P., Rummakko, P., Sipila, J., Brunow, G., The formation of dibenzodioxocin structures by oxidative coupling. a model reaction for lignin biosynthesis. *Tetrahedron Lett.*, 36, 4501, 1995.
6. Couto, S.R., Toca-Herrera, J.L., Laccase production at reactor scale by filamentous fungi. *Biotechnol. Adv.*, 25, 558, 2007.
7. Madhavi, V., Lele, S.S., Laccase: properties and applications. *BioResources.*, 4, 1694, 2009.
8. Malmstrom, B.G., Reinhammar, B., Vanngard, T., The state of copper in stellacyanin and laccase from the lacquer tree *Rhus vernicifera*. *Biochem. Biophys. Acta.*, 205, 48, 1970.
9. De'Souza, T.M., Boomminathan, K., Reddy, C.A., Isolation of laccase gene-specific sequences from white rot and brown rot fungi by PCR. *Appl. Environ. Microbiol.*, 62, 3739, 1996.

10. An, H., Xiao, T., Huan, F., Wei, D., Molecular characterization of a novel thermostable laccase PPLCC2 from the brown rot fungus *Postia placenta* MAD-698-R. *Electron. J. Biotechnol.*, 18, 451, 2015.
11. Alexandre, G., Zhulin, I.B., Laccases are wide spread in bacteria. *Trends Biotechnol.*, 18, 41, 2000.
12. Enguita, F.J., Martins, L.O., Henriques, A.O., Carrondo, M.A., Crystal structure of a bacterial endospore coat component: a laccase with enhanced thermostability properties. *J. Biol. Chem.*, 278, 19416, 2003.
13. Galai, S., Touhammi, Y., Marzouki, M.N., Response surface methodology applied to laccase activities exhibited by *Stenotrophomonas maltophilia* AAP56 in different growth conditions. *BioResources.*, 7, 706, 2012.
14. Senthivelan, T., Kanagaraj, J., Panda, R.C., Recent trends in fungal laccase for various industrial applications: an eco-friendly approach – a review. *Biotechnol. Bioprocess Engg.*, 21, 19, 2016.
15. Sadhasivam, S., Savitha, S., Swaminathan, K., Lin, F.H., Production, purification and characterization of midredoxpotential laccase from a newly isolated TrichodermaharzianumWL1. *Process Biochem.*,43, 736, 2008.
16. Xu, F., Laccase. In M. C. Flickinger & S. W. Drew (Eds.), *The encyclopedia of bioprocessing technology: fermentation, biocatalysis and bioseparation.* New York: Wiley, 1545, 1999.
17. Shraddha, R.S., Sehgal, S., Kamthania, M., Kumar, A., Laccase: microbial sources, production, purification, and potential biotechnological applications. *Enzyme Res.*, 2011.
18. Dittmer, N.T., Suderman, R.J., Jiang, H., Zhu, Y.C., Gorman, M.J., Kramer, K.J., Kanost, M.R., Characterization of cDNA encoding putative laccase-like multicopper oxidases and developmental expression in the tobacco hornworm, Manducasexta, and the malaria mosquito. *Anopheles gambiae. Insect Biochem. Mol. Biol.*, 34, 29, 2004.
19. Sharma, K.K., Kuhad, R.C., Laccase: enzyme revisited and function refined. *Ind. J.Microb.*, 48, 309, 2008.
20. Chhaya, R., Modi, H.A., Statistical optimization of laccase producing *Streptomyces chartreusis* by solid state fermentation. *CIBTech J. Microbiol.*, 3, 8, 2013.
21. Revankar, M.S., Lele, S.S., Increased production of extracellular laccase by the white rot fungus *Coriolus versicolor* MTCC 138. *World J. Microbiol. Biotechnol.*, 22, 921, 2006.
22. Saravanakumar, K., Saranya, R., Arathi, S., Kaviyarasan, V., Statistical design and response surface technique for the optimization of extra cellular laccase enzyme production by using *Pleurotus sp. Rec. Res. Sci. Technol.*, 2, 104, 2010.
23. Saravanakumar, K., Kaviyarasan, V., Response Surface methodological approach to optimize the nutritional parameters for extracellular laccase production by *Lentinus* sp. *J. Biosci. Res.*, 1, 40, 2010.
24. Bettin, F., Rosa, L.O., Montanari, Q., Calloni, R., Gaio, T.A., Malvessi, E., Silveira, M.M., Dillon, A.J.P., Growth kinetics, production, and

characterization of extracellular laccases from *Pleurotus sajor-caju* PS-2001. *Process Biochem.*, 46, 758, 2011.

25. Junior, N.L., Gern, R.M.M., Furlan, S.A., Schlosser, D., Laccase production by the aquatic *Ascomycete Phoma sp.* UHH 5-1-03 and the white rot *Basidiomycete Pleurotus ostreatus* DSM 1833 during submerged cultivation on banana peels and enzyme applicability for the removal of endocrine-disrupting chemicals. *Appl. Biochem. Biotechnol.*, 167, 1144, 2012.
26. Revankar, M.S., Lele, S.S., Increased production of extracellular laccase by the white rot fungus *Coriolus versicolor* MTCC 138. *World J. Microbiol. Biotechnol.*, 22, 921, 2006.
27. Hatvani, N., Mecs, I., Production of laccase and manganese peroxidase by *Lentinus edodes* on malt-containing by-product of the brewing process. *Process Biochem.*, 37, 491, 2001.
28. Johnsy, G., Sargunam, S., Dinesh, M.G., Kaviyarasan, V., Nutritive value of edible wild mushrooms collected from the western Ghats of Kanyakumari district. *Botany Res. Int.*, 69, 2011.
29. Vivekanand, V., Dwivedi, P., Pareek, N., Singh, R.P., Banana peel: a potential substrate for laccase production by *Aspergillus fumigates* VkJ2.4.5 in solid-state fermentation. *Appl. Biochem. Biotechnol.*, 165, 204, 2011.
30. Galhaup, C., Haltrich, D., Enhanced formation of laccase activity by the white-rot fungus *Trametes pubescens* in the presence of copper. *Appl. Microbiol. Biotechnol.*, 56, 225, 2001.
31. Ding, Z., Peng, L., Chen, Y., Zhang, L., Gu, Z., Shi, G., Zhang, K., Production and characterization of thermostable laccase from the mushroom, *Ganoderma lucidum*, using submerged fermentation. *Afri. J. Microbiol. Res.*, 6, 1147, 2012.
32. Patel, H., Gupte, A., Optimization of different culture conditions for enhanced laccase production and its purification from *Tricholoma giganteum* AGHP. *Biores. Bioprocessing.*, 3, 11, 2016.
33. Lee, K.H., Wi, S.G., Singh, A.P., Kim, Y.S., Micromorphological characteristics of decayed wood and laccase produced by the brown-rot fungus *Coniophora puteana*. *J. Wood Sci.*, 50, 281, 2004.
34. Patrick, F., Mtui, G., Mshandete, A.M., Kivaisi, A., Optimization of laccase and manganese peroxidase production in submerged culture of *Pleurotus sajorcaju*. *Afri. J. Biotechnol.*, 10, 10166, 2011.
35. Patel, H., Gupte, A., Gupte, S., Effect of different culture conditions and inducers on production of laccase by a basidiomycete fungal isolate *Pleurotus ostreatus* HP-1 under solid state fermentation. *BioResources.*, 4, 268, 2009.
36. Prasher, I.B., Chauhan, R., Effect of carbon and nitrogen sources on the growth, reproduction and ligninolytic enzymes activity of *Dictyoarthrinium synnematicum somrith*. *Adv. Zool. Bot.*, 3, 24, 2015.
37. Levin, L., Melignani, E., Ramos, A.M., Effect of nitrogen sources and vitamins on ligninolytic enzyme production by some white-rot fungi. Dye decolorization by selected culture filtrates. *Bioresour. Technol.*, 101, 4554, 2010.

38. Eugenio, M.E., Carbajo, J.M., Martín, J.A., González, A.E., Villar, J.C., Laccase production by *Pycnoporus sanguineus* under different culture conditions. *J. Basic Microbiol.*, 49, 433, 2009.
39. Kenkebashvili, N., Elisashvili, V., Wasser, S.P., Effect of carbon, nitrogen sources, and copper concentration on the ligninolytic enzyme production by *Coriolopsis gallica*. *J. Waste Conversion, Bioproducts Biotechnol.*, 1, 22, 2012.
40. Collins, P.J., Dobson, A.D.W., Regulation of laccase gene transcription in *Trametes versicolor*. *Appl. Environ. Microbiol.*, 63, 3444, 1997.
41. Elisashvili, V., Kachlishvili, E., Physiological regulation of laccase and manganese peroxidase production by white-rot basidiomycetes. *J. Biotechnol.*, 144, 37, 2009.
42. Monteiro, M.C., Carvalho, M.E.A.D., Pulp bleaching using laccase from *Trametes versicolor* under high temperature and alkaline conditions. *Appl. Biochem. Biotechnol.*, 70, 983, 1998.
43. Patel, S.S., Parekh, S.N., Laccase–production and it's application. *World J. Pharma. Biotechnol.*, 2, 57, 2015.
44. Buswell, J.A., Cai, Y.J., Chang, S.T., Effect of nutrient and manganese on manganese peroxidase and laccase production by *Lentinula (Lentinus) edodes*. *Microbiol. Lett.*, 128, 81, 1995.
45. Dong, J.L., Zhang, Y.W., Zhang, R.H., Huang, W.Z., Zhang, Y.Z., Influence of culture conditions on laccase production and isozyme patterns in the white-rot fungus *Trametes gallica*. *J. Basic Microbiol.*, 45, 190, 2005.
46. Pandey, A., Soccol, C.R., Laroche, C., *Current development in solid state fermentation*. Springer, Asiatech Publisher Inc., New Delhi, 2008.
47. Ravikumar, G., Kalaiselvi, M., Gomathi, D., Vidhya, B., Devaki, K., Uma, C., Effect of laccase from *Hypsizygus ulmarius* in decolorization of different dyes. *J. Appl. Pharma. Sci.*, 3, 150, 2013.
48. Chhaya, U., Gupte, A., Effect of different cultivation conditions and inducers on the production of laccase by the litter-dwelling fungal isolate *Fusarium incarnatum* LD-3 under solid substrate fermentation. *Annal. Microbiol.*, 63, 215, 2013.
49. Elsayed, M.A., Mohamed, M.H., Ali, M., ElshafeiBakry, M.H., Abdelmageed, M.O., Optimization of cultural and nutritional parameters for the production of laccase by *Pleurotus ostreatus* ARC280. *Br. Biotechnol J.*, 2, 115, 2012.
50. Nandal, P., Ravella, S.R., Kuhad, R.C., Laccase production by *Coriolopsis caperata* RCK2011: optimization under solid state fermentation by Taguchi DOE methodology. *Sci. Rep.*, 3, 1386, 2013.
51. Nasreen, Z., Usman, S., Yasmeen, A., Nazir, S., Yaseen, T., Ali, S., Ahmad, S., Production of laccase enzyme by basidomycetes *Coriolus versicolor* through solid state fermentation. *Int. J. Curr. Microbiol. Appl. Sci.*, 4, 1069, 2015.
52. Vantamuri, A.B., Kaliwal, B.B., Production of laccase by newly isolated *Marasmius* sp. BBKAV79 in solid state fermentation and its anti-proliferative activity. *Int. J. Pharma Sci. Res.*, 7, 4978, 2016.

53. Xin, F.X., Geng, A.L., Utilization of horticultural waste for laccase production by *Trametes versicolor* under solid-state fermentation. *Appl. Biochem. Biotechnol.*, 163, 235, 2011.
54. Sun, W., Xu, M., Xia, C., Li, A., Sun, G., Enhanced production of laccase by *Coriolus hirsutus* using molasses distillery wastewater. *Front. Environ. Sci. Engg.*, 7, 200, 2013.
55. Youshuang, Z., Haibo, Z., Mingle, C., Zhenzhen, W., Feng, H., Peiji, G., Production of a thermostable metal-tolerant laccase from *Trametes versicolor* and its application in dye decolorization. *Biotechnol. Bioprocess Engg.*, 16, 1027, 2011.
56. Ding, Z., Chen, Y., Xu, Z., Peng, L., Xu, G., Gu, Z., Zhang, L., Shi, G., Zhang, K., Production and characterization of laccase from *Pleurotus ferulae* in submerged fermentation. *Annal. Microbiol.*, 64, 121, 2014.
57. Niladevi, K.N., Prema, P., Effect of inducers and process parameters on laccase production by *Streptomyces psammoticus* and its application in dye decolourization. *Bioresour.Technol.*, 99, 4583, 2008.
58. Meza, J.C., Lomascolo, A., Casalot, L., Sigoillot, J.C., Auria, R., Laccase production by Pycnoporus cinnabarinus grown on sugar-cane bagasse: influence of ethanol vapours as inducer. *Process Biochem.*, 40, 3365–3371, 2005.
59. Strong, P.J., Improved laccase production by *Trametes pubescens* MB89 in distillery wastewaters. *Enzyme Res.*, 2011.
60. Birhanli, E., Yesilada, O., The effect of various inducers and their combinations with copper on laccase production of *Trametes versicolor* pellets in a repeated-batch process. *Turkish J. Biol.*, 41, 587, 2017.
61. Valle, J.S., Vandenberghe, L.P.S., Oliveira, A.C.C., Tavares, M.F., Linde, G.A., Colauto, N.B., Soccol, C.R., Effect of different compounds on the induction of laccase production by *Agaricus blazei*. *Genet. Mol. Res.*, 14, 15882, 2015.
62. Barbosa, A.M., Dekker, R.F.H., Hardy, S.G.E., Veratryl alcohol as an inducer of laccase by an ascomycete, *Botryosphaeria* sp., when screened on the polymeric dye Poly R-478. *Lett. Appl. Microbiol.*, 23, 93, 1996.
63. Manavalan, T., Manavalan, A., Thangavelu, K.P., Heese, K., Characterization of optimized production, purification and application of laccase from *Ganoderma lucidum*. *Biochem. Eng. J.*, 70, 106–114, 2013.
64. Mann, J.J.L., Paul, M.P.R.N., Paul, S.-H.H., Nair (Tan), N.G., Use of olive mill wastewater as a suitable substrate for the production of laccase by *Cerrena consors*. *Int. Biodeterior. Biodegradation.*, 99, 138, 2015.
65. Schliephake, K., Mainwaring, D.E., Lonergan, G.T., Jones, I.K., Baker, W.L., Transformation and degradation of the disazo dye Chicago Sky Blue by a purified laccase from *Pycnoporus cinnabarinus*. *Enzyme Microbial Technol.*, 27, 100, 2000.
66. Vasconcelos, A.F.D., Barbosa, A.M., Dekker, R.F.H., Scarminio, I.S., Rezende, M.I., Optimization of laccase production by *Botryosphaeria* sp. in the presence of veratryl alcohol by the response-surface method. *Process Biochem.*, 35, 1131, 2000.

67. Bollag, J.M., Leonowicz, A., Comparative studies of extracellular laccases. *Appl. Environ. Microbiol.*, 48, 849, 1984.
68. Robene-Soustrade I., Lung-Escarmant, B., Laccase isoenzyme patterns of European Armillaria species from culture filtrates and infected woody plant tissues. *Eur. J. Forest Pathol.*, 27, 105, 1997.
69. Luke, A.K., Burton, S.G., A novel application for *Neurospora crassa:* Progress from batch culture to a membrane bioreactor for the bioremediation of phenols. *Enzyme Microbial Technol.*, 29, 348, 2001.
70. Sedarati, M.R., Keshavarz, T., Leontievsky, A.A., Evans, C.S., Transformation of high concentrations of chlorophenols by the white-rot basidiomycete *Trametes versicolor* immobilized on nylon mesh. *Electron. J. Biotechnol.*, 6, 27, 2003.
71. Park, C., Lee, B., Han, E.-J., Lee, J., Kim, S., Decolorization of acid black 52 by fungal immobilization. *Enzyme Microbial Technol.*, 39, 371, 2006.
72. Mohorcic, M., Friedrich, J., Pavko, A., Decoloration of the diazo dye reactive black 5 by immobilized *Bjerkundera adusta* in a stirred tank bioreactor. *Acta. Chim. Slov.*, 51, 619, 2004.
73. Ghosh, P., Ghosh, U., Statistical optimization of laccase production by *Aspergillus flavus* PUF5 through submerged fermentation using agro-waste as cheap substrate. *Acta Biol. Szeged.*, 61, 25, 2017.
74. Sarnthima, R., Khammuang, S., Laccase Production by *Pycnoporus sanguineus* grown under liquid state culture and its potential in remazol brilliant blue R decolorization. *Int. J. Agri. Biol.*, 15, 215, 2013.
75. Johnsy, G., Kaviyarasan, V., Effect of physico-chemical parameters on ligninolytic enzyme production of an indigenous isolate of *Neolentinus kauffmanii* -under submerged culture condition. *World J. Pharm. Sci.*, 2(5), 464, 2014.
76. Prasad, K.K., Chaganti, S.R., Venkata Mohan, S., Sarma, P.N., Solid state fermentation of laccase from new pulse husks: process optimization and bioprocess study. *Int. J. Innovations Biol. Chem. Sci.*, 2, 22, 2011.
77. Mathur, G., Nigam, R., Jaiswal, A., Kumar, C., Bioprocess parameter optimization for laccase production in solid state fermentation. *Int. J. Biotechnol. Bioengg. Res.*, 4, 521, 2013.
78. Kalra, K., Shavez, M., Production of laccase from banana skin by *Trametes hirsuta* in solid state and submerged fermentation. *Asian J. Biochem. Pharm. Res.*, 4, 2014.
79. Chiranjeevi, P.V., Sathish, T., Pandian, M.R., Harmonizing various culture conditions and inducers for hyper laccase production by *Pleurotus ostreatus* PVCRSP-7 in solid state fermentation. *J. Pharm. Res.*, 8, 526, 2014.
80. Niladevi, K.N., Sukumaran, R.K., Prema, P., Utilization of rice straw for laccase production by *Streptomyces psammoticus* in solid-state fermentation. *J. Indus. Microbiol. Biotechnol.*, 34, 665, 2007.
81. Zilly, A., Bazanella, G.C.S., Helm, C.V., Araújo, C.A.V., Souza, C.G.M., Rosane, A.B., Peralta, M., Solid-state bioconversion of passion fruit waste by white-rot fungi for production of oxidative and hydrolytic enzymes. *Food Bioprocess Technol.*, 5, 1573, 2012.

82. Sun, Q.Y., Hong, Y.Z., Xiao, Y., Fang, W., Fang, J., Decolorization of textile reactive dyes by the crude laccase produced from solid-state fermentation of agro-byproducts. *World J. Microbiol. Biotechnol.*, 25, 1153, 2009.
83. Couto, S.R., Laccase from *Trametes hirsute* grown on paper cuttings: application to synthetic dye decolorization at different pH values. *Engg. Life Sci.*, 7, 229, 2007.
84. Arora, D.S., Sharma, R.K., Ligninolytic fungal laccases and their biotechnological applications. *Appl. Biochem. Biotechnol.*, 160, 1760, 2010.
85. Renzetti, S., Courtin, C.M., Delcour, J.A., Arendt, E.K., Oxidative and proteolytic enzyme preparations as promising improvers for oat bread formulations: rheological, biochemical and microstructural background. *Food Chem.*, 119, 1465, 2010.
86. Oktem, H.A., Senyurt, O., Eyidogan, F.I., Bayrac, C., Yilmaz, R., Development of a laccase based paper biosensor for the detection of phenolic compounds. *J. Food Agric. Environ.*, 10, 1030, 2012.
87. Ali, H., Ahmad, W., Haq, T., Decolorization and degradation of malachite green by *Aspergillus flavus* and *Alternaria solani. Afric. J. Biotechnol.*, 8, 1574, 2009.
88. Brondani, D., Souza, B.D., Souza, B.S., Neves, A., Vieira, L.C., PEI-coated gold nanoparticles decorated with laccase: a new platform for direct electrochemistry of enzymes and biosensing applications. *Biosen. Bioelect.*, 42, 242, 2013.
89. Faramarzi, M.A., Forootanfar, H., Biosynthesis and characterization of gold nanoparticles produced by laccase from *Paraconiothyrium variabile. Colloids Surf. B: Biointerfaces.*, 87, 23, 2011.
90. Lee, K.M., Kalyani, D., Tiwari, M.K., Kim, T.S., Dhiman, S.S., Lee, J.K., Kima, I.W., Enhanced enzymatic hydrolysis of rice straw by removal of Phenolic compounds using a novel laccase from yeast *Yarrowia lipolytica. Bioresour. Technol.*, 123, 636, 2012.
91. Rocasalbas, G., Francesko, A., Tourino, S., Fernandez-Francos, X., Guebitzc, G.M., Tzanov, T., Laccase-assisted formation of bioactive chitosan/gelatin hydrogel stabilized with plant polyphenols. *Carbohyd. Polym.*, 92, 989, 2013.
92. Minussi, R.C., Miranda, M.A., Silva, J.A., Ferreira, C.V., Aoyama, H., Marangoni, S., Rotilio, D., Pastore, G.M., Duran, N., Purification, characterization and application of laccase from *Trametes versicolor* for colour and phenolic removal of olive mill wastewater in the presence of 1- hydroxybenzotriazole. *Afric. J. Biotechnol.*, 6, 1248, 2007.
93. Luo, H., Jin, S., Fallgren, P.H., Park, H.J., Johnson, P.A., A novel laccase-catalyzed cathode for microbial fuel cells. *Chem. Engg.* J. 165, 524, 2010.
94. Ardhaoui, M., Bhatt, S., Zheng, M., Dowling, D., Jolivalt, C., Khonsari, F.A., Biosensor based on laccase immobilized on plasma polymerized allyl amine/carbon electrode. *Mat. Sci. Engin. C*, 33, 3197, 2013.
95. Wong, K.S., Huang, Q., Au, C.H., Wang, J., Kwan, H.S., Biodegradation of dyes and polyaromatic hydrocarbons by two allelic forms of Lentinulaedodes laccase expressed from *Pichia pastoris. Bioresour. Technol.*, 104, 157, 2012.

96. Gómez-Toribio, V., García-Martín, A.B., Martínez, M.J., Martínez, Á.T., Guillén, F., Enhancing the production of hydroxyl radicals by *Pleurotus eryngii* via Quinone Redox Cycling for pollutant removal. *Appl. Environ. Microbiol.*, 75, 3954, 2009.
97. Jang, M.Y., Ryu, W.R., Cho, M.H., Laccase production from repeated batch cultures using free mycelia of *Trametes* sp. *Enzyme Microbial. Technol.*, 30, 741, 2002.
98. Gaitan, I.J., Medina, S.C., González, J.C., Rodríguez, A., Espejo, A.J., Osma, J.F., Sarria, V., Alméciga-Díaz, C.J., Sánchez, O.F., Evaluation of toxicity and degradation of a chlorophenol mixture by the laccase produced by *Trametes pubescens*. *Bioresour. Technol.*, 102, 3632, 2011.
99. Moya, R., Hernández, M., García-Martín, A., Ball, A.S., Arias, M.E., Contributions to a better comprehension of redox-mediated decolouration and detoxification of azo dyes by a laccase produced by *Streptomyces cyaneus* CECT 3335. *Bioresour. Technol.*, 101, 2224, 2010.
100. Sanghi, R., Verma, P., Biomimetic synthesis and characterisation of protein capped silver nanoparticles. *Bioresour. Technol.*, 100, 501, 2009.
101. Gnanasalomi, D.V., Gnanadoss, J.J., Laccases from fungi and their applications: recent developments. *Asian J. Exp. Biol. Sci.*, 4, 581, 2013.
102. Aracri, E., Colom, J.F., Vidal, T., Application of laccase-natural mediator systems to sisal pulp: An effective approach to bio bleaching or functionalizing pulp fibres. *Bioresour. Technol.*, 100, 5911, 2009.
103. Kuuva, T., Lantto, R., Reinikainen, T., Buchert, J., Autio, K., Rheological properties of laccase-induced sugar beet pectin gels. *Food Hydrocoll.*, 17, 679, 2003.
104. Claus, H., Faber, G., Konig, H., Redox-mediated decolorization of synthetic dyes by fungal laccases. *Appl. Microbiol. Biotechnol.*, 59, 672, 2002.
105. Vijaykumar, M.H., Veeranagouda, Y., Neelakanteshwar, K., Karegoudar, T.B., Decolorization of 1:2 metal complex dye Acid blue 193 by a newly isolated fungus, *Cladosporium cladosporioides*. *World J. Microbiol. Biotechnol.*, 22, 157, 2006.
106. Kirby, N., Marchant, R., Geoffrey M., Decolourisation of synthetic textile dyes by *Phlebia tremellosa*. *FEMS: Microbiol. Lett.*, 188, 93, 2000.
107. Pazarloglu, N.K., Sariisik, M., Telefoncu, A., Laccase: production by *Trametes versicolor* and application to denim washing. *Process Biochem.*, 40, 1673, 2005.
108. Montazer, M., and Maryan, A.S., Application of laccases with cellulases on denim for clean effluent and repeatable biowashing. *J. Appl. Polym. Sci.*, 110, 3121, 2008.
109. Pointing, S.B., Feasibility of bioremediation by white-rot fungi. *Appl. Microbiol. Biotechnol.*, 57, 20, 2001.

8

Use of Fermentation Technology for Value Added Industrial Research

Biva Ghosh[1], Debalina Bhattacharya[2] and Mainak Mukhopadhyay[1]*

[1]Department of Biotechnology, JIS University, Kolkata, West Bengal, India
[2]Department of Biochemistry, University of Calcutta, Kolkata, West Bengal, India

Abstract

Fermentation is a biochemical process of microorganism for the production of different enzymes, antibiotics, etc. which are used in different industry for the energy production, food processing industries, antibiotic production, protein synthesis, hormone synthesis, and many more industrially important enzymatic productions. It is also used for waste management such as biofuels production from lignocellulosic biomass that includes mainly agricultural wastes, which are produced in high amount from household, industries and agricultural fields. Other than this is also used to produce bio-surfactant, polymers production such as bacterial cellulose production. There are different types of fermentation process for enzyme production which are submerged and solid state production system. Except this generalized fermentation process includes batch fermentation, fed-batch fermentation and continuous culture. In submerged fermentation microorganisms are grown in liquid medium where as in solid state fermentation microorganism are grown on solid substrate. The microorganism used the nutrients present in the substrate such as carbon, nitrogen, essential salts, amino acids, etc. to synthesize compounds which are needed for their survival. Change in the fermentation condition leads to different compound production which can be economically important or useless. Thus, for the production of precise and faultless production, fermentation condition should be perfect which will enhance the production rate of fermentation product and increase its economic value. In this chapter, different type of microorganisms, their application, and their recent developments are discussed. Beside this, industries depending on fermentation of plant based processing substrate has lots of unwanted wastes due to easy decay of the substrates.

**Corresponding author*: m.mukhopadhyay85@gmail.com

Arindam Kuila and Vinay Sharma (eds.) Principles and Applications of Fermentation Technology, (141–162) © 2018 Scrivener Publishing LLC

Thus many researcher has suggested different strategies for zero waste or to reuse waste for different other product formation. This leads to waste management as well as it is eco-friendly and also prevents loss of valuable products. This chapter thus also put some light on different enhancement of fermentation technology for the production of zero waste and value added products.

Keywords: Fermentation, value added products, zero waste

8.1 Introduction

Fermentation is a biochemical process which converts sugar molecules into different components such as acid, alcohol, gases and more compounds using different types of microorganisms. It is used for the production of many value added products such as enzymes, biofuels as renewable energy source, food products, medicines etc. Fermentation of lignocellulosic biomass such as agricultural waste leads to waste management by the conversion of these waste into biofuels or important chemical products such as lactic acid succinic acid etc. and many more chemical such as 1,3-Propanediol, Polyhydroxyalkanoates (PHA), Exopolysaccharides, Vanillin etc. Other than this many more valuable products such as biofuels like biodiesel, ethanol, biogas, biohydrogen, etc. are produced from lignocellulosic waste. These chemical has high economic importance. Singh *et al.* (2011) has already discussed about the zero waste management of lignocellulosic wastes by using fermentation technology and synthesizing important products such as enzymes, biofuels, biogas, organic acids, and polyhydroxybutyrate (PHB) and the left over can be converted to biomanure. Erickson *et al.* (2012) has also discussed about many industrially important chemicals which are used for many other synthesis purpose such as bioplastics, pharmaceuticals use, surfactants, biopolymers, etc. Lactic acid and vanillin are highly used for milk product formation and flavor respectively. In recent days, food additives have become very popular due to different reasons such as food colors, flavors, acid regulators, essence, preservatives, etc. Other than this, many chemical used in cosmetic products such as scents are also synthesized by fermentation process. This help to reduce the load of synthetic products over nature and leads to eco-friendlier process for the production of needful things. These products have high value in market and are unavoidable needs of this century. However, the production cost of these valuable products is costly and researches are already done to lower the cost of production and optimize the fermentation system to zero waste. Thus, this chapter focuses on the production of different valuable products using fermentation technology on different waste sources or renewable sources for the economically important high yield products.

8.2 Fermentation

Fermentation process is a biochemical activity of microorganism during their life process such as growth and development, senescence and death. Fermentation process mainly converts sugar molecules into industrially valuable compounds. In this process the microorganisms are grown under suitable condition on suitable substrate and the end product formed by the microorganism are extracted from the media. The container in which the fermentation is done is called bioreactor or fermenter. There are different types of bioreactors which are used to control fermentation condition and produce different end products. Some of the types of bioreactors are as follows.

- External recycle airlift bioreactor
- Internal recycle airlift bioreactor
- Tubular tower bioreactor
- Nathan bioreactor
- Stirred bioreactor

Depending on the bioreactor different type of fermentation process are involved. There are mainly three types of fermentation process.

1. Batch fermentation
In bath fermentation, sterile nutrient broth is inoculated with microorganism and cultured in a closed bioreactor for specified time and in specific conditions. Nothing is added in the mid of the fermentation process.

2. Fed batch fermentation
In fed batch fermentation, initially small concentration of substances are added and then these substances are added in small quantities continuously throughout the whole process of fermentation.

3. Continuous fermentation
In continuous fermentation process, equal amount of sterile nutrient broth is added to the open bioreactor and simultaneously equal amount of cultured broth with the products is taken out for purification of products.

Depending on the type of fermentation different products such as proteins, enzymes, alcohol, acids, etc. are produced. Some of the industrially valuable fermentation products are discussed below.

8.3 Biofuel Production

In recent years increase in need for biofuels in replace of fossil fuels are in high demands. Fermentation is one of the most important step followed to generate fossil fuels from lignocellulosic waste or other biodegradable sources such as microalgae etc. Biofuels such as biodiesels, bioethanol, butanol, biohydrogen etc. are one of the most important value added products which are produced from biodegradable and waste materials leading to zero waste concept.

8.3.1 Biohydrogen

Since, hydrogen has unique qualities such as high gravimetric based energy combustion and absence of oxidation or greenhouse gases production, is considered as green energy source. Since decades, production of biohydrogen from natural biodegradable sources has been studied. Present scenario of global need for green energy in replacement of fossil fuels has gained focus on biohydrogen production from different natural sources. Biohydrogen production was first reported by fermentation process during 1970s [1]. Though during fermentation process different by-products are formed which reduces the yield of biohydrogen production but still continuous fermentation using simple sugars as substrates dominates the industry for biohydrogen formation [2]. One of the simple polysaccharide, glycerol has high potential to produce high yield of biohydrogen by fermentation process. Glycerol itself is produces as waste during pretreatment and production of organic alcohols and organic acids from lignocellulose wastes. Other than this, it is cheaper to be produced and on fermentation of glycerol, only 10% is converted to biodiesel as by-product [2]. Thus, it is most suitable substrate for biohydrogen formation and leads to zero waste strategy. Different studied on rate of biohydrogen generation depending on glycerol concentration are studied in recent years [2]. The maximum biohydrogen produced by fermentation of glycerol was estimated as 0.8 mol/mol glycerol by using *T. maritima* in batch chemostat fermenter [3]. Dark fermentation of glycerol produces hydrogen acetate, butyrate, ethanol, and 1,3-propanediol as by-products along with biohydrogen production. These are used for Nicotinamide adenine dinucleotide (NADH) production and adversely affect the yield of hydrogen formation. Other than this direct fermentation of lignocellulosic biomass after pretreatment also produces biohydrogen [2]. Pretreatment of lignocellulose yields reducing sugar and polysaccharides which on fermentation produces biohydrogen. Plant resources itself acts as a storage of biohydrogen. Photosynthesis is another process through which biohydrogen is produced. Photosynthetic process occurs in two stages, that is, light dependent and light independent [4]. In light dependent pathways, the light energy is directly absorbed by the chlorophyll and converted to Adenosine Triphosphate (ATP) and Nicotinamide adenine dinucleotide phosphate with hydrogen (NADPH). During this process electron transport chain is activated by the electron carriers present in thylakoids such as Ferredoxin, Plastoquinone, Plastocyanin, Cytochrome C. Ferredoxin Nicotinamide adenine dinucleotide phosphate (NADP) oxidoreductase reduces $NADP^+$ to NADPH in presence of water. In oxygenic condition of photosynthesis water molecules are broken in electrons and protons. These protons are sometimes converted to molecular hydrogen by hydrogenase rather than accepted by Ferredoxin. Thus, oxygenic condition of photosynthesis produces bio-hydrogen biofuel. On the other hand, light independent photosynthesis produces ATP and NADPH, which further converts to sugar molecules which help to synthesize bio-alcohol, biodiesel and on further fermentation produce bio-hydrogen as biofuels [4, 5]. Thus, production of biohydrogen from optimizing the photosynthesis pathway as well as reusing the glycerol rich waste make this process clean and zero waste process. In this way, even the waste can be converted to valuable industrial product.

8.3.2 Biodiesel

Vegetable oil or animal fats or any natural form of oil after transesterification produces biodiesel. Transesterification is a process of exchanging organic group of an ester with the organic group of an alcohol under catalysis of acid or base. Fatty acids, triglycerides or oils are found from many plant animal or microbial resources [6]. Many process are involved to extract oils from lignocellulosic biomass which on further trans-esterified to form biodiesel. Though till now plant based oils are most abundantly used for biodiesel production but it has some limitation. Plant based oils are also used as edible sources thus, raising conflict on providing priority on production of biofuels or used as foods. Other than this some non-edible oil rich crops are present but less abundance of lands for non-edible crops raise the problem [7, 8]. Microalgae in such case is the solution. Many researches are done on microalgae for the production of biodiesel. Depending on species 70% of the cell consist of triglyceride in case of microalgae [8]. Microalgae can be easily cultured and maintained even in hash condition and in large quantity. Fermentation of microalgae produces bio-oil which on further transesterification produces biodiesel. As microalgae grows very fast thus cultivating microalgae in fermenter solve the problem of space as well as conflict on food crop. Heterotrophic fermentation of *Chlorella protothecoides* on glycerol as carbon source using semi continuous fermentation process produced more bio-oil than fed-batch fermentation thus, yielding high bio-oil for biodiesel production [9]. Being scarcity of fossil fuel and its high need, production of biodiesel from these renewable sources is a relief for the world, leading to rapidly growing industries in this sector.

8.3.3 Bioethanol

Another biofuel widely used in worldwide is ethanol. Ethanol is produced by scarification and fermentation of reducing sugars. Lignocellulosic wastes are the best source for generating reducing sugars. Different pretreatments are performed to hydrolyze lignocellulose into simple sugar moieties. Bioethanol from secondary waste can be produced by fermentation process. Yeast are generally used to produce bioethanol [10]. *Saccharomyces cerevisiae* are the most commonly used yeast for production of ethanol. But *Saccharomyces cerevisiae* has some limitation such as, it only ferments hexose into ethanol but is inefficient to grow on pentose sugar [11]. Some other yeasts such as *Candida shehatae*, *Candida tropicalis*, and *Pichia stipitis* can synthesize ethanol from both pentose and hexose sugar but have low tolerance towards ethanol [10]. Thus, genetically modified bacteria are produced to combat this problem. Some of the genetically modified bacteria synthesizing ethanol are *Escherichia coli* KO11, *Klebsiellaoxytoca* P2, and *Erwiniachrysanthemi* EC16 [12]. Berlowska *et al.* (2016) has used two different strains of yeast for synthesis of ethanol by fermentation. They are *Saccharomyces cerevisiae* for hexose fermentation and *Pichia stipites* for pentose fermentation from sugar beet pulp. Among them *Saccharomyces cerevisiae* has higher yields (1 g/L) than *Pichia stipitis* (0.5 g/L) [10]. Thus, researches are still needed to generate robust ethanogenic bacteria whose tolerance level is high for ethanol as well as for inhibitors and has high productive yield [13].

8.4 1,3-Propanediol

1,3-Propanediol is an important monomer produce by fermentation of glycerol. It is used in the production of polyethers, polyesters, polyurethanes by polycondensation of this monomer. Most of the fermentation process leading to alcohol production such as bioethanol production leads to glycerol production as by-products along with other products. Glycerol on further fermentation produces 1,3-propanediol. Some of the bacterium responsible for fermentation of glycerol to 1,3-propanediol are *Klebsiella pneumoniae*, *Citrobacter freundii*, and *Clostridium pastwureunum* [14]. During fermentation first 3-hydroxypropionaldehyde is produced then further reduced by $NADH_2$ to 1,3-propanediol. This $NADH_2$ is the result of oxidative pathway of glycolysis reaction of glycerol and produces many by-products other than 1,3-Propanediol such as acetate, lactate, succinate, butyrate, ethanol, butanol, and 2,3-butanediol. Production 1,3-propanediol is high as 67% mol/mol with acetic acid as maximum by-product of oxidative pathway [14]. As the yield of 1,3-propanediol depends on the reductive and oxidative reaction pathway during fermentation; fermentation process needs to be monitored. Most of the loss of yield of 1,3-propanediol production is due to the conversion of glycerol into cell mass by microbes. Thus, a variety of fermentation process and their optimization are done to maximize the 1,3-propanediol yield. Even product tolerant mutant species of microbes such as *C. butyricum* and *K. pneummoniae* are known to produce 70 g/L and 78 g/L of 1,3-propanediol in fed-batch fermentation system with controlled pH [13]. Another approach of conversion of glucose to 1,3-propanediol was done by two step fermentation. Firstly, conversion of glucose to glycerol by fermentation and then again ferment the glycerol to 1,3-propanediol [15]. *S. cerevisiae* produces glycerol from glycolytic intermediate that is dihydroxyacetone 3-phosphate by using two enzymes-dihydroxyacetone 3-phosphate dehydrogenase and glycerol-3-phosphate phosphatase. On the other hand, *K. pneumonia* converts 1,3-propanediol from glycerol by using two enzymes that are glycerol dehydratase and 1,3-propanediol dehydrogenase [16]. Thus, an *E. coli* strain was created by inserting genes for glycerol synthesis and genes for conversion of glycerol to 1,3-propanediol from *S. cerevisiae* and *K. pneumonia*, respectively. This mutated strain of *E. coli* can synthesize glycerol from sugar molecules as well as can convert glycerol to 1,3-propanediol as efficiently as in their native forms [16]. Thus, this not only reduces cost of production but also can be used to treatment of leftover waste biomass of lignocellulose with mixture of sugar monomers and by-products such as glycerol for 1,3-propanediol production and leads to zero waste technology.

8.5 Lactic Acid

Lactic acid (2-hydroxypropionic acid) is one of the most important organic acid used abundantly by industries [13]. It has many applications in food, pharmaceutical, and textile industries. It is used as preservative, acidulant, and flavoring in food industries and in chemical industries it is used as raw material for synthesis of lactate ester, propylene glycol, 2,3-pentanedione, propanoic acid, acrylic acid, acetaldehyde, and dilactide [17, 18]. In recent days, its demand has increased globally due to its ability to form poly lactic acid (PLA) which is used in formation of different biodegradable bioplastics or polymers with modifications. Lactic acid has two optical isomers L(+)-lactic acid and D(−)-lactic acid [19]. For high quality PLA production optical quality of lactic acid is crucial [17]. Lactic acid production can be done by chemical process or by fermentation process [20]. Fermentation process is more favorable than chemical due to many reasons such as fermentation process produces optically pure form of lactic acid whereas, chemical process produces racemic mixture of DL-lactic acid [21]. Chemical process uses petrochemical resources as substrate for lactic acid production whereas, in biological process microorganism ferment sugar molecules present in lignocellulosic biomass or other biodegradable material into lactic acid [22]. As in fermentation process all substrates are renewable and biodegradable thus this process is eco-friendlier. In this process waste products such as agricultural waste [23], food industries waste [24], Cassava starch wastewater [25], and many more lignocellulose waste biomass can be used for lactic acid synthesis. Lactic acid producing bacteria can be classified into two groups, homofermentative and heterofermentative [17]. Homofermentor bacteria convert sugar completely into lactic acid whereas heterofermentor bacteria catabolize glucose into ethanol and carbon dioxide as by-products and lactic acid as main product. Homofermentor bacteria synthesize lactic acid by Embden-Meyerhof pathway or also known as glycolysis pathway [26]. Some of the microorganism well known for lactic acid production are *Rhizopus oryzae* ATCC 52311, *Rhizopus oryzae* NRRL 395, *Enterococcus faecalis* RKY1, *Lactobacillus rhamnosus* ATCC 10863, *Lactobacillus helveticus* ATCC 15009, *Lactobacillus bulgaricus* NRRL B-548, *Lactobacillus casei* NRRL B-441, *Lactobacillus plantarum* ATCC 21028, *Lactobacillus pentosus* ATCC 8041, *Lactobacillus amylophilus* GV6, *Lactobacillus delbrueckii* NCIMB 8130, *Lactococcus lactis sp.* IFO 12007 [17]. Chaisu *et al.* (2014) has optimized the lactic acid production from Sugarcane molasses, which is an agricultural waste by using *Lactobacillus casei* M-15 [23]. The optimized the maximum yield of lactic acid by response surface methodology (RSM) by culturing the bacteria under 3.82% of molasses and 8.02% of inoculum level within 24 hours and at 37 °C. This process relatively increased the lactic acid production and follows sustainable development. It also reuses the agricultural waste and help in waste management. This process is a pollution free method for lactic acid production [23]. Tosungnoen *et al.* (2014) treated synthetic cassava starch wastewater (SCW) with amylolytic *Lactobacillus plantarum* MSUL 702 bacteria by repeated-batch and simultaneous saccharification and fermentation for lactic acid production

[25]. In this process, not only 28.71 g/L of lactic acid was produced in first batch of culture but also 98% of chemical oxygen demand (COD) and 85% of total kjeldahl nitrogen (TKN) was removed in second batch of culture. This bacterium not only retains high lactic acid production but also retains treatment efficiency for consecutive four batches. Thus, waste is also reused for value added product formation [25]. Pleissner *et al.* (2016) used mixed restaurant food waste as raw material for lactic acid fermentation [24]. Food contains lots of sugar and nitrogen sources which on simultaneous saccharification and fermentation by Lactobacillus sp. or Streptococcus sp. strains produces L(+)-lactic acid. In laboratory under technical scale and sterile condition 702 g L^{-1} of L(+)-lactic acid is produced after purification [24]. Thus, even the restaurant waste with other household or industrial waste can be used to produce valuable industrial products.

8.6 Polyhydroxyalkanoates

PHA are natural homo and hetero polysaccharides such as poly 3-hydroxybutyric acid (PHB) and poly 3-hydroxybutyric-co-3-hydroxyvaleric acid (PHB-V) [13]. These are produced by many microorganism such as *Ralstonia eutropha, Alcaligenes latus, Azotobacter vinelandii, Chromobacteruium violaceum, methylotrophs*, and *pseudomonads* in limited nutrient condition [27]. These microorganisms when grown in media containing limitation in one of the nutrient element N, P, S, 0_2, or Mg and excess of carbon sources synthesizes PHA intracellularly as stored carbon and energy source. PHA readily breakdown to carbon dioxide and water by microorganism [27]. It is a green polymer for bioplastic formation like PLA. PLA is a polymer of lactic acid and produced by microorganisms. PHA synthesized by bacteria are of two types short-chain-length PHAs (scl-PHAs) consist of 3–5 carbon atoms and medium-chain-length PHAs (mcl-PHAs) consist of 6–14 carbon atoms. PHB is a type of scl-PHA which is stiff and brittle whereas, PHB-V which is a combination of PHB and 3-hydroxyvalerate (HV) is tougher, flexible, and has more thermal stability [28]. Sci-PHB is generally used in disposable food packaging whereas mcl-PHA is used as elastomers. As biodegradable plastics and polymers is in high demand, thus, PHA has become one of the most wanted product of industries. But due to its high cost of production wide commercialization and industrialization of PHA is struggling [29]. Most of the bacteria synthesizing PHA are known to survive in stress conditions. Bacterial strains such as genus of *Sphingobacterium, Bacillus, Pseudomonas*, and *Rhodococcus* are known to produce PHA and degrade environmental pollutants [29]. Most of the environment pollution includes volatiles aromatic hydrocarbon that causes mutagenic, carcinogenic, allergic, and cytotoxic threats to life. Due to extensive use of petrochemical products these environmental pollution has increased. As PHA synthesizing bacteria has the ability to degrade these hydrocarbon and produce PHA which is a biodegradable valuable polymer is a great relief to the society [30]. Some of the *Pseudomonas* species: *P. putida* F1, mt-2, and CA-3 together are known to synthesize PHA from toluene, benzene, ethylbenzene, xylene, and styrene mixture [31]. Another example of substrate for PHA synthesis are vegetable oils. Vegetable oils are rich in carbon sources and can act as good substrate for PHA synthesis. Hwan *et al.* (2008) has also discussed the synthesis of PHA on vegetable oils by *Pseudomonas sp.* strain DR2. This bacterium has successfully synthesized PHA on waste vegetable oils [32]. Rhu *et al.* (2003) has discussed about waste treatment of sewage waste and PHA production. As discussed earlier PHA synthesizing bacteria are stress tolerant thus, it can be used to treat sewage waste, which are rich in carbon and hydrocarbon sources [33]. Colombo *et al.* (2017) has discussed about PHA synthesis from municipality waste using mixture of PHA producing microbes in Europe [34]. PHA synthesized by bacteria is done by an enzyme called PHA synthase. *P. pseudoflava* and *P. palleronii* are known to degrade synthetic waste and synthesis of PHA. PHA synthase enzyme was isolated by Reddy *et al.* (2017) from these bacteria and can be used for waste treatment and PHA synthesis in a cost effective manner [35]. Thus, production of PHA from waste sources lowers the cost of production which is major problem and help in bioremediation. In this way, waste is also converted to value added industrial products.

8.7 Exopolysaccharides

Exopolysaccharides (EPS) are synthesized by different bacteria from sugar molecules. They are of two types, homopolysaccharides and heteropolysaccharides. Homopolysaccharides such as dextran, alternan, pullulan, levan, β-D-glucansare synthesized by bacteria *Leu. mesenteroides subsp. mesenteroides, Leu. mesenteroides, Aureobasidium pullulons, levan, Z. mobilis, Streptococcus sp.* respectively. Hetropolysaccharides such as alginate, gellan, xanthan are synthesized by bacteria *Pseudomonas aeruginosa, Sphingomonas paucimobilis, Xanthomonas campe* respectively. Other than these EPS are also synthesized by lactic acid bacteria of different chemical structures [13]. EPS synthesized by lactic acid bacteria provides consistency, texture, and rheology of the fermented milk products. Synthesis of microbial EPS is very complex process and involves lots of genes.

Generally, synthesis of EPS involved four major steps which are: import of sugar into the cytoplasm, synthesis of sugar-1-phosphates, activation and coupling of sugar molecules, export of EPS from cytoplasm to media. EPS has numerous application in food industries, medical, and pharmaceuticals [36]. EPS producing bacteria are also known to grow on waste water and acts as waste treatment. Some of the incidences are discussed here. Taskin *et al.* (2011) has reported about synthesis of EPS by *Morchella esculenta* on extract of waste loquat (*Eriobotrya japonica l.*) kernels. EPS production on an average of 5 g/L was recovered [37]. Shochu is popular beverage in Japan which is produced from rice, sweet potatoes, barley, buckwheat, or sugar beets, shochukasu which is the distillery waste of shochu production is prohibited to dump in river or fields. Thus, it is dumped in ocean which is costlier as well as causes marine pollution. As this waste is rich in sugar contained, thus, Yuliani *et al.* (2011) used this wastewater to culture *Lactobacillus sakei* CY1 and produced EPS of 25.5% glucose and 13.2% galactose along with other sugars [38]. Morillo *et al.* (2007) isolated a bacterial strain *Paenibacillus jamilae* from olive mill wastewater which produces EPS. This bacterium produces 5.1 g/L of EPS from this wastewater using batch culture fermentation method [39]. Sellami *et al.* (2015) synthesized EPS by cultivating *Rhizobium leguminosarum* on waste water of oil company and wastewater of fish processing company. The highest EPS production of 42.4 g/L, after 96 h of culture was found when both oil and fish processing company wastewater was mixed. Thus, the discussion emphasizes that EPS which is one of the valuable industrial product can be synthesized efficiently from waste biomass and also help in bioremediation of wastes [40].

8.8 Succinic Acid

Succinic acid or butanedioic acid is an aliphatic C4 dicarboxylic acid. It is a very important value added product for industrial purpose. It is a co-product of biorefineries as it displaces petroleum based chemical and polymers for different product formation [41–43]. Succinic acid can be converted to many industrially important chemicals such as 1,4-butanediol, tetrahydrofuran, and γ-butyrolactone etc. [44]. The salt form of succinic acid that is succinate is a primary constituent of tricarboxylic acid cycle, thus, it can be produced from lignocellulosic sugar at high carbon efficiency [45, 46]. *A. succinogenes* and *Mannheimia succiniciproducens* are some of the well-known microorganism used for the production succinic acid. *A. succinogenes* is well-known for the production succinic acid from soft wood lignocellulosic biomass such as corn stalk and sugarcane hydrolysate where as *Mannheimia succiniciproducens* is known to produce succinic acid from wood hydrolysate. *A. succinogenes* is partly high acid tolerant and also fix CO_2 and consume broad range of sugar substrates such as C6 and C5 sugar molecules [47]. Thus, *Anaerobiospirillum succiniciproducens* is an engineered strain of *A. succinogenes* and *Mannheimia succiniciproducens* with *E. coli* which produces high yield and productivity of succinic acid [48, 49]. Production of succinic acid from other waste sources are also discussed such as Zhang *et al.* (2013) has used pre-treated bakery waste for the production of succinic acid by *Actinobacillus succinogenes* and yield 24.8 and 31.7 g L^{-1} of succinic acid from cake and pastry respectively [50]. Leung *et al.* (2012) used another common food waste that is bread as substrate for the production of succinic acid by *Actinobacillus succinogenes.* It yields 47.3 g/L of succinic acid [51]. Some other lignocellulosic waste sources for succinic acid production are crop stalk wastes [52], corncob hydrolysate [53], fruit, and vegetable wastes [54] by *Actinobacillus succinogenes.* Thus, succinic acid production can be done from different sugar rich waste biomass, which also reduces the cost of production.

8.9 Flavoring and Fragrance Substances

There are varieties of flavoring and fragrance substances produced by chemical process and are important for industrial use especially in food and cosmetic industries. Flavoring substances are generally volatile and non-volatile chemicals which are extracted from plant sources or synthesized chemically [55]. Volatile chemical only contribute to the aroma but non-volatile chemicals contribute both to the aroma and taste [56]. Flavors are generally extracted from plant sources and then their chemical structure is studied and chemical flavors are synthesized. The aroma of the food is responsible for some compounds such as alcohols, aldehydes, esters, dicarbonyls, short to medium- chain free fatty acids, methyl ketones, lactones, phenolic compounds, and sulphur compounds [56]. Though flavors can be extracted from plant sources but the downstream process for such extraction are very costly and also very small quantities of compound can be extracted from a large biomass. Thus, flavors and fragrances are chemically mimicked. But this chemical process of flavoring substance causes environmentally unhealthy production and lacks substrate selectivity which causes undesirable racemic mixtures, thus increasing the cost of processing [55]. As natural fragrances and flavoring substances are in high demand thus, microbial technology are used for the production of safe and natural fragrance and flavoring substances. Microorganism can synthesize these compounds as secondary metabolites on nutrients such as sugars and amino acids. This can be done by two process, one by *in situ* production of aroma and flavor as integral part of food and beverage production process and another is by synthesizing aroma or flavors from specifically designed microbial system [55]. Some of the example of flavors and fragrance synthesis from waste biomass as substrate are discussed. Phenylethyl alcohol which is aroma of rose can be synthesized by *Kluyveromyces marxianus* from cheese waste [57]. Terpenoid flavor and fragrance compounds are one of the most industrial valuable compounds which could be synthesized by microorganism fermentation of house- hold vegetable wastes [58]. *Rhizopus oryzae* and *Candida tropicalis* has the ability to synthesize D-limonene a flavoring substance from olive oil mill waste [59]. Some of the flavoring and fragrance substance are listed in the Table 8.1.

8.10 Hormones and Enzymes

Hormones, enzymes, and antibiotics are of one most important industrial products. Different hormones such as plant hormones and human hormones are produced by fermentation process. Plant hormones such as auxins, cytokinins, ethylene, gibberellins (GAs), abscisic acid (ABA), brassinosteroid, and salicylates are produced by fermentation process [68]. Recombinant human growth hormones (rHGH) are produced by genetically engineered *E. coli* through fermentation process [69]. Other than this, economically important enzymes such as ligninolytic, protease, lipase, chitinolytic, cellulase, amylase, etc. are produced by fermentation process by various microorganisms. They are used in various industrial process such as in paper industry, textile industries, pharmaceutical industries, food industries, chemical industries, cosmetic industries, etc. These enzymes have multiple function such as hydrolysis, degradation of polysaccharides, used in biofuels cells and many more. Ligninolytic enzymes such as laccase, manganese peroxidase, lignin peroxidase, and aryl alcohol oxidase are produced by *Pleurotus ostreatus* on potato peel waste by solid state fermentation process. *Pleurotus ostreatus* also secreted amylase and protease enzyme under same condition [70]. These enzymes are significant for their efficiency in bioconversion of plant waste and they have promising biotechnological applications in pulp and paper, food, textile and dye industries, bioremediation, cosmetics, and many others industrial applications [70]. Another example of environment pollution are municipality wastes dumped in land pits, which contaminate soil as well as spreads pathogens

Table 8.1 List of some flavonoids and their active component synthesize by microorganism using waste materials.

Sl no.	Flavonoids	Active compounds	Microorganism	Substrate	References
1	Almond flavor	Benzaldehyde	*Rhizopus oligosporus* USM R	Soy bean meal and rice husks	[60]
2	Apple and pineapple	Butyric acid	*Clostridium tyrobutyricum* strain	Wheat straw	[61]
3	Flavor component of dairy products	Acetaldehyde, acetone, 2-butanone, dimethyl disulfide, acetoin, 2,3-butanedione, 2,3-pentanedione, and acetic, hexanoic and butanoic acids	*L. bulgaricus* strains and *streptococci* cultures	Fermented milk (laban)	[62]
4	Lilac	R-(+)-α-terpineol	*Penicillium sp.* 2025 and *Fusarium oxysporum* 152B	Liquid cassava waste and orange essential oil	[63]

5	Vanilla	Vanillin	*Aspergillus niger* CC-MCC0774 and *Pycnoporus cinnabarinus* CGMCC1115	Waste residue of rice bran oil	[64]
6	Rosary	2-Phenylethanol	*Saccharomyces cerevisiae*	Cassava wastewater	[65]
7	Coconut aroma	Saturated lactones, δ-octalactone, γ-nonalactone, γ-undecalactone, γ-dodecalactone and δ-dodecalactone	*Trichoderma viride* EMCC-107	Sugarcane bagasse	[66]
8	Cinnamon essential oil	Cinnamic acid and p-hydroxycinnamic acid	Engineered *Escherichia coli*	Lignocellulosic hydrolysates	[67]

in the surrounding areas. Municipality wastes mainly consist of lignocellulose biomass. And lignocellulose biomass has high opportunities to be converted into different value added products. But the robust nature of lignocellulose requires a cocktail of enzymes for pretreatment and hydrolysis into simple sugar. Cellulase is one of the most important enzymes involved in this cocktail of enzymes. Abdullah *et al.* (2016) has optimized cellulase production from municipality solid waste by *Trichoderma reesei* and *Aspergillus niger* using solid state fermentation with a yield of 24.7% of cellulose [71]. Another example of cellulase production by *Aspergillus niger* is using coir waste [72]. Fishery processing industries produces a large number of by-products or waste whose disposal is a big issue. These waste leads to environment and health problems. The by-products include mostly heads, viscera, chitinous material, wastewater that are a good substrate for microorganism to grow for enzymes production. Though to avoid wasting the by-products various disposal methods are used such as including, ensilation, fermentation, hydrolysate, and fish oil production etc. But, using fermentation technology for enzymes production using these wastes as substrate is an interesting issue. Rebah and Miled (2012) has discussed in a review the possibility of production of various microbial enzymes such as protease, lipase, chitinolytic, and ligninolytic enzymes. Thus, even these wastes can be converted to such as valuable products. It not only decreases the cost of production of enzymes but also clean the environment.

8.11 Conclusion

Fermentation technology is use of microbes for the production of various industrially valuable products. Optimization of fermentation technology provide green route for production of various chemicals, enzymes, proteins, biofuels, polymers, etc. with low cost of production and no pollution. Now day, many synthetic products are used derived from petrochemical industries in our day to lives which are harmful for human health and environment. But these products are need of day to day life. Therefore, to replace petrochemical products fermentation technology is the best option. Through fermentation technology, we can produce many chemicals which mimics the chemical produced by petrochemical process. But, the by-products formed and hazard to the environment are not done by fermentation technology. Even with fermentation technology waste products of the industries can be converted to valuable products leading to zero waste technology. Biofuels can be easily produced from different industrial waste and similarly different chemicals, polymers, enzymes, flavoring substance are produced by fermentation process and using waste materials as substrate. This not only reduces the cost of production but also help in bioremediation. Some of the polymers such as PHA and PLA are used for the production of bioplastics and replace the petrochemical-based plastics. Other than this hormones and enzymes production were costlier earlier, but with the help of recombinant DNA technology, genetically modified microorganism are capable of producing those enzymes and hormones by fermentation technology. Some of the plant hormones, ligninolytic enzymes, cellulase, amylase, etc. can be produce by fermentation technology using waste products as substrate. Exopolysaccharides are secreted by bacteria which are highly valuable for pharmaceutical industries. Whereas ligninolytic enzymes like laccase and cellulase are valuable for paper, textile, and biofuel industries. This biofuel itself is a big relief to the environment as it is supplement to fossil fuels. Thus, industrially value added product formation and converting waste as valuable raw material has been discussed in this chapter. More research is furnishing with this enormous microbial diversity for green technology development and producing every possible product naturally and safe for the environment.

References

1. Karube, I., *et al.*, Continous hydrogen production by immobilized whole cells of Clostridium butyricum. *Biochim. Biophys. Acta.*, 444(2), 338–343, 1976.
2. Sivagurunathan, P., *et al.*, Fermentative hydrogen production using lignocellulose biomass: an overview of pre-treatment methods, inhibitor effects and detoxification experiences. *Renewable Sustainable Energy Rev.*, 77(Supplement C), 28–42, 2017.
3. Maru, B.T., *et al.*, Glycerol fermentation to hydrogen by Thermotoga maritima: proposed pathway and bioenergetic considerations. *Int. J. Hydrogen Energy*, 38(14), 5563–5572, 2013.
4. Voloshin, R.A., *et al.*, Photoelectrochemical cells based on photosynthetic systems: a review. *Biofuel Res. J.*, 2(2), 227–235, 2015.
5. Roy, L., Garlapati, V., Banerjee, R., Challenges in harnessing the potential of Lignocellulosic biofuels and the probably combating strategies. *Bioenergy*, Apple Academic Press, 171–203, 2015.
6. Demirbas, A., Competitive liquid biofuels from biomass. *Appl. Energy*, 88(1), 17–28, 2011.
7. Arazo, R.O., de Luna, M.D.G., Capareda, S.C., Assessing biodiesel production from sewage sludge-derived bio-oil. *Biocatal. Agric. Biotechnol.*, 10(Supplement C), 189–196, 2017.
8. Milano, J., *et al.*, Microalgae biofuels as an alternative to fossil fuel for power generation. *Renewable Sustainable Energy Rev.*, 58(Supplement C), 180–197, 2016.

9. Cerón-García, M.C., *et al.*, A process for biodiesel production involving the heterotrophic fermentation of *Chlorella protothecoides* with glycerol as the carbon source. *Appl. Energy*, 103(Supplement C), 341–349, 2013.
10. Ber, *et al.*, Simultaneous saccharification and fermentation of sugar beet pulp for efficient bioethanol production. *BioMed Res. Int.*, 10, 2016.
11. Bai, F.W., W.A. Anderson, and M. Moo-Young, Ethanol fermentation technologies from sugar and starch feedstocks. *Biotechnology Advances*, 26(1): p. 89–105, 2008.
12. Zheng, Y., *et al.*, Integrating sugar beet pulp storage, hydrolysis and fermentation for fuel ethanol production. *Applied Energy*, 93(Supplement C): p. 168–175, 2012.
13. Saha, B.C., Commodity Chemicals Production by Fermentation: An Overview, in Fermentation Biotechnology. *American Chemical Society*. p. 3–17, 2003.
14. Zeng, A.-P. and H. Biebl, *Bulk Chemicals from Biotechnology: The Case of 1,3-Propanediol Production and the New Trends*, in *Tools and Applications of Biochemical Engineering Science*, K. Schügerl, *et al.*, Editors. Springer Berlin Heidelberg: Berlin, Heidelberg. p. 239–259, 2002.
15. Cameron, D.C., *et al.*, Metabolic engineering of propanediol pathways. *Biotechnol Prog*, 14(1): pp. 116–125, 1998.
16. Chotani, G., *et al.*, The commercial production of chemicals using pathway engineering. *Biochim Biophys Acta*, 1543(2): pp. 434–455, 2000.
17. Wee, Y., Kim, J., Ryu, H., Biotechnological Production of Lactic Acid and Its Recent Applications. *Food Technology and Biotechnology*, 44(2): p. 163–172, 2006.
18. Varadarajan, S. and D.J. Miller, Catalytic Upgrading of Fermentation-Derived Organic Acids. *Biotechnol Prog*, 15(5): p. 845–854, 1999.
19. Datta, R., *et al.*, Technological and economic potential of poly(lactic acid) and lactic acid derivatives. *FEMS Microbiology Reviews*, 16(2): pp. 221–231, 1995.
20. Lunt, J., Large-scale production, properties and commercial applications of polylactic acid polymers. *Polymer Degradation and Stability*, 59(1): pp. 145–152, 1998.
21. Ghaffar, T., *et al.*, Recent trends in lactic acid biotechnology: A brief review on production to purification. *Journal of Radiation Research and Applied Sciences*, 7(2): p. 222–229, 2014.
22. Åkerberg, C. and G. Zacchi, An economic evaluation of the fermentative production of lactic acid from wheat flour. *Bioresource Technology*, 75(2): p. 119–126, 2000.
23. Chaisu, K., *et al.*, Optimization Lactic Acid Production from Molasses Renewable Raw Material through Response Surface Methodology with Lactobacillus Casei M-15. *APCBEE Procedia*, 8(Supplement C): p. 194–198, 2014.
24. Pleissner, D., *et al.*, Direct production of lactic acid based on simultaneous saccharification and fermentation of mixed restaurant food waste. *Journal of Cleaner Production*, 143(Supplement C): p. 615–623, 2017.
25. Tosungnoen, S., K. Chookietwattana, and S. Dararat, Lactic Acid Production from Repeated-Batch and Simultaneous Saccharification and Fermentation

of Cassava Starch Wastewater by Amylolytic Lactobacillus Plantarum MSUL 702. *APCBEE Procedia*, 8(Supplement C): p. 204–209, 2014.

26. Thomas, T.D., D.C. Ellwood, and V.M. Longyear, Change from homo- to heterolactic fermentation by Streptococcus lactis resulting from glucose limitation in anaerobic chemostat cultures. *J. Bacteriol*, 138(1): p. 109–17, 1979.
27. Madison, L.L. and G.W. Huisman, Metabolic engineering of poly(3-hydroxyalkanoates): from DNA to plastic. *Microbiol Mol. Biol. Rev.*, 63(1): p. 21–53, 1999.
28. Lopez, N.I., *et al.*, Polyhydroxyalkanoates: Much More than Biodegradable Plastics. *Adv Appl Microbiol*, 93: p. 73–106, 2015.
29. Kourmentza, C., *et al.*, Recent Advances and Challenges towards Sustainable Polyhydroxyalkanoate (PHA) Production. *Bioengineering*, 4(2): p. 55, 2017.
30. Khandare, R.V. and S.P. Govindwar, Phytoremediation of textile dyes and effluents: Current scenario and future prospects. *Biotechnology Advances*, 33(8): p. 1697–1714, 2015.
31. Nikodinovic, J., *et al.*, The conversion of BTEX compounds by single and defined mixed cultures to medium-chain-length polyhydroxyalkanoate. *Appl. Microbiol. Biotechnol.*, 80(4): p. 665–73, 2008.
32. Song, J.H., *et al.*, Polyhydroxyalkanoate (PHA) production using waste vegetable oil by Pseudomonas sp. strain DR2. *J. Microbiol Biotechnol*, 18(8): p. 1408–15, 2008.
33. Rhu, D.H., *et al.*, Polyhydroxyalkanoate (PHA) production from waste. *Water Sci. Technol.*, 48(8): p. 221–8, 2003.
34. Colombo, B., *et al.*, Enhanced polyhydroxyalkanoate (PHA) production from the organic fraction of municipal solid waste by using mixed microbial culture. *Biotechnology for Biofuels*, 10(1): p. 201, 2017.
35. Venkateswar Reddy, M., *et al.*, Polyhydroxyalkanoates (PHA) production from synthetic waste using Pseudomonas pseudoflava: PHA synthase enzyme activity analysis from P. pseudoflava and P. palleronii. *Bioresource Technology*, 234(Supplement C): p. 99–105, 2017.
36. Laws, A., Y. Gu, and V. Marshall, Biosynthesis, characterisation, and design of bacterial exopolysaccharides from lactic acid bacteria. *Biotechnol. Adv.*, 19(8): p. 597–625, 2001.
37. Taskin, M., S. Erdal, and M. Genisel, Biomass and Exopolysaccharide Production by Morchella Esculenta in Submerged Culture Using the Extract from Waste Loquat (Eriobotrya Japonica L.) Kernels. *Journal of Food Processing and Preservation*, 35(5): p. 623–630, 2011.
38. Yuliani, E., *et al.*, Exopolysaccharide Production from Sweet Potato-Shochu Distillery Wastewater by Lactobacillus Sakei CY1. *Biotechnology & Biotechnological Equipment*, 25(2): p. 2329–2333, 2011.
39. Morillo, J.A., *et al.*, Production and characterization of the exopolysaccharide produced by Paenibacillus jamilae grown on olive mill-waste waters. *World Journal of Microbiology and Biotechnology*, 23(12): p. 1705, 2007.
40. Sellami, M., *et al.*, Industrial wastewater as raw material for exopolysaccharide production by Rhizobium leguminosarum. *Brazilian Journal of Microbiology*, 46(2): p. 407–413, 2015.

41. Werpy T, Petersen G. Top value added chemicals from biomass. Results of screening for potential candidates from sugars and synthesis gas. US. Department of Energy. http://www.osti.gov/bridge, 2004.
42. Gallezot, P., Conversion of biomass to selected chemical products. *Chem. Soc. Rev.*, 41, 2012.
43. Gallezot, P., Process options for converting renewable feedstocks to bioproducts. *Green Chem*, 9, 2007.
44. Song, H. and S.Y. Lee, Production of succinic acid by bacterial fermentation. *Enzyme Microb. Technol.*, 39, 2006.
45. Salvachúa, D., *et al.*, Succinic acid production on xylose-enriched biorefinery streams by Actinobacillus succinogenes in batch fermentation. *Biotechnology for Biofuels*, 9(1): p. 28, 2016.
46. Jansen, M.L. and W.M. Gulik, Towards large scale fermentative production of succinic acid. *Curr. Opin. Biotechnol.*, 30, 2014.
47. Guettler, M., D. Rumler, and M. Jain, Actinobacillus succinogenes sp. nov., a novel succinic-acid-producing strain from the bovine rumen. *Int. J. Syst. Bact.*, 49, 1999.
48. McKinlay, J.B., C. Vieille, and J.G. Zeikus, Prospects for a bio-based succinate industry. *Appl. Microb. Biotechnol.*, 76, 2007.
49. Lee, S.Y., P. Cheon, and H.N. Chang, Kinetic study of organic acid formations and growth of Anaerobiospirillum succiniciproducens during continuous cultures. *J. Microbiol. Biotechnol.*, 19, 2009.
50. Zhang, A.Y.-z., *et al.*, Valorisation of bakery waste for succinic acid production. *Green Chemistry*, 15(3): p. 690–695, 2013.
51. Leung, C.C.J., *et al.*, Utilisation of waste bread for fermentative succinic acid production. *Biochemical Engineering Journal*, 65(Supplement C): p. 10–15, 2012.
52. Li, Q., *et al.*, Efficient conversion of crop stalk wastes into succinic acid production by Actinobacillus succinogenes. *Bioresource Technology*, 101(9): p. 3292–3294, 2010.
53. Yu, J., *et al.*, Development of succinic acid production from corncob hydrolysate by Actinobacillus succinogenes. *J. Ind. Microbiol. Biotechnol.*, 37(10): p. 1033–40, 2010.
54. Dessie, W., *et al.*, Succinic acid production from fruit and vegetable wastes hydrolyzed by on-site enzyme mixtures through solid state fermentation. *Bioresour. Technol.*, 2017.
55. Sanromán, M.A.L.a.M.A., Production of Food Aroma Compounds: Microbial and Enzymatic Methodologies. *Food Technol. Biotechnol.*, 44(3): p. 335–353, 2006.
56. Urbach, G., The flavour of milk and dairy products: II. Cheese: contribution of volatile compounds. *International Journal of Dairy Technology*, 50(3): p. 79–89, 1997.
57. Conde-Báez, L., *et al.*, Evaluation of Waste of the Cheese Industry for the Production of Aroma of Roses (Phenylethyl Alcohol). *Waste and Biomass Valorization*, 8(4): p. 1343–1350, 2017.

58. Schempp, F.M., *et al.*, Microbial Cell Factories for the Production of Terpenoid Flavor and Fragrance Compounds. *Journal of Agricultural and Food Chemistry*, 2017.
59. Guneser, O., *et al.*, Production of flavor compounds from olive mill waste by Rhizopus oryzae and Candida tropicalis. *Brazilian Journal of Microbiology*, 48(2): p. 275–285, 2017.
60. Norliza, A.W.a.I., C.O., The production of Benzaldehyde by Rhizopus oligosporus USM R1 in a Solid State Fermentation (SSF) System of Soy Bean Meal: rice husks. *Malaysian Journal of Microbiology*, 1(2): p. 17–24, 2005.
61. Baroi, G.N., *et al.*, Butyric acid fermentation from pretreated and hydrolysed wheat straw by an adapted Clostridium tyrobutyricum strain. *Microbial Biotechnology*, 8(5): p. 874–882, 2015.
62. Chammas, G.I., *et al.*, Characterisation of lactic acid bacteria isolated from fermented milk "laban". *Int. J. Food Microbiol.*, 110(1): p. 52–61, 2006.
63. Maróstica, M.R. and G.M. Pastore, Production of R-(+)-α-terpineol by the biotransformation of limonene from orange essential oil, using cassava waste water as medium. *Food Chemistry*, 101(1): p. 345–350, 2007.
64. Zheng, L., *et al.*, Production of vanillin from waste residue of rice bran oil by Aspergillus niger and Pycnoporus cinnabarinus. *Bioresour. Technol.*, 98(5): p. 1115–9, 2007.
65. Oliveira, S.M.M., *et al.*, Production of natural aroma by yeast in wastewater of cassava starch industry. *Engenharia Agrícola*, 35: p. 721–732, 2015.
66. Fadel, H.H.M., *et al.*, Characterization and evaluation of coconut aroma produced by Trichoderma viride EMCC-107 in solid state fermentation on sugarcane bagasse. *Electronic Journal of Biotechnology*, 18(1): p. 5–9, 2015.
67. Vargas-Tah, A., *et al.*, Production of cinnamic and p-hydroxycinnamic acid from sugar mixtures with engineered Escherichia coli. *Microbial Cell Factories*, 14(1): p. 6, 2015.
68. Shi, T.-Q., *et al.*, Microbial production of plant hormones: Opportunities and challenges. *Bioengineered*, 8(2): p. 124–128, 2017.
69. Zhang, X.-W., *et al.*, Human growth hormone production by high cell density fermentation of recombinant Escherichia coli. *Process Biochemistry*, 33(6): p. 683–686, 1998.
70. Ozcirak Ergun, S. and R. Ozturk Urek, Production of ligninolytic enzymes by solid state fermentation using Pleurotus ostreatus. *Annals of Agrarian Science*, 15(2): p. 273–277, 2017.
71. Abdullah, J.J., G.D., Pensupa N, Tucker GA, Du C, Optimizing Cellulase Production from Municipal Solid Waste (MSW) using Solid State Fermentation (SSF). *J. Fundam. Renewable Energy Appl*, 6(206), 2016.
72. Mrudula, S. and R. Murugammal, Production of cellulase by Aspergillus niger under submerged and solid state fermentation using coir waste as a substrate. *Brazilian Journal of Microbiology*, 42: p. 1119–1127, 2011.

9

Valorization of Lignin: Emerging Technologies and Limitations in Biorefineries

Gourav Dhiman[1], Nadeem Akhtar[2] and Gunjan Mukherjee[1,3,*]

[1]*Department of Biotechnology, Chandigarh University, Punjab, India*
[2]*Department of Animal Biosciences, University of Guelph, Guelph, ON, Canada*
[3]*The Energy and Resources Institute, New Delhi, India*

Abstract

The depletion of fossil fuel reserves coupled with the increase in demands for fuels across the globe has raised concerns for improved utilization of renewable energy resources. Among three main components of the lignocellulosic biomass, lignin holds approximately 15–40% of the organic carbon sequestered in the biosphere. Being rich in aromatic carbons, lignin has the potential to act as a raw material for the production of valuable materials, chemicals, polymers, and bio-fuels through bio-engineering routes. Efficient utilization of lignocellulosic biomass is a pre-requisite for reduction of pre-treatment cost due to recalcitrant matrix of cell wall-associated polysaccharides. Structural modification and/or alteration in lignin structure have shown promising result for optimal recovery and chemical transformation of lignin in a biorefinery. In addition, bioengineering approaches reduce recalcitrance in crops with desired physico–chemical properties which may help to overcome the limitations. However, valorization of lignin-based value-added products poses challenges for its depolymerization using a cost-effective sustainable technology. Realization of biorefinery approaches with fully integrated bioprocesses technology requires appropriate fractionation of biomass to reduce our reliance on petroleum-based products. This chapter highlights recent developments and their limitations in lignin biosynthesis, lignin genetic engineering, and depolymerization strategies used for the production of lignin-based bio-fuels and bio-products.

Keywords: Lignocellulosic biomass, biorefinery, valorization, biodegradation, biosynthesis, bio-fuels

**Corresponding author*: gunjanmukherjee@gmail.com

Arindam Kuila and Vinay Sharma (eds.) Principles and Applications of Fermentation Technology, (163–180) © 2018 Scrivener Publishing LLC

9.1 Introduction

The depleting fossil fuel reserves and their serious effect on the environmental have generated a strong will for finding alternative renewable and clean sources of energy. The abundance of the lignocellulosic biomass provides the most suitable alternative for cleaner fuel and other value-added product development. Several technologies have either been developed or are in their advanced stages of development for efficiently converting lignocellulosic biomass into a variety of fuels and other products. However, the research and technological development in efficient utilization of the lignocellulosic biomass have taken a new turn in the direction of lignin utilization, which was earlier primarily focused on the cellulosic component. Lignin being abundant and making up 15–40% of the lignocellulosic component indeed acts as an important substrate for various industrial applications [1]. Lignin, despite being a complex aromatic polymer, has been successfully converted into various value-added products. A variety of polymeric foams, thermoplastics, membranes, and fuels have been developed from the lignin, which can be a vital alternative to the similar products currently produced from the petroleum [2].

Since lignin assists the cementing of cellulose and hemicelluloses in the cell wall and is not easily separable. Yet the separation of the lignin from the other components of a cell wall, that is, cellulose is essential in various industries, mainly the paper and pulp industry. Therefore, lignin is primary waste product in the paper and pulp industry, and a significant amount of it remains unutilized. Therefore, its use for commercial production can give rise to a range of chemicals and value-added products. Hence, lignin valorization has become the need of the hour for its effective commercial exploitation. It is estimated that the demand for various plant-based chemical and food resources will increase by 50% in next three decades [3]. Lignin, being organic polymers reasonably abundant in oxygen compared to carbon and hydrogen of the petroleum-based hydrocarbons. A wide variety of products can be obtained from such chemical compositions based on different C5 and C6 sugar production [4]. Syngas products (methanol, DME, ethanol, etc.), hydrocarbons (benzene, toluene, xylene, etc.), phenols (cresols, eugenols, etc.), and oxidized products(vanillin, vanillic acid, etc.) are amongst the major categories of the chemicals which can be produced using lignin as a raw material [5].

There are arrays of chemical, biological, and engineering developments underway to convert understand, and modify the structure of lignin for effective technological development for its industrial applications and producing value-added products. Also, the commercial production of bio-fuels from the lignocellulosic biomass will further enhance the usage of unutilized lignocellulosic biomass waste or industrial lignin which otherwise remains unutilized. In this chapter, we shall explain the structural, biological, and engineering advances in lignin valorization.

9.2 Lignocellulosic Material: Focus on Second Generation Biofuel

The production of bio-ethanol from carbohydrate substrate is a well-established process. Mainly first generation bio-fuels, which are produced from fruits, vegetables, and cereal crops, is a well-established technology. Since the use of primary food materials for fuel production can pose serious threats to the food security worldwide, the use of food crops for fuel production has been criticized [3]. Therefore, use of lignocellulosic biomass is a viable option for producing biofuel, that is, second generation biofuel. At the same time growing bio-fuel crops on marginal agricultural land can also further fulfill the demand of raw materials for the second generation biofuel production without posing any serious threats to conventional cultivation and food security [6]. Hence, dedicated raw material resources can be available for the uninterrupted supply in the years to come. The lignocellulosic biomass is mainly composed of 30–55% cellulose, 20–35% hemicellulose, and 15–40% lignin [7]. However, the distribution of these component's chemical and molecular characteristics may vary amongst different plant species, tissues, and ages of the biomass [4, 8]. Currently, in available technology, cellulose is the main component of lignocellulosic biomass which is widely exploited for the bio-fuel production. This mainly consists of glucose units linearly linked by β-1,4 glycosidic bonds [9] further organized into microfibrils via weak interactions [7]. Since the microfibrils are tightly packed with hemicelluloses and lignin [10], it requires a pre-treatment prior to its subjection to hydrolysis, that is, the removal of lignin from the substrate to make it available for enzymatic digestion [11]. There are various different pre-treatment technologies like physical, microwave, acid, ionic liquid, ultrasonication, and alkaline hydrolysis for lignin removal from biomass [12]. However, the commercial applications of them are not very common except for physical and acid hydrolysis, which is the most widely, used pre-treatment methods. The pre-treatment of lignocellulosic biomass separates the lignin from the cellulosic microfibers and hemicelluloses fibres. As result of which cellulose is available for their breakdown into fermentable sugars from cellulose, that is, glucose, and then into bio-ethanol. Also, hemicellulose can give rise to other pentose sugars such as xylose, arabinose, hexoses like mannose, galactose, and rhamnose, which can be subjected to fermentation for the production of bio-ethanol and other value added products [13]. Second generation bio-fuel production is more acceptable compared to first generation bio-fuel production. However, the byproduct lignin, which is often produced in abundance during the pre-treatment process, carries huge potential for its conversion into biodegradable chemical products.

9.3 Composition and Biosynthesis of Lignin

Lignin is among the most abundant polymer after cellulose, and like cellulose, it is readily available for commercial usage. Lignin is a polymer mainly made of its monomeric units called monolignols, which make lignin H-, G- and S-units [14]. It is a complex molecular structure made from the polymerization of substituted phenyl propylene units [2]. It is the most important component of the plant cell wall system for providing rigidity to the plant. It also helps in the plant's defence system [15] and seed dispersal mechanisms [16]. The three monomeric units mainly involved in lignin synthesis are *p*-coumaryl (H), coniferyl (G), and sinapyl (S), which are produced via the phenylpropanoid pathway [17]. However, the monomeric units may vary in the degree of methoxylation on the aromatic rings depending upon the type of lignin and plant species [2]. Phenylalanine is the main precursor for lignin biosynthesis derived from the shikimate pathway, which is converted into cinnamic acid through the action of the enzyme phenylalanine ammonia lyase (PAL). Subsequently, various different enzymes such as*p*-coumarate-3-hydroxylase (C3H), cinnamate-4-hydroxylase (C4H), ferulate-5-hydroxylase (F5H), caffeic acid-O-methyltransferase (COMT), caffeoyl-CoA-O-methyltransferase (CCoAOMT), cinnamoyl-CoA-reductase (CCR),cinnamyl alcohol dehydrogenase (CAD), and sinapyl alcohol dehydrogenase (SAD) produce different alcohols (monolignols), acids, and aldehydes which are used for the lignin biosynthesis and for the production of other aromatic compounds [17]. As soon as the monomeric units of lignin (monolignols) are formed they are transported into the apoplastic space where they are converted into free radicals and undergo the polymerization process. As a result, the middle lamella of the plant cell wall is lignified, while the composition and structural integrity will depend upon the type of bond formed between the monomeric units [18]. Slight modifications in the biosynthetic pathways can result in varied compositions of lignin and structural integrity.

This can be exploited for generation of plant species with different lignin content for industrial applications.

9.3.1 Structure Analysis of Lignin

As discussed above, lignin composition varies amongst different plant species, which depends on the enzymes, polymerization, and availability of monolignol. Therefore, the exact lignin composition will vary among different cell types and will require detailed gene study to deduce the composition and structural variations. For instance, the lignin content in the coconut shell is different from the coconut wood [3].

In order to find the structural components of lignin, it can be broken down into its monomeric units and then subjected to degradative techniques such as oxidation, reduction, hydrolysis, and acidolysis. Also, derivatization techniques like thioglycolic acid (TGA) and acetyl bromide (ACBR) can be employed for analyzing the structural features. In degradation techniques, the end product obtained is based on the monomeric subunits of the lignin degraded. These end products can be further studied by applying high throughput techniques like nuclear magnetic resonance (NMR), gas chromatography (GC), and mass spectroscopy [3]. On the other hand, neutron scattering is also one of the emerging techniques that can be employed for the structural analysis of lignin. Particularly, small-angle neutron scattering (SANS) techniques have a precision that ranges from nm to μm. Therefore it can give a deeper structural analysis and provide atomic level predictions [19]. The integration of these sophisticated techniques with computer programs can further enhance the insights to structural integrity. On the other hand, microscopic techniques like scanning electron and confocal Raman microscopies can also be employed for the structural and morphological analysis of the lignins [20].

9.3.2 Degradative Analytical Techniques (Oxidation, Reduction, Hydrolysis, and Acidolysis)

There are various oxidizing agents like nitrobenzene, cupric oxide, and potassium permanganate used for the lignin oxidation, which break down the side chain of lignin, liberating the aromatic rings for analysis. The aromatic rings released as a result of oxidation can be further analyzed for the presence of available aromatic species. For instance, the oxidation with nitrobenzene results in guaiacol, vanillic acid, and 5-carboxyvanillin [21]. Since the oxidation method does not give any information about the side chain, other methods can be employed for side chain analysis. For example, the when the lignin is subjected to ozonolysis, it will target the unsaturated bonds in the ring structure liberating the side chain as carboxylic acid products, like formic and oxalic acids [22]. Therefore, the two techniques can be collectively employed for the complete lignin structure deduction [23]. Similarly, derivatization followed by reductive cleavage (DFRC) of lignin is another important structure deduction technique widely employed, where lignin is subjected to ACBR which forms the benzyl bromide, which on treatment with zinc dust gives cinnamyl alcohol derivatives [23]. As discussed, there are multiple methods for the lignin structure deduction but the selective targeting of these methods still has some drawbacks for complete structural analysis.

9.3.3 Non-Degradative Analytical Techniques

In addition to the degradative methods, non-degradative techniques are used for complete structural analysis. Non-degradative analysis provides a good method for the analysis of size, structure, and degree of polymerization in the lignin molecule. The most widely used non-degradative techniques are the TGA and ACBR techniques [24, 25]. In these techniques the functional group is added on a lignin molecule and on solubilizing in a suitable solvent it is subjected to size exclusion chromatography for determining the size of the lignin polymer. However, the native structure of the lignin molecule remains intact. On the other hand, the dynamic light scattering method can be used for molecular mass and shape determination [26].

9.4 Bioengineering of Lignin

It has been discussed in earlier segments that the alteration of lignin content is important in various chemical processes. A reduction or increase in the lignin content or altering the lignin structure may be needed for various industrial applications. Therefore, engineering a lignin source for multiple purposes could be a great approach in future lignin biorefinery. In this segment bioengineering approach for the improving lignin characteristics will be discussed.

9.4.1 Reducing the Recalcitrance Nature of Biomass

The recalcitrance nature of biomass is contributed to by various factors, that is, tissue cuticle, inhibitory molecules, and complex cell wall structure. The nature of lignin content, cell type, cell wall, and other chemical components makes it difficult for commercial applications. Since lignin is the main component of cell walls that contribute in recalcitration, there are several genes encoding these enzymes are involved in its biosynthesis. These genes can be altered for reduced lignin polymerization. Using T-DNA for the knockout and knockdown of genes can result in reduced expression of genes associated with lignin biosynthesis [27]. However, it affects plant health as reduced lignin content disables the plant against various biotic and abiotic stresses. Other than altering the lignin biosynthesis genes, phenol oxidase genes can also be manipulated for reduced lignin content. Since the lignin polymerizing enzymes reside in the apoplast and generate free radicals for the polymerization process, a slight change can have a significant effect on the lignin content [28]. At the same time reduced laccase content can also cause a significant reduction in the lignin content [29]. It has also been found that brown midrib lines in chemically mutated maize and sorgum were due to a mutation in the catechol-o-methyltransferase (COMT)gene, a lignin biosynthetic pathway gene which resulted in a significant lignin reduction and brown midribs [30]. Therefore, it is evident that a reduction in lignin content is very much possible with the help of the genetic engineering where a simple gene knockout or mutation can profoundly reduce the lignin content and thereby reducing the recalcitrance. However, the genetics described above can reduce the recalcitrance along with the serious effect on plant heath. To overcome this problem, tissue-specific promoters can be used instead of constitutive gene promoters, so that a selective reduction in the lignin content can be achieved and deleterious effects on the plant can be reduced [31]. The tissue specific expression of lignin synthesis genes can also be regulated by using different TF factors which can regulate the lignin content in different tissues [32]. The usage of new bioengineering techniques like gene editing with point mutations can also edit lignin biosynthesis in specific cells, thereby increasing the digestibility of the biomass and reducing the recalcitrance [33].

9.4.2 Improving Lignin Content for Production of High Energy Feedstock

Lignin, being a highly polymerized and reduced polymer, carries a great calorific value [34]. Lignin biomass can be used for high energy fuels, biofuels, and fuel electricity generation [35]. Therefore, an increased lignin deposition on the other hand may be desired for the conversion of lignins directly into bio-fuels, as pre-treatment methods in bio-ethanol production are still not economically viable. Likewise, the genetic manipulations for decreasing the lignin content can also be employed for increasing the same. Mutations in the genes can trigger increased lignin content [36]. Ectopic lignin depositions in plants that otherwise do not have any lignin in them appear to be a scientifically promising alternative for producing high lignin plants, which can be achieved by inducing genetic mutations [36]. A mutation in the ELI 1 gene (ectopic lignin gene 1) and the cellulose synthase 3 gene resulted in reduced ectopic lignin deposition and cellulose content in mutant plants [37]. In the mutant plants with an increased lignin content, their defence responses were enhanced and their growth reduced [38], which is a common phenomenon in plants growing under stressful conditions.

At the same time, by controlling the regulatory factors associated with the lignin biosynthesis, like negative micro RNA regulator of lignin biosynthesis gene, can elevate the lignin content [39]. Several biotic and abiotic factors can also trigger the higher lignin biosynthesis by up-regulating the transcription factors, for example, *MYB58* and *MYB63I*,in *Arabidopsis* sp. [40]. Also, elevated production of plant hormones such as auxin and indole-3-acetic acid (IAA) can result in increased lignin content [41].

Therefore, it can be inferred from the above discussion that up-regulation and down-regulation of lignin biosynthesis is advantageous for industrial applications. On the one hand, where reduced lignin content can enhance cellulose utilization for bio-ethanol production, and on the other hand an increased content can result in more lignin based fuels. Innovation in these sectors can not only provide clean and green energy but also an efficient management of biomass produced from genetically modified crops.

9.5 Lignin Separation and Recovery

Proper access and availability of substrate is the key to any chemical and biological reaction. Therefore, an efficient lignin conversion requires an efficient separation of lignin from the cell wall materials so that they can be made available for their proper conversion into valuable products. There are different chemical and biological methods that can be employed for this separation. In this segment, we will discuss these methods in detail and various implications associated with them.

9.5.1 Chemical- and Physical-Based Lignin Separations

There have been several studies pertaining to the successful isolation of lignin from the lignocellulosic biomass. However, the chemical treatment usually leads an alteration in lignin structure thereby reducing its value for the conversion into valuable products [42]. There are many different techniques developed for lignin isolation and conversion from lignocellulosic biomass, such aspyrolysis, steam explosion, and organosolv-based lignin separation [43].

The most common method for lignin separation is kraft process, where lignin is precipitated at a low pH in the presence of minerals. The lignin produced out of this process is called kraft's lignin, and is of low quality and impure. This method is commonly employed in the pulping process in the paper industry [44]. However, other processes like organosolv, ionic liquids, and membrane-based separation may be employed yet are relatively expensive processes compared tokraft's process [11].

Pyrolysis, that is, thermal destruction of biomass in the absence of oxygen usually results in three main components: oil, char, and gas. The component yield in lignin pyrolysis often differs from other pyrolysed biomass due to its structural complexity. Usually, the production of bio-oils is less in lignin pyrolysis while the process conditions may hugely affect the component yield. For instance, corn stalk yields G, H phenols and oils when pyrolysed at different temperatures for a specific time duration. Therefore, a variety of products may be produced via lignin pyrolysis depending upon the process and the type of raw material [45].

The isolation, polymerization, depolymerization, and chemical conversion of lignin require well defined chemical reactions. Therefore, effective chemical reactions with suitable catalysts have been widely explored for their adequate use on the commercial scale. Various catalysts have been studied for catalytic lignin transformations. Oxidation, polymerization, depolymerization, gasification, and pyrolysisin the presence of organometallics, acids, bases, noble mineral salts, and zeolites have been studied for their catalytic property and feasibility in suitable commercial processes [46].

Lignin polymers usually depolymerize before they can be converted into any valuable chemical products. The depolymerization can be carried out via combustion, pyrolysis, or oxidation depending upon the solubilization process [47]. Since all these processes are carried out at a higher temperature, sometimes many aromatic byproducts are formed as a result, which makes their recovery difficult.

Thermal depolymeraization of lignin has been practiced for a long time and is one of the most widely used techniques. Thermal cracking of lignin at a higher temperature (400–1100 °C), thermal treatment of lignocellulosic biomass with organosolv, salts and acids has been used for lignin depolymerization and separation for further commercial utilization of lignin in a biorefinery [48].

9.5.2 Biological Degradation of Lignin

The biological recovery of lignin is a more eco-friendly method for delignification of lignocellulosic biomass. A variety of fungal and bacterial strains are used for delignification purposes. Usually, thermotolerant mould are used for the delignification of lignocellulosic biomass [49]. A *Saccharomyces cerevisiae*strain with a transgene for laccase from *Trametes versicolor* has been found to be efficient in the delignification of lignocellulosic biomass [50]. Other than genetically modified organisms, some naturally occurring bacteria also have a higher delignification activity. Thermo-tolerant laccases from *Thermus thermophilus* have been isolated with enhanced activity and half-life [51]. Although many microorganisms have been isolated for their commercial advantages in lignin isolation, not much success has been obtained in this sector. Still, not many of the microbes are available for the commercial applications. When it comes to the delignification of lignocellulosic biomass, chemical processes are still more economic and commercially applicable when compared to the biological process. However, significant progress is being made in the biological delignification process which may bring some more efficient outcomes in the near future.

9.6 Lignin-Based Materials and Polymers

A higher degree of aromatic polymerization in lignin makes it one of the most desired precursors for biodegradable polymers and chemicals hydro gels. There are many approaches to develop the technology for harnessing the full potential of lignin for producing commercially viable products [13]. Vanillin, dimethyl sulphate, and dimethyl sulfoxide are amongst the few successfully commercialized lignin derived products [52].

There are a variety of polymerization techniques for the production of a variety of lignin based polymers, that is, reversible addition fragmentation chain transfer (RAFT) polymerization, nitroxide mediated polymerization (NMP), metal-mediated living radical polymerization, cobalt catalyzed chain transfer polymerization (CCCTP), and ring opening polymerization (ROP), with selected click reactions including copper catalyzed azide-alkyne cycloaddition (CuAAC), p-fluorothiol click, thiol-ene click, thio-halogen click, and thiol-yne click. A glycopolymer is synthesized and further modified via a click reaction by adding functional moieties on them. The above mentioned polymeric techniques can give rise to better biodegradable polymers for medical and food packaging applications [53].

Similarly, alkaline oxidation is used to make vanillin from lignin, which is used as a flavouring agent in various foods as well as in pharmaceutical preparation [54]. Likewise, vanillin based resins are made using polymerization of methacrylate and glycerol dimethacrylate, having high thermostable properties, and which can be used as a substitute for styrene [54]. Similarly, cyanide-based resin [55], epoxide resin [56], and benaoxazine resins [57] can be produced from lignin with excellent polymeric properties, and can serve as a vital alternative to petroleum-based polymers. Another important usage of lignin is the production of polyester, where lignin condensation with sebacoyl chloride yielded highly thermostable polyester [58]. Similarly, the production of polyurethanes from a lignin has been investigated and resulted in a high strength polymer with greater biodegradability [59].

9.7 Lignin-Based Fuels and Chemicals

Lignin, being one of the most abundant available substrates for different chemical conversions carries huge untapped applications (Table 9.1). It is one of the most important byproducts of the paper and pulp industry which remains unused. A high calorific value of the lignin (27 KJ/g) can therefore act as a potential source of energy [60], and also lignin-based oils can be of great value for their application as fuel [3]. Other than as an energy source, lignin is an antimicrobial agent and an antioxidant source [67]. As explained above, lignin is an important contributor for generation of various bio-based polymers, but at the same time it acts as a chelating agent, absorbent, emulsifier, and dispersant reagent in cement manufacturing gypsum blend [3]. Also, lignin is used for the manufacturing of energy storage devices [68] and slow release fertilizers [3].

Polymers based on C5 and C6 sugars from biomass have been successfully produced via fermentation of sugars, which produces 1,4-diacids (succinic acid, fumaric acid, malic acid). These are further used for a variety of chemical conversions for polymer synthesis, for instance, polyamides, polyesters and polyester amides, polybutylene terephthalate polyamides, polyesters, and polyester amides are made using the 1,4-diacids as synthesis platforms [69]. Similarly,2,5-furan dicarboxylic acid (2,5-FDCA), 5-HMF, 3- hydroxy propionic acid (3-HPA), aspartic acid, glucaric acid, glutamic

Table 9.1 Lignin-based polymers/products and their economic importance.

Product(s)	Economic importance	Reference
Polyurethanes	Polymeric backbone with enhanced thermo-stable properties and biodegradability.	[61]
Epoxide resins	Important material used in coatings, adhesives, composites, and electronic materials.	[62]
Phenolic Resins	Used as plywood adhesive for providing high shear strength.	[63]
Hydrogels	Used as delivery systems for proteins or drugs, tissue engineering scaffolds, or absorbents.	[64]
Vinyl-based graft copolymers	Enhances the mechanical properties of composites and reduces their flammability.	[65]
Vanillin	Used as a flavoring agent in various foods and pharmaceutical products.	[66]
Vanillin-based Resins	Possesses high thermal stability over styrene, and therefore can be used as a replacement for styrene.	[54]
Polyesters	Thermoplastic synthesis.	[58]

acid, itaconic acid, levulinic acid, 3-hydroxybutyrolactone (3-HBL), glycerol, sorbitol, and xylitol/arabinitol are used as the basis for conversion into a variety of valuable chemicals and polymers [4].

9.8 Concluding Remarks and Future Prospects

A very small fraction of the lignin produced annually is properly channelled for various commercial applications. There is a huge potential for further research for producing newer chemicals and improving the existing technologies which are otherwise not very practical commercially. There is a huge untapped potential for lignin-based chemicals like thermo-stable polymers, natural flavors, and fuels. A race for developing the chemical or biological process for the successful conversion of lignin into value added chemicals has already started. Using lignin fermented products for chemical modification, subjecting the same directly to different chemical processes (e.g., pyrolysis), gives different valuable products and precursors. Also, the advanced separation techniques, effective catalytic processes, and purification techniques can further be advantageous assets for creating an economically viable technology for bringing in a new era of bio-based polymers which is currently far from reality. A sustainable biorefinery approach for lignin-based products is already being developed, which can provide an alternative to more environmentally friendly biodegradable products. It is appropriate to say that in future we may see many bio-based products, fuels, and chemicals in the market where a significant share will be lignocellulosic-based, and a greater part of which will be derived from lignin.

References

1. Li, S.H., Liu, S., Colmenares, J.C., Xu, Y.J., A sustainable approach for lignin valorization by heterogeneous photocatalysis. *Green Chem.*, 18, 594, 2016.
2. Ragauskas, A.J., Beckham, G.T., Biddy, M.J., Chandra, R., Chen, F., Davis, M.F., Davison, B.H., Dixon, R.A., Gilna, P., Keller, M., Langan, P., Naskar, A.K., Saddler, J.N., Tschaplinski, T.J., Tuskan, G.A., Wyman C.E., Lignin valorization: Improving lignin processing in the biorefinery. *Science*, 344, 1246843, 2014.
3. Welker, C.M., Balasubramanian, V.K., Petti, C., Rai, K.M., DeBolt, S., Mendu,V., Engineering plant biomass lignin content and composition for biofuels and bioproducts. *Energies*, 8, 7654–7676, 2015.
4. Isikgora, F.H., Becer, C.R., Lignocellulosic biomass: Asustainable platform for the production of bio-based chemicals and polymers. *Polym. Chem.*, 6, 4497–4559, 2015.
5. Jong, E.D., Higson, A., Walsh, P., Wellisch, M., Barbosa, M., Biobased chemicals value added products from biorefiniaries. IEA Bioenergy Task 42 Biorefinery (report).
6. Qin, Z., Zhuang, Q., Zhu, X., Cai, X., Zhang, X., Carbon consequences and agricultural implications of growing biofuel crops on marginal agricultural lands in China. *Environ. Sci. Technol.*, 45, 10765–10772, 2011.
7. Akhtar, N., Gupta, K., Goyal, D., Recent advances in pre-treatment technologies for efficient hydrolysis of lignocellulosic biomass. *Environ. Prog. Sustainable Energy*, 35, 489–511, 2016.
8. Barakat, A., de Vries, H., Rouau, X., Dry fractionation process as an important step in current and future lignocellulose biorefineries: A review. *Bioresour. Technol.*, 134, 362–373, 2013.

9. Cherubini, F., Strømman, A.H., Chemicals from lignocellulosic biomass: Opportunities, perspectives and potentials of biorefinery systems. *Biofuels, Bioprod. Biorefin.*, 5, 548–561, 2011.
10. Agbor, V.B., Cicek, N., Sparling, R., Berlin, A., Levin, D.B., Biomass pretreatment: Fundamentals toward application. *Biotechnol. Adv.*,29, 675–685, 2011.
11. Mukherjee, G., Dhiman, G., Akhtar, N., Efficient hydrolysis of lignocellulosic biomass: Potential challenges and future perspectives for bio refinery. In: *Bioremediation and Sustainable Environmental Technologies for Cleaner Environment*, pp. 213–237. Springer, Switzerland, 2017.
12. Alvira, P., Tomás-Pejó, E., Ballesteros, M., Negro, M. J., Pre-treatment technologies for an efficient bio-ethanol production process based on enzymatic hydrolysis: A review. *Bioresour. Technol.*, 101, 4851–4861, 2010.
13. Mukherjee, G., Dhiman, G., Akhtar N., Lignocellulosic biomass utilization for the production of sustainable chemicals and polymers. In: *Lignocellulosic Biomass Production and Industrial Applications*, 215–245. John Wiley & Sons, Inc., 2017a.
14. Simon, C., Lion, C., Huss, B., Blervacq, A.S., Spriet, C., Guérardel, Y., Biot, C., Hawkins S., BLISS: shining a light on lignification in plants. *Plant Signaling Behav.*, 2017 (In press).
15. Bhuiyan, N.H., Selvaraj, G., Wei, Y., King, J., Role of lignification in plant defense. *Plant Signaling Behav.*, 4, 158–159, 2009.
16. Liljegren, S.J., Ditta, G.S., Eshed, Y., Savidge, B., Bowman, J.L., Yanofsky, M.F., Shatterproof Mads-Box genes control seed dispersal in Arabidopsis. *Nature*, 404, 766–770, 2000.
17. Boerjan, W., Ralph, J., Baucher, M., Lignin biosynthesis. *Annu. Rev. Plant Biol.*, 54, 519–546, 2003.
18. Terashima, N., Yoshida, M., Hafrén, J., Fukushima, K., Westermark, U., Proposed supramolecular structure of lignin in softwood tracheid compound middle lamella regions. In: *Holzforschung*, Volume 66, pp. 907–915. Walter de Gruyter, Berlin, Germany, 2012.
19. Pingali, S.V., Urban, V.S., Heller, W.T., McGaughey, J., Neill, H., Foston, M., Myles D.A., Ragauskas, A., Evans, B.R., Breakdown of cell wall nanostructure in dilute acid pretreated biomass. *Biomacromol.*, 11, 2329–2335, 2010.
20. Wang, C., Li, H., Li, M., Bian, J., Sun, R., Revealing thestructureand distribution changes of Eucalyptusligninduring the hydrothermal and alkaline pretreatments. *Sci. Rep.*,7, 593, 2017.
21. Chen, C.L., Nitrobenzene and cupric oxide oxidations. In: *Methods in Lignin Chemistry*, S. Lin, C. Dence (Eds.), pp. 301–321, Springer, Berlin/Heidelberg, Germany, 1992.
22. Sarkanen, K.V., Islam, A., Anderson, C.D., Ozonation. In: *Methods in Lignin Chemistry*, S. Lin, C. Dence, (Eds.), pp. 387–406. Springer, Berlin/Heidelberg, Germany, 1992.

23. Fachuang, L., John, R., Efficient ether cleavage in lignins: The derivatization followed by reductive cleavage procedure as a basis for new analytical methods. In Lignin and Lignan Biosynthesis, *American Chemical Society*, Washington, DC, USA, pp. 294–322, 1998.
24. Morrison, I.M., Improvements in the acetyl bromide technique to determine lignin and digestibility and its application to legumes. *J. Sci. Food Agri.*, 23, 1463–1469, 1971.
25. Bruce, R.J., West C.A., Elicitation of lignin biosynthesis and isoperoxidase activity by pectic fragments in suspension cultures of castor bean. *Plant Physiol.*, 91, 889–897, 1989.
26. Gidh, A.V., Decker, S.R., Vinzant, T.B., Himmel, M.E., Williford, C., Determination of lignin by size exclusion chromatography using multi angle laser light scattering. *J. Chromatogr. A*, 1114, 102–110, 2006.
27. Poovaiah, C.R., Nageswara-Rao, M., Soneji, J.R., Baxter, H.L., Stewart C.N., Jr., Altered lignin biosynthesis using biotechnology to improve lignocellulosic biofuel feedstocks. *Plant Biotechnol. J.*, 12, 1163–1173, 2014.
28. Reiss, R., Ihssen, J., Richter, M., Eichhorn, E., Schilling, B., Thöny-Meyer L., Laccase versus laccase-like multi-copper oxidase: A comparative study of similar enzymes with diverse substrate spectra. *PLoS ONE*, 8(6), e65633, 2013.
29. Cai, X., Davis, E.J., Ballif, J., Liang, M., Bushman, E., Haroldsen, V., Torabinejad, J., Wu, Y., Mutant identification and characterization of the laccase gene family in Arabidopsis. *J. Exp. Bot.*, 57, 2563–2569, 2006.
30. Saballos, A., Vermerris, W., Rivera, L., Ejeta, G., Allelic Association, Chemical characterization and saccharification properties of brown midrib mutants of sorghum (*Sorghum bicolor (L.) Moench*). *Bioenergy Res.*, 1, 193–204, 2008.
31. Ko, J.H., Kim, H.T., Hwang, I., Han, K.H., Tissue-type-specific transcriptome analysis identifies developing xylem-specific promoters in poplar. *Plant Biotechnol. J.*, 10, 587–596, 2012.
32. Ralph, J., Hatfield, R.D., Quideau, S., Helm, R.F., Grabber, J.H., Jung H.-J.G., Pathway of p-Coumaric acid incorporation into maize lignin as revealed by NMR, *Journal of American Chemical Society*, 116, 9448–9456,1994.
33. Gaj, T., Gersbach, C.A., Barbas, C.F., 3rd, ZFN, TALEN, and CRISPR/Cas-based methods forgenome engineering. *Trends Biotechnol.*, 31, 397–405, 2013.
34. Mendu, V., Harman-Ware, A.E., Crocker, M., Jae, J., Stork, J., Morton, S., 3rd., Placido, A., Huber, G., Debolt, S., Identification and thermochemical analysis of high-lignin feedstocks forbiofuel and biochemical production. *Biotechnol. Biofuels*, 2011, 4, 2016.
35. Marsh, G., Biofuels: Aviation alternative? *Renew. Energy Focus*, 9, 48–51, 2008.
36. Regalbuto, J.R., Cellulosic biofuels—got gasoline? *Science*, 325, 822–824, 2009.
37. Cano-Delgado, A., Penfield, S., Smith, C., Catley, M., Bevan M., Reduced cellulose synthesis invokes lignification and defense responses in *Arabidopsis thaliana*. *Plant J.*, 34, 351–362, 2003.

38. Moura, J.C., Bonine, C.A., de Oliveira Fernandes Viana, J., Dornelas, M.C., Mazzafera, P., Abiotic and biotic stresses and changes in the lignin content and composition in plants. *J. Integr. Plant Biol.*, 52, 360–376, 2010.
39. Ong, S.S., Wickneswari, R., Characterization of microRNAs expressed during secondary wall biosynthesis in *Acacia mangium. PLoS ONE*, 7, 2012.
40. Zhou, J., Lee, C., Zhong, R., Ye, Z.H., MYB58 and MYB63 are transcriptional activators of the lignin biosynthetic pathway during secondary cell wall formation in Arabidopsis. *Plant Cell*, 21, 248–266, 2009.
41. Sitbon, F., Hennion, S., Little, C.H.A., Sundberg, B., Enhanced ethylene production and peroxidase activity in IAA-overproducing transgenic tobacco plants is associated with increased lignin content and altered lignin composition. *Plant Sci.*, 141, 165–173, 1999.
42. Hématy, K., Sado, P.-E., Van Tuinen, A., Rochange, S., Desnos, T., Balzergue, S., Pelletier, S., Renou J.-P., Höfte, H., A receptor-like kinase mediates the response of arabidopsis cells to the inhibition of cellulose synthesis. *Curr. Biol.*, 17, 922–931, 2007.
43. Wang, H., Tucker, M., Ji, Y., Recent development in chemical depolymerization of lignin: A review. *J. Appl. Chem.*, 838645, 2013.
44. Toledano, A., García, A., Mondragon, I., Labidi, J., Lignin separation and fractionation by ultrafiltration. *Sep. Purif. Technol.*, 71, 38–43, 2010.
45. Obydenkova, S.V., Kouris, P.D., Hensen, E.J.M., Heeres, H.J., Boot, M.D., Environmental economics of lignin derived transport fuels. *Bioresour. Technol.*, 243, 589–599,2017.
46. Li, C., Zhao, X., Wang, A., Huber, G.W., Zhang, T., Catalytic transformation of lignin for the production of chemicals and fuels. *Chem. Rev.*, 115(21), 11559–11624, 2015.
47. Pandey M.P., Kim C.S., Lignin depolymerization and conversion: A review of thermochemical methods. *Chem. Eng. Technol.*, 34, 29–41, 2011.
48. Sannigrahi, P., Pu, Y., Ragauskas A., Cellulosic biorefineries—unleashing lignin opportunities. *Curr. Opin. Environ. Sustain.*, 2, 383–393, 2010.
49. López-Abelairas, M., ÁlvarezPallín, M., Salvachúa, D., Lú-Chau, T., Martínez, M.J., Lema J.M., Optimisation of the biological pre-treatment of wheat straw with white-rot fungi for ethanol production. *Bioprocess Biosyst. Eng.*, 36, 1251–1260, 2013.
50. Larsson, S., Cassland, P., Jönsson L.J., Development of *Saccharomyces cerevisiae* with enhanced resistance to phenolic fermentation inhibitors in lignocelluloses hydrolysates by heterologous expression of laccase. *Appl. Environ. Microbiol.*, 67, 1163–1170, 2011.
51. Miyazaki, T., Miyazaki, J., Yamane, H., Nishiyama, M., alpha-Amino-adipate aminotransferase from an extremely thermophilic bacterium, Thermusthermophilus. *Microbiol.*, 150, 2327–2334, 2004.
52. Laurichesse, S., Avérous, L., Chemical modification of lignins: Towards bio-based polymers. *Prog. Polym. Sci.*, 39, 1266–1290, 2014.

53. Munoz-Guerra, S., Carbohydrate-based polyamides and polyesters: An overview illustrated with two selected examples. *High Perform. Polym.*, 24, 9–23, 2012.
54. Stanzione, J.F., Sadler, J.M., La Scala, J.J., Wool, R.P., Lignin model compounds as bio-based reactive diluents for liquid molding resins. *ChemSusChem.*, 5(7), 1291–1297, 2012.
55. Meylemans, H.A., Harvey, B.G., Reams, J.T., Guenthner, A.J., Cambrea, L.R., Groshens, T.J., Baldwin, L.C., Garrison, M.D., Mabry, J.M., Synthesis, characterization, and cure chemistry of renewable bis(cyanate) esters derived from 2-methoxy-4-methylphenol. *Biomacromol.*, 14, 771–780, 2013.
56. Fache, M., Darroman, E., Besse, V., Auvergne, R., Caillol, S., Boutevin, B., Vanillin, a promising biobased building-block for monomer synthesis. *Green Chem.*, 16, 1987–1998, 2014.
57. Wang, C., Sun, J., Liu, X., Sudo, A., Endo, T., Synthesis and copolymerization of fully bio-based benzoxazines from guaiacol, furfurylamine and stearylamine. *Green Chem.*, 14, 2799–2806, 2012.
58. Binh, N.T.T., Luong, N.D., Kim, D.O., Lee, S.H., Kim, B.J., Lee, Y.S., Nam, J.D., Synthesis of lignin-based thermoplastic copolyester using kraft lignin as a macromonomer. *Compos. Interfaces*, 16, 923–935, 2009.
59. Chung, H., Washburn N.R., Improved lignin polyurethane properties with Lewis acid treatment. *ACS Appl. Mater. Interfaces*, 4, 2840–2846, 2012.
60. Mendu, V., Shearin, T., Campbell, J.E., Stork, J., Jae, J., Crocker, M., Huber, G., DeBolt, S., Global bioenergy potential from high-lignin agricultural residue. *Proc. Natl. Acad. Sci.*, 109, 4014–4019, 2012.
61. Gandini, A., Belgacem, M.N., Guo, Z.-X., Montanari, S., Lignins as macromonomers for polyesters and polyurethanes. In: *Chemical Modification, Properties and Usage of Lignin*, T.Q. Hu (Ed.), pp 57–80. Kluwer Academic/ Plenum: New York, 2002.
62. Aurvergne, R., Caillol, S., David, G., Boutevin, B., Pascault J.-P., Bio based thermosetting epoxy: Present and future. *Chem. Rev.*, 114, 1082–1115, 2014.
63. Danielson. B., Simonson, R., Kraft lignin in phenol formaldehyde resin. Part 1. Partial replacement of phenol by kraft lignin in phenol-formaldehyde adhesives for plywood. *J. Adhes. Sci. Technol.*, 12, 923–939, 1998.
64. Passauer, L., Highly swellable lignin hydrogels: Novel materials with interesting properties. In: *Functional Materials from Renewable Sources*, pp. 211–228. American Chemical Society, Washington, DC, 2012.
65. Li, H., Zhang, Q., Gao, P., Wang, L., Preparation and characterization of graft copolymer from dealkaline lignin and styrene. *J. Appl. Polym. Sci.*, 132, 41900, 2015.
66. Strassberger, Z., Tanase, S., Rothenberg, G., The pros and cons of lignin valorisation in an integrated biorefinery. *RSC Adv.*, 4, 25310, 2014.
67. Thakur, V.K., Thakur, M.K., Recent advances in green hydrogels from lignin: Areview. *Int. J. Biol. Macromol.*, 72, 834–847, 2015.

68. Hu, S., Zhang, S., Pan, N., Hsieh, Y.-L., High energy density supercapacitors from lignin derivedsubmicron activated carbon fibers in aqueous electrolytes. *J. Power Sources*, 270, 106–112, 2014.
69. Kabasci, S., Bretz, I., Succinic acid: Synthesis of biobased polymers from renewable resources. In: *Renewable Polymers*, V. Mittal (Ed.), pp. 355–379, John Wiley & Sons, Inc., 2011.

10

Exploring the Fermentation Technology for Biocatalysts Production

Ronivaldo Rodrigues da Silva

Instituto de Biociências, Letras e Ciências Exatas, Universidade Estadual Paulista, São José do Rio Preto, São Paulo, Brazil

Abstract

Investigating enzymes to degrade complex compounds constitute to a sustainable tool for depolymerization of organic matter and xenobiotic compounds. In this sense, microorganisms are the main sources of enzyme production with recognized capacity for application in various market segments. By considering the enzyme production, submerged and solid state bioprocesses have gained great popularity in the scientific world and have been extensively exploited. Using the fermentation technology, the metabolic potential of microorganisms, especially bacteria and fungi, has been exploited for many biotechnological studies. These microorganisms can serve as promising candidates for enzymes production, particularly oxidative and hydrolytic enzymes, with applications in many industrial sectors, such as food processing, pharmaceuticals, leather treatment, cosmetics, fine chemicals, among others.

Keywords: Bioprocess, enzyme technology, fermentation, microorganisms

10.1 Introduction

Microbiology is an invaluable resource for enzyme technology. Since long time, microorganisms are used to produce biocatalysts gaining position in the industrial scenario [12, 13, 15]. Some characteristics justify the interest in microorganisms for the production of enzymes, such as ease of

Corresponding author: rds.roni@yahoo.com.br

Arindam Kuila and Vinay Sharma (eds.) Principles and Applications of Fermentation Technology, (181–188) © 2018 Scrivener Publishing LLC

handling, possibility of genetic manipulation, diversity in production and secretion, and great profitability in industrial production.

Enzymes are biocatalysts involved in chemical reactions in all living organisms. They exhibit specificity to the substrate, in which it accelerates the chemical reaction for its conversion into product. Each enzyme has particular functional and structural biochemical properties, thus being interesting tools for *in vitro* reactions and, therefore, targets of constant biotechnological studies for application in several commercial sectors [12, 13, 15, 17, 18].

The literature reports the importance of enzymes in the transformation of matter from ancient times [5, 13]. For a long time, microorganisms have been used to obtain products of human interest, such as beverages (fermented) and foods (bakery and cheeses) [13]. With the increase in the demand for these products, the improvement of cultivation techniques that propitiated greater profitability became necessary. The exploitation of microbial bioprocesses allowed the production of specific and large-scale metabolites, ensuring the use of enzymes as a commercial product in several industrial sectors [13].

In this context, investigating into new enzymes and their biochemical properties have been the object of many studies in the enzymatic technology. For the commercial demand in the use of biocatalysts, thermostability and thermoactivity of the enzyme are factors that gain prominence in industrial processes due to greater stability in the use and storage of these molecules [5].

To understand better the production of enzymes, we need to deepen our discussion about microbial bioprocesses.

10.2 Biotechnology Fermentation

Industrial fermentation is understood as the process of microbial growth in an aerobic or anaerobic condition under determined physical–chemical parameters. Several products of industrial interest are generated as, amino acids, vitamins, organic acids, alcohol, antibiotics, and enzymes [3]. The production of enzymes is basically carried out by two fermentation systems: submerged fermentation (SmF) and solid state fermentation (SSF), whose differences in bioprocesses and composition of the medium are determinant in the production of enzymes [3, 14].

10.2.1 Submerged Fermentation

In this process, the microorganism is grown in liquid culture medium under agitation. This method, in an industrial fermenter, allows the control of physicochemical factors such as pH, temperature, and oxygenation, besides guaranteeing the homogenization of nutrients [13, 21]. It presents advantages such as control of fermentative parameters and greater efficiency of nutrient absorption and metabolic excretion by the microbial cell [21] and disadvantages such as higher energy and water requirement, with high aeration and agitation costs [13].

10.2.2 Solid State Fermentation

Many definitions have been proposed to define this type of microbial growth. However, it is a consensus that in this bioprocess the microbial growth, in solid medium, occurs in the absence or little presence of free water. In this bioprocess it is necessary that the substrate is sufficiently moistened to allow microbial metabolism [9].

SSF is characterized by advantages such as low energy requirements, lower water consumption and simplicity of the culture medium, allowing the use of agro-industrial residues as raw material in the production of enzymes [9]; and disadvantages such as lower accessibility to the substrate, difficulty in controlling physicochemical variables (pH, temperature, oxygenation), difficulty in nutrient transfer, and separation of metabolites by chromatographic processes [13].

10.3 Production of Enzymes

The production of enzymes varies according to the source of prospection (animal, vegetable, microorganisms) and according to culture conditions. Many studies have corroborated this assertion and thus, the fermentative parameters are still widely studied in order to improve yield and variety in the production of biocatalysts [13].

By exploiting microbial biodiversity, enzymatic technology has enabled the substitution of many chemical reactions that have a complicating effect on the environment or that are more costly for commercialization. An overview of chemical reactions substituted by the use of enzymes is shown in Table 10.1.

The success of filamentous fungi in the generation of products of industrial application is guaranteed by the wide metabolic versatility of these microorganisms. Some filamentous fungi are known to produce organic acids, polysaccharides, enzymes, alkaloids, and antibiotics [4]. The largest industrial exploitation fungi to-date include species of *Aspergillus*,

Table 10.1 Examples of enzyme application.

Enzyme	Chemical process previously employed	Enzyme application	References
Aspartic peptidase	Calf chymosin	Cheese manufacturing	[15, 18]
Peptidases	Sodium sulfide and Chromium salts	Leather treatment	[11, 13, 20]
Chitinolytic enzymes	Pesticides	Agriculture	[10]
Amylases, peptidases	Manual scrubbing using chemicals	Biofilm removal	[20]
Amylases, lipases, peptidases	Detergent formulation without enzymes	Detergent	[20]
Xylanases	Chlorinated compounds	Paper biobleaching	[23]

such as *A. awamori*, *A. niger*, *A. oryzae*, and *A. nidulans*, *Rhizomucor* and *Penicillium* species, and *Trichoderma reesei* [1, 13, 22].

White-rot fungi are also highly exploited because of their large capacity for oxidoreductases production, especially to degrade lignocellulolytic biomass and bioremediation processes [12].

Bacteria and yeasts are also extensively exploited for the production of metabolites, such as enzymes, antibiotics, and ethanol. In general, these microorganisms present a better enzyme production in submerged bioprocesses [13]. The production of enzymes by bacteria has been much evidenced in species of the genera *Bacillus* and *Streptomyces* [13] and for yeasts, species such as *Aureobasidium pullulans*, *Aureobasidium leucospermi*, and *Yarrowia lipolytica* [7, 8].

Many commercial sectors have been contemplated with the use of enzymes, such as the pharmaceutical industry (production of biopharmaceuticals, e.g., L-Asparaginase and collagenases), leather industry (peptidases) degradation of plant biomass (cellulases, esterases, xylanases, and ligninases) bioremediation (oxidoreductases) [2, 16, 17, 18, 19].

Bioprospecting of microorganisms for the production of enzymes extends to a wide spectrum of biocatalysts contemplating the different classes of enzymes: Oxidoreductases (e.g., Catalases, Glucose oxidases,

Laccases), Transferases (e.g., Fructosyltransferases, Glucosyltransferases), Hydrolases (e.g., Proteases, Glycoside hydrolases, Lipases, Esterases), Lyases (Pectate lyases, Alpha-acetolactate, Decarboxylases), Isomerases (e.g., Glucose isomerases, Epimerases, Mutases), and Ligases (e.g., Argininosuccinate, Glutathione synthase) [5, 12, 17].

To date, the hydrolases correspond to the main class of enzymes commercialized, with emphasis on proteolytic enzymes, glycosidases, and lipases. Applications such as detergents, leather, textiles, and food are examples of commercial segments that employ these biocatalysts in the processing of macromolecules.

Proteases or peptidases are hydrolases that catalyze the cleavage of peptide bonds in proteins and peptides. These enzymes are widely prospected from microbial sources and used in industrial sectors such as food industry (cheese production, protein hydrolysates), detergents [6], leather treatment, basic research (trypsin and proteinase K) and pharmaceuticals (collagenases) [13, 14].

By considering the production of proteolytic enzymes, the catalytic specificity study is a valuable information to determine the potential application of these enzymes [13, 19]. According to the catalytic properties of proteolytic enzymes, it is possible to evaluate the application scenario, such as application in detergents, whose action of proteases with broad spectrum of action on different proteins and stability at alkaline pH are very required [6]; high specificity, such as trypsin, for studies in basic research among others [13].

The glycoside hydrolases are enzymes that catalyze the cleavage of glycosidic bonds in disaccharides, oligosaccharides, and polysaccharides. In this group of enzymes, the most important are amylases, cellulolytic, and hemicellulolytic enzymes, with a large number of published articles and many segments for industrial application, such as food, textile, and paper industries, among others [17].

It is important to emphasize that the microbial source for enzyme production is an invaluable natural resource for biotechnology, either by investigating the biochemical potential for expression and secretion of enzymes by microorganisms that are already well studied or by exploiting the microbial diversity in the discovery of new unknown enzymes.

References

1. Ângelo, T., Silva, R.R., Cabral, H., Concomitant production of peptidases and lipases by fungus using agroindustrial residue in solid-state fermentation. *Int. J. Cur. Microbiol. Appl. Sci.*, 3, 810–823, 2014.

2. Biaggio, R.T., Silva, R.R., Rosa, N.G., Leite, R.S.R., Arantes, E.C., Cabral, T.P.F., Juliano, M.A., Juliano, L., Cabral, H., Purification and biochemical characterization of an extracellular serine peptidase from *Aspergillus terreus*. *Prep. Biochem. Biotechnol.*, 46, 298–304, 2015.
3. Demain, A.L., The business of biotechnology. *Ind. Biotechnol.*, 3(3), 269–283, 2007.
4. El-Enshasy, H.A., Filamentous fungal cultures–process characteristics, products, and applications. *Bioproc. Value-Added Prod. Renew. Resour.*, 225–262, 2007.
5. Gurung, N., Ray, S., Bose, S., Rai, V., A broader view: Microbial enzymes and their relevance in industries, medicine, and beyond. *BioMed Res. Int.*, 1–18, 2013. http://dx.doi.org/10.1155/2013/329121.
6. Ida, E.L., Silva, R.R., Oliveira, T.B., Souto, T.B., Leite, J.A., Rodrigues, A., Cabral, H., Biochemical properties and evaluation of washing performance in commercial detergent compatibility of two collagenolytic serine peptidases secreted by *Aspergillus fischeri* and *Penicillium citrinum*. *Prep. Biochem. Biotechnol.*, 1–9, 2016.
7. Kumar, C.G., Takagi, H., Microbial alkaline proteases: From a bioindustrial viewpoint. *Biotechnol. Adv.*, 17, 561–594, 1999.
8. Ni, X., Yue, L., Chi, Z., Li, J., Wang, X., Madzak, C., Alkaline protease gene cloning from the marine yeast *Aureobasidium pullulans* HN2-3 and the protease surface display on *Yarrowia lipolytica* for bioactive peptide production. *Mar. Biotechnol.*, 11, 81–89, 2009.
9. Pandey, A., Solid-state fermentation. *Biochem. Eng. J.*, 13, 81–84, 2003.
10. Patil, R.S., Ghormade, V., Deshpande, M.V., Chitinolytic enzymes: An exploration. *Enzyme Microb. Technol.*, 26, 473–483, 2000.
11. Rao, M.B., Tanksale, A.M., Ghatge, M.S., Deshpande, V.V., Molecular and biotechnological aspects of microbial proteases. *Microbiol. Mol. Biol. Rev.*, 62(3), 597–635, 1998.
12. Silva, R.R., Potential of white-rot fungi for bioremediation. *Revista Brasileira de Gestão Ambiental e Sustentabilidade, Brazilian Journal of Environmental Management and Sustainability*, 4, 229–232, 2017.
13. Silva, R.R., Bacterial and fungal proteolytic enzymes: Production, catalysis and potential applications. *Appl. Biochem. Biotechnol.*, 183, 1–19, 2017.
14. Silva, R.R., Cabral, T.P.F., Rodrigues, A., Cabral, H., Production and partial characterization of serine and metallo peptidases secreted by *Aspergillus fumigatus* Fresenius in submerged and solid state fermentation. *Braz. J. Microbiol.*, 44, 235–243, 2013.
15. Silva, R.R., Oliveira, L.C.G., Juliano, M.A., Juliano, L., Oliveira, A.H.C., Rosa, J.C., Cabral, H., Biochemical and milk-clotting properties and mapping of catalytic subsites of an extracellular aspartic peptidase from basidiomycete fungus *Phanerochaete chrysosporium*, *Food Chemistry*, 225, 45–54, 2017.
16. Silva, R. R., Oliveira, L. C. G., Juliano, M. A., Juliano, L., Rosa, J. C., Cabral, H., Activity of a peptidase secreted by *Phanerochaete chrysosporium* depends on lysine to subsite S'1. *Int. J. Biol. Macromol.*, 94, 474–483, 2017.

17. Silva, R.R., Pedezzi, R., Souto, T.B., Exploring the bioprospecting and biotechnological potential of *white-rot* and anaerobic *Neocallimastigomycota* fungi: Peptidases, esterases, and lignocellulolytic enzymes. *Appl. Microbiol. Biotechnol.*, 1–13, 2017.
18. Silva, R.R., Souto, T.B., Oliveira, T.B., Oliveira, L.C.G., Karcher, D., Juliano, M.A., Juliano, L., Oliveira, A.H.C., Rodrigues, A., Rosa, J.C., Cabral, H., Evaluation of the catalytic specificity, biochemical properties, and milk clotting abilities of an aspartic peptidase from *Rhizomucor miehei*. *J. Ind. Microbiol. Biotechnol.*, 43, 1059–1069, 2016.
19. Silva, R., Caetano, R., Okamoto, D., Oliveira, L., Bertolin, T., Juliano, M.A., Juliano, L., Oliveira, A.H., Rosa, J., Cabral, H., The identification and biochemical properties of the catalytic specificity of a serine peptidase secreted by *Aspergillus fumigatus* Fresenius. *Protein Pept. Lett.*, 21, 663–671, 2014.
20. Singh, S., Bajaj, B.K., Potential application spectrum of microbial proteases for clean and green industrial production. *Energ. Ecol. Environ.* http://dx.doi.org/10.1007/s40974-017-0076-5.
21. Sinha, S., Sinha, S. Studies on the production of acid protease by submerged fermentation. *Int. J. Food Eng.*, 5, 2009.
22. Van Den Homberg, J.P.T.W., Van De Vondervoort, P.J.I., Fraissinettachet, L., Visser, J., Aspergillus as a host for heterologous protein production: The problem of proteases. *Trends Biotechnol.*, 15, 256–263, 1997.
23. Walia, A., Guleria, S., Mehta, P., Chauhan, A., Parkash, J., Microbial xylanases and their industrial application in pulp and paper biobleaching: A review. *3 Biotech.*, 2017. http://dx.doi.org/10.1007/s13205-016-0584-6.

11

Microbial CYP450: An Insight into Its Molecular/Catalytic Mechanism, Production and Industrial Application

Abhilek Kumar Nautiyal[1,2], Arijit Jana[1,2], Sourya Bhattacharya[3], Tripti Sharma[1,2], Neha Bansal[1,2], Sree Sai Ogetiammini[3], Debashish Ghosh[1,2], Saugata Hazra[3,4,*] and Diptarka Dasgupta[1,2,*]

[1]*Biotechnology Conversion Area, Bio Fuels Division; CSIR-Indian Institute of Petroleum, Mohkampur, Uttarakhand, India*
[2]*Academy of Scientific and Innovative Research (AcSIR)*
[3]*Department of Biotechnology & [4]Center of Nanotechnology, Indian Institute of Technology, Roorkee, Uttarakhand, India*

Abstract

Selective C-H bond activation at inactivated carbon by Cytochrome P450 monooxygenase (CYP450) mediated oxidation under mild condition always remains challenging to the scientific community. CYPs are heme-containing enzymes that use molecular oxygen and the hydride donor Nicotinamide adenine dinucleotide phosphate (NAD[P]H) (coupled via redox partners) to insert atomic oxygen into the organic substrate. Oxidation is manifested as hydroxylation, epoxidation, dealkylation, and other transformations and is carried out in a regio- and stereoselective manner. Distributed within various species, microbial CYP450 plays a significant role in environment bioremediation or particular medicinal applications. Requisite supply of reductant during biotransformation hinders its cell-free application and thus engineering this protein for its use as whole cell biocatalyst is a major question in this area. The present chapter deals with the development of understanding the role of microbial CYP450, and its structure, function, active sites, and mode of action. Production strategies by submerged culturing with various cultivation aspects, strain engineering for improved catalytic activity and

**Corresponding authors*: saugata.iitk@gmail.com; ddgupta@iip.res.in

Arindam Kuila and Vinay Sharma (eds.) Principles and Applications of Fermentation Technology, (189–216) © 2018 Scrivener Publishing LLC

down-streaming of the enzyme have been highlighted. The chapter concludes with various applications of CYP450s in environmental bioremediation or medicinal biotransformations.

Keywords: Cytochrome P450, microbial genome, catalytic cycle, CYPBM3

11.1 Introduction

Cytochrome P450 enzymes (P450s or CYPs) are membrane-bound hemoproteins with diverse oxygenation functions and widely distributed in living organisms like archaea, bacteria as well as human beings [37]. They are a part of multienzymatic systems called the monooxygenases and catalyze a variety of oxidation reactions such as hydroxylation, N and O-dealkylation reactions, oxidation of thiols, and epoxidation of double bonds [73]. During catalysis, P450 carries out the oxidation of non-activated C–H bonds through the transfer of a singlet oxygen atom in a regio- and/or stereospecific manner [27], by utilizing electrons provided through one NAD(P)H molecule. P450-dependent metabolism requires two specific protein components, an active P450 catalytic domain which catalyzes the oxidation reaction assisted by an oxidoreductase domain which shuttles electrons from NADPH to P450 for its functional catalytic sphere [76]. The enzyme complex is anchored to the membrane of endoplasmic reticulum (ER) by a transmembrane linker with the catalytic domain directed towards the cytosol [26].

The first experimental evidence connecting CYP450 was discovered as early as in the fifties by Brodie and Axelrod (1955). They identified the enzyme system in the ER of rat liver cells that were capable of oxidizing xenobiotic compounds. CYP450 derives its name from the characteristic absorption peak at 450 nm wavelength with their carbon monoxide bound form. Electron spin resonance spectroscopy suggested that P450 is a low spin ferric hemoprotein [46, 81] with a thiol residue as an axial haem ligand [8, 9]. The absorption peak at 450 nm corresponds to the Soret π-π^* band of the (Iron)-CO chromophore. During catalysis, P450 undergoes a conformational change regarding charge transfer upon substrate binding that is modulated by a covalently bonded Cysteine residue linked with the Fe ligand [18].

CYP450 plays a pivotal role in detoxification of xenobiotics, cellular metabolism, and homeostasis [114]. CYP enzymes can be transcriptionally activated by various chemicals and endogenous substrates which include fatty acids, steroids, lipids hydroperoxides, retinoids, etc. through receptor-dependent mechanisms [91, 115]. Many chemotherapeutic agents can cause drug interactions with P450 due to their ability to either inhibit or induce the CYP enzyme system [74, 94]. Nucleotide polymorphisms and epigenetic changes in CYP genes may be responsible for the variation in therapeutic efficacy of drugs between individuals and belonging to different origins [113]. Careful knowledge about the substrates, inducers, and inhibitors of CYP isoforms and its polymorphic varieties may be used as a clinical tool to either determine a precise therapeutic strategy or more specifically treatment doses for drugs that are metabolized by these enzymes [80]. One of the few examples of the industrial application of a P450 is the 6β-hydroxylation of Compactin to produce a compound Mevastatin, used as an active precursor in the synthesis of drug pravastatin to regulate high lipid levels in the blood [110]. A further example of a commercially used CYP450 includes the production of 1α, 25-dihydroxyvitamin D_3 from vitamin D_3 used in the treatment of both hyperthyroidism and osteoporosis [116]. Apart from their plausible role in the pharmaceutical sector, CYPs are of great interest to the chemical industry owing to their potential as catalysts for synthetic transformations such as the hydroxylation of aliphatic and aromatic hydrocarbons, epoxidation of carbon-carbon double bonds and heteroatom oxygenation [40].

Despite several advantages, a few crucial limitations associated with these enzymes have hindered their application on a large scale. The enzymes have a relatively low turnover number, less stable, and being intracellular trans-membrane protein requires expensive purification for recovery [59, 88]. Protein engineering studies guided by structural insight have been envisaged to improve the catalytic efficiency and expand the inventory of substrates, and recombinant expression systems are being designed for heterologous expression of soluble microbial P450s in a suitable host to alleviate the issues associated with protein recovery [36, 49].

The present chapter focuses on a specific class of CYP450, the microbial system and discusses the enzymatic properties in light of their structure, function, active sites, and mode of action. The different production strategies of P450 with cultivation aspects, possible strategies for the catalytic improvement, and methods for enzyme recovery have been covered. The chapter concludes with the plausible scope of applications of CYP450s in environmental bioremediation or medicinal biotransformation.

11.2 Microbial Cytochrome P450

Nearly 250 CYP450s across different species have been identified, purified, and characterized in detail to infer the structural-functional properties

[87]. Each CYP450 gene is named with CYP, to signify it as a part of the CYP450 gene family. The gene is also given a number associated with a specific group within the gene family, a letter representing the gene's subfamily, and a number assigned to the specific gene within the subfamily. For example, the CYP450 gene that is in group 102, subfamily A, gene 1 is written as *CYP102A1*. Based on the electron transfer partner during catalysis, P450 varieties have primarily been classified into two major categories. First group comprises of P450s with a single flavoprotein reductase containing both flavin adenine dinucleotide (FAD) and flavin mononucleotide (FMN). Examples include P450s of the ER of higher organisms. Only one member of this group which has been carefully studied in a prokaryote and that is the soluble, catalytically self-sufficient, belongs to *Bacillus megaterium,* P450BM-3. The second group comprises of P450s which have two electron transfer partners: (1) a FAD-containing reduced pyridine nucleotide dehydrogenase and (2) an iron-sulfur protein. The members of this group are typically localized in mitochondria.

The very first P450 discovered as a fusion to its redox partner was CYP102A1, a mammalian-like diflavin reductase. This fusion of the two enzymatic units makes soluble CYP102A1 an ideal model for their mammalian counterpart, particularly human, P450 enzymes to study. CYP102A1 is called to be a versatile monooxygenase with a recognized, diverse applicability and an established biotechnological relevance. Wild-type CYP102A1 catalyzes the hydroxylation of long-chain fatty acids at subterminal positions (ω-1 to ω-3). Based on the crystal structure of CYP102A1 available, 20 amino acid residues of CYP102A1 have been suggested to be involved in a 'substrate channel' or 'active site'. These recommended amino acids have been targeted in studies characterizing the structure and enzymatic mechanism of CYP102A1. Bacterial CYP are soluble and metabolize only a partial number of natural substrates, such as fatty acids, vitamins, styrene, erythromycin, and terpenes, and mammalian CYP are membrane-bound and metabolize group of substrates, such as steroids, fatty acids, drugs, prodrugs, carcinogens, pesticides, and herbicides [25]. Mammalian P450s, show a low turnover rate and stability and requires additional redox partners, such as CYP450 reductase (CPR) and often cytochrome b_5, an expensive cofactor NADPH, and lipids. In contrast to the bacterial system, mammalian CYP enzymes have large and flexible active sites, which are capable of adapting their conformation based on the size, shape, and geometry of substrates [55, 98]. Bacterial $P450_{BM3}$ (CYP102) is catalytically self-sufficient, that is, the electron transfer from NADPH to FMN/FAD reductase domain and the P450 monooxygenase domain are in a single peptide [54, 77]. Moreover, bacterial CYP shows a much higher turnover, expression in *E. coli*, and coupling efficiency (utilization of NADPH in substrate metabolism vs formation of superoxide, peroxide, and water molecules) compared to that of their mammalian counterparts [2, 47]. Furthermore, bacterial CYP biocatalysts can function or can be tailored to operate under extreme conditions of temperature, pH, buffer system, or solvent. Bacterial CYP enzymes have been recently engineered for the metabolism of several non-natural substrates, such as smaller alkanes and drugs [59]. However, it remains to be seen whether bacterial enzymes can be tailored for the regio- or stereoselective synthesis of drugs and drug metabolites of variable size, shape, and geometry [71, 94]. These contrasting features of bacterial and mammalian CYP suggest that bacterial CYP are easy to use as biocatalysts, whereas mammalian CYP has much broader applications in pharmaceuticals/biotechnological industries and green chemistry. Across nature, CYP450 enzymes have an extraordinarily diverse substrate range. To expand the inventory of substrates, mutagenesis strategies play a crucial role in the case of soluble microbial P450s for which structural data help to guide protein engineering and thus replace synthetic chemistry for production of specialty chemicals. P450s can catalyse the specific addition of oxygen atoms at positions on chemical scaffolds, while this remains very challenging through traditional synthetic methods [59].

11.3 Extent of P450s in Microbial Genome

CYP genes present in microbial organisms differ broadly, even between species of the same genus. The novel P450s genes can be revealed by genome sequencing to identify unique classes of these enzymes from the bacteria and archaea and for the fungal kingdom of the Eukarya domain. Archaea and bacteria, however, contain fairly few P450s (in comparison to the most eukaryotes) with few organisms like *E. coli* [13], *Helicobacter pylori* [106] having no *CYP* genes. Moderate numbers of P450 (2–8 P450s/genome) and extents of genetic diversity are observed in *Bacillus* species genomes across a limited number of *CYP* gene families (*CYP102, CYP106, CYP107, CYP109, CYP134, CYP152,* and *CYP197)*, with 51 genes identified [70]. Most of the *Bacillus* P450s have undefined or unclear physiological roles, although numerous studies have been done on particular *Bacillus* sp P450s, including intensive characterization of the CYP102A1 (P450 BM3) P450-cytochrome, CPR fusion enzyme. P450 BM3 which is found in some *Bacillus* sp as well as in other bacteria is a highly efficient fatty acid monooxygenase, whose definitive physiological function remains elusive to date

[70]. Other characterised *Bacillus* P450s include isoforms that possess fatty acid or steroid hydroxylating activities. In contrast to *Bacillus* and many other bacteria, actinobacteria often contain large numbers of P450 genes. For example, there are 17, 20, 21, 39, and 47 *CYP* genes in *Mycobacterium bovis* AF2122/97, *M. tuberculosis* H37Rv, *M. ulcerans Agy99, M. smegmatis* MC2155, and *M. marinum,* respectively [60].

11.4 Structure, Function and Catalytic Cycle

Knowledge of the 3D structure has been imperative to understand the functional properties featuring the catalytic mechanism of CYP450. The first report elucidating structural information was reported by [89] for a class I *Pseudomonas putida* CYP. Subsequently, the crystal structure of other class varieties of CYP450 was reported. P450s comprise of nearly 400 and 500 amino acids containing a single heme prosthetic group. The structural fold is conserved throughout different P450 classes despite less than 20% sequence homology across the gene superfamily [27] (Figure 11.1). There are regions of variability across primary, secondary, and tertiary sequences which is attributed to the recognition and binding of diverse substrates, cofactors and their location within the cell [90].

The heme binding region essentially conserved in all microbial CYP450 is the core of the catalytic reaction process. The Fe atom (in the heme moiety) in the Fe^{3+} state is covalently bonded to a cysteine residue (Figure 11.2) in co-ordination with two water molecules. The residues adjacent to this Cys are highly conserved and regulate the hydrophobicity of the environment which helps to maintain the redox potential of the heme. A reaction is triggered when a substrate binds to the active site cavity within the heme protein in a region close to the Fe center resulting in excretion of water. Substrate binding lowers the redox potential by approximately 100mV resulting in a conformation that facilitates interaction of active P450 with its redox counterpart. This alteration allows Fe^{3+} to be reduced to Fe^{2+} by accepting an electron transferred from NAD(P)H via an electron transfer chain. An O_2 molecule binds rapidly to the Fe^{2+} ion forming ferrous dioxy complex Fe^{2+}-O_2 (Figure 11.3). This complex then slowly undergoes a transition to a more stable complex Fe^{3+}-O_2^- [37]. A second electron is then accepted by the complex whereby Fe^3+-O_2- is reduced to Fe^{3+}-O_2^2- [41]. This is the rate-determining step of the overall reaction process. O_2^{2-} then reacts with two protons from the surrounding solvent, cleaving the O-O bond, forming water and leaving the complex $(Fe\text{-}O)^{3+}$. The O atom ligated

Substrate binding site and mutations studied		R-->Q, S				Palmitic acid binding site					
Residue position	47	48	49	50	51	52	53	54	55	56	57
Bacillus megatarium	G	R	V	T	R	Y	L	S	S	Q	R
Bacillus aryabhattai	G	R	V	T	R	Y	L	S	S	Q	R
Bacillus cereus	S	D	T	I	I	V	V	S	G	H	E
Bacillus flexus	G	R	V	T	R	Y	L	S	S	Q	R
Bacillus mycoides	S	R	V	T	R	Y	V	S	S	Q	R
Rhodococcus jostii	G	N	R	F	V	F	A	S	G	A	D
Jeotgalicoccus sp.	N	R	L	N	T	S	L	G	G	K	P
Bacillus subtilis	G	K	N	F	I	C	M	T	G	A	E
Ehodococcus rhodochrous	-	-	-	-	-	-	-	-	-	-	-

Substrate binding site and mutations studied	Palmitic acid binding site	Palmitic acid binding site A-->G							A-->F, I, W
Residue position	74	75	76	77	78	79	80	81	82	83
Bacillus megatarium	Q	A	L	K	F	V	R	D	F	A
Bacillus aryabhattai	Q	A	L	K	F	V	R	D	F	A
Bacillus cereus	G	A	L	A	K	V	R	A	F	A
Bacillus flexus	Q	A	L	K	F	V	R	D	F	A
Bacillus mycoides	Q	A	L	K	N	V	R	S	F	T
Rhodococcus jostii	P	G	V	A	S	L	R	E	V	G
Jeotgalicoccus sp.	Q	P	K	R	I	V	N	T	L	F
Bacillus subtilis	L	P	K	R	V	Q	K	S	L	F
Ehodococcus rhodochrous	-	K	S	P	D	V	G	G	F	R

Substrate binding site and mutations studied			L-->E	F-->V								
Residue position	85	86	87	88	89	90	91	92	93	94	95	96
Bacillus megatarium	D	G	L	F	T	S	W	T	H	E	K	N
Bacillus aryabhattai	D	G	L	F	T	S	W	T	H	E	K	N
Bacillus cereus	D	G	L	F	T	S	E	T	H	E	P	N
Bacillus flexus	D	G	L	F	T	S	W	T	H	E	K	N
Bacillus mycoides	D	G	L	F	T	S	W	S	Y	E	K	K
Rhodococcus jostii	D	G	L	F	T	A	Y	N	H	E	P	N
Jeotgalicoccus sp.	K	G	A	F	M	S	L	M	T	E	G	N
Bacillus subtilis	V	N	A	I	Q	G	M	D	G	S	A	H
Ehodococcus rhodochrous	E	G	F	W	V	V	S	R	H	E	S	N

Figure 11.1 Sequence alignment between different cytochromes P450s from different bacteria; Effect of different mutational study in *Bacillus megaterium* CypP450 and residual diversity in other bacteria.

to Fe is subsequently transferred to the bound substrate converting it to its hydroxylated form. The product is then released from the active site of the enzyme while the Fe^{+3} atom returns to its initial state by co-ordinating back with a water molecule. The overall reaction is depicted as below.

$$RH + O_2 + 2H^+ + 2e^- \rightarrow ROH + H_2O$$

Substrate binding site and mutations studied		L-->Q										
Residue position	188	189	190	191	192	193	194	195	196	197	198	199
Bacillus megatarium	K	L	Q	R	A	N	P	D	D	P	A	Y
Bacillus aryabhattai	K	L	Q	R	A	N	P	D	D	P	A	Y
Bacillus cereus	Q	L	Q	R	L	D	I	E	D	K	L	M
Bacillus flexus	K	L	Q	R	A	N	P	D	D	P	A	Y
Bacillus mycoides	R	L	Q	R	A	N	P	N	D	S	L	Y
Rhodococcus jostii	T	T	F	V	K	S	L	G	R	L	L	M
Jeotgalicoccus sp.	-	-	-	-	-	-	-	-	-	-	G	Y
Bacillus subtilis	K	G	R	R	A	R	P	R	A	E	-	-
Ehodococcus rhodochrous	K	I	F	E	W	S	N	Q	M	T	G	Y

Substrate binding site and mutations studied		F-->E			A-->C, H, K, M, Q				T-->A, N		
Residue position	261	262	263	264	265	266	267	268	269	270	271
Bacillus megatarium	T	F	L	I	A	G	H	E	T	T	S
Bacillus aryabhattai	T	F	L	I	A	G	H	E	T	T	S
Bacillus cereus	T	F	L	I	A	G	H	E	T	T	S
Bacillus flexus	T	F	L	I	A	G	H	E	T	T	S
Bacillus mycoides	T	F	L	I	A	G	H	E	T	T	S
Rhodococcus jostii	T	F	L	V	A	G	H	E	T	T	S
Jeotgalicoccus sp.	N	T	F	R	-	P	L	I	A	I	N
Bacillus subtilis	N	V	L	R	-	-	-	-	-	-	-
Ehodococcus rhodochrous	V	L	A	V	A	G	N	E	T	T	R

Substrate binding site and mutations studied		A-->V		A-->P								
Residue position	**328**	**329**	**330**	**331**	**332**	**333**	**334**	**335**	**336**	**337**	**338**	**339**
Bacillus megatarium	T	A	P	A	F	S	L	Y	A	K	E	D
Bacillus aryabhattai	T	A	P	A	F	S	L	Y	A	K	E	D
Bacillus cereus	T	A	P	A	F	S	L	Y	A	K	E	D
Bacillus flexus	T	A	P	A	F	S	L	Y	A	K	E	D
Bacillus mycoides	T	A	P	A	F	S	L	Y	P	K	E	D
Rhodococcus jostii	T	A	P	A	Y	G	R	E	A	T	V	D
Jeotgalicoccus sp.	F	V	P	F	L	P	G	K	A	K	V	D
Bacillus subtilis	F	G	P	F	L	G	A	L	V	K	K	D
Ehodococcus rhodochrous	P	V	T	S	F	Q	R	T	A	L	E	D

Substrate binding site and mutations studied		F-->H, W							Fe binding site	I-->E, P	
Residue position	**393**	**394**	**395**	**396**	**397**	**398**	**399**	**400**	**401**	**402**	**403**
Bacillus megatarium	P	F	G	N	G	Q	R	A	C	I	G
Bacillus aryabhattai	P	F	G	N	G	Q	R	A	C	I	G
Bacillus cereus	P	F	G	N	G	Q	R	A	C	I	G
Bacillus flexus	P	F	G	N	G	Q	R	A	C	I	G
Bacillus mycoides	P	F	G	N	G	Q	R	A	C	I	G
Rhodococcus jostii	P	F	G	T	G	E	R	A	C	I	G
Jeotgalicoccus sp.	P	Q	G	G	G	D	H	R	C	A	G
Bacillus subtilis	P	Q	G	G	G	G	H	R	C	P	G
Ehodococcus rhodochrous	F	G	G	T	G	A	H	Y	C	L	G

Figure 11.1 (*Continued*).

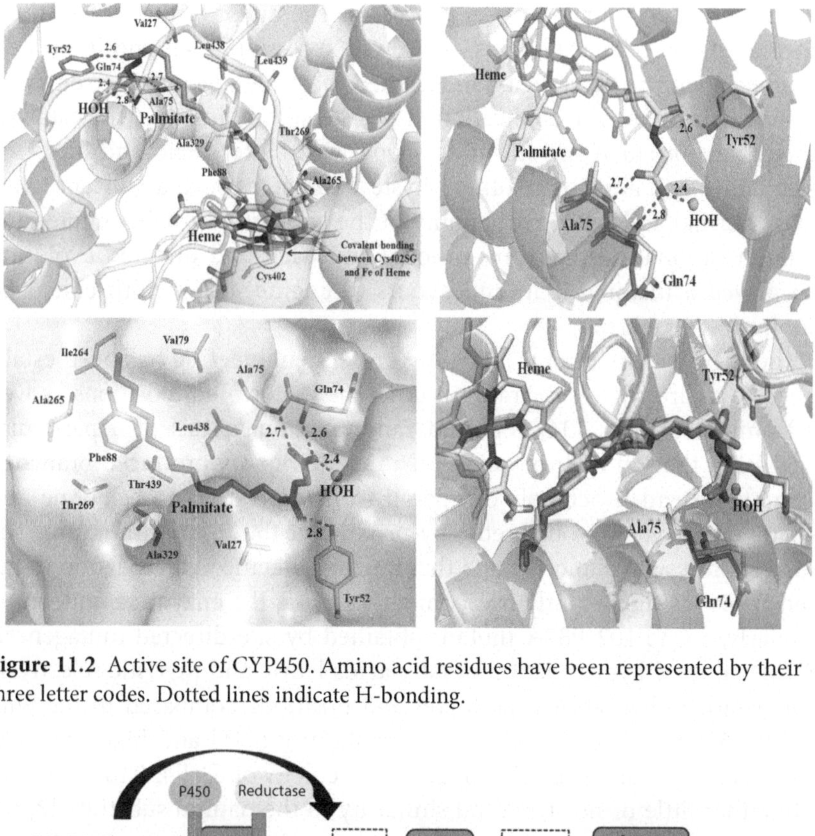

Figure 11.2 Active site of CYP450. Amino acid residues have been represented by their three letter codes. Dotted lines indicate H-bonding.

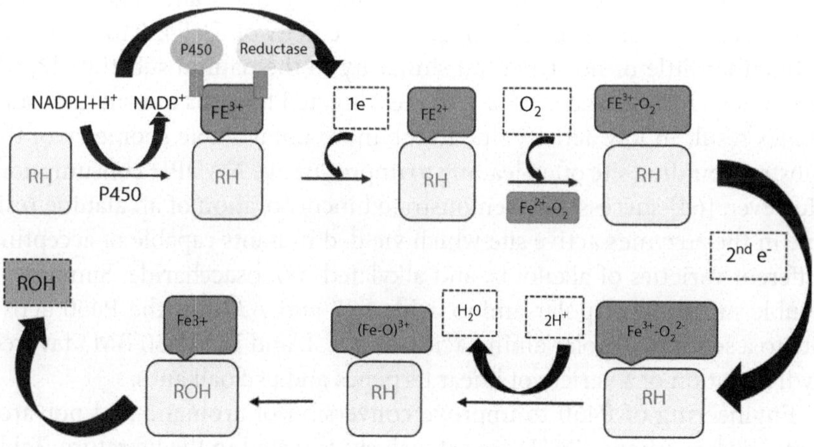

Figure 11.3 Catalytic cycle of CYP450.

11.5 Strain Engineering for Improved Activity

Apart from unique capabilities of CYP450 catalysing important chemical reactions, very few of these have been utilized in either preparative chemical reactions or industrial processes. The major limitations of these biocatalysts are primarily due to their low expression levels within the cell, insufficient enzymatic activity, and limited substrate specificity. To enhance their practical scope of application, it is necessary to improve the catalytic efficiencies, broaden the substrate scope with better regio and/or stereo selectivities and ameliorate other physical attributes such as solvent tolerance, thermal and oxidative stability of the enzymes. Protein engineering through rational design, semi-rational approach and directed evolution has played a major role in endeavours to engineer CYP with enhanced biophysical properties.

The regioselectivity of CYP450 by protein engineering has been established to hydroxylate substrates in different subterminal positions as well as terminal carbon. [34] reported random mutagenesis of a particular P450 BM3 that on hydroxylation of n-octane rapidly produced branched octanols instead n-octanols compared to native variety. 4-octanone and 3-octanone have also obtained a by-product of the reaction mixture. The conversion results demonstrated that rate of n-octane hydroxylation in the wild type was significantly overcome by altering the enzyme architecture. Similarly, a CYP102 F87A mutant obtained by site-directed mutagenesis (SDM) allowed hydroxylation of fatty acids lauric and myristic selectively at ω-4 rather than at ω-1, ω-2, and ω-3 positions compared to the wild variety [59, 61]. Alternatively, a combination of SDM and high throughput screening could be used to alter the specificity of CYP450 to substrates with either little or no structural similarity to the natural substrate [3, 63, 66]. In general, mutagenesis of enzymes directed towards non-natural substrates result in low activity due to the more unfavorable geometry of the substrate binding site often leading to unproductive NADPH consumption. However, [64] successfully demonstrated incorporation of an alanine residue in the enzyme's active site which yielded mutants capable of accepting different varieties of alkaloids, and alkylated monosaccharide. Similarly, a double mutation of polar amino acids F87 and A328 in the P450 active site to a set of non-polar amino acids A, V, F, L and I in P450 BM3 favored hydroxylation of a variety of linear terpenes and cycloalkanes.

Engineering of P450 to improve conversion of aromatic and polyaromatic hydrocarbons (PAH) are extensively reported in the literature. Table 11.1 and Figure 11.4 illustrate the different amino acid substitutions that have been selectively targeted to better the rate kinetics of the reaction above. [33] reported oxidation of naphthalene and pyrene by an engineered *P. putida* CYP450, which was initially involved in camphor oxidation. Replacement of a tyrosine residue at 96th position with either residue Ala or Phe improved the oxidation rate by ~10 to 20 fold. The results

Table 11.1 Effect of different mutational study in *Bacillus megaterium* CYP450.

Functional residue	Role in native strain	Altered residue	Strategy	Promoter	Host organism	Remarks	Ref
Ala_{75}	PAH hydroxylation	Gly	Site-directed mutagenesis	P_RP_L- of pCYT-EXP1	Catalase deficient *E. coli* UM2	higher activity towards PAH, increasing NADPH consumption rates	[66]
Ala83	Conversion of laurate as a substrate	Phe	Site-directed mutagenesis	pGLWBM3	*E. coli* JM109	800 fold increase in binding affinity	[52]
Phe87	PAH binding in the vicinity of active site	Val	Site-directed mutagenesis	P_RP_L- of pCYT-EXP1	Catalase deficient *E. coli* UM2	Increase in PAH-induced heme spin-state shift and coupling efficiencies of NADPH utilization by ~50 fold	[66]
Leu189	PAH hydroxylation	Gln	Site-directed mutagenesis	P_RP_L- of pCYT-EXP1	Catalase deficient *E. coli* UM2	High enzyme activities towards all three-ring PAHs by more than 30-fold	[66]

(*Continued*)

Table 11.1 Cont.

Functional residue	Role in native strain	Altered residue	Strategy	Promoter	Host organism	Remarks	Ref
Ala265	NADPH-dependent fatty acid hydroxylation	Cys/His/Lys/Met/Gln	Site-directed mutagenesis	NR	*E. coli* strain TG1	No substantial change in fatty acid oxidation or electron transportfrom NADPH to FMN	[38]
Thr268	Cyclopropanation of styrene.	Ala/Asn	Site-directed mutagenesis	pCWori plasmid under control of *Lac* promoter	E. coli DH5α	High substrate free turnover rate constant	[23]

Ala328	Palmitate conversion	Val	Site-directed mutagenesis	pT7BM-3 plasmid with T7 gene 10 promoter	E. coli DH5αFαIQ	5-10 fold increase in substrate binding affinity; 2-8 increase in turnover number for palmitate conversion	[44]
Ala330	PAH hydroxylation	Pro	Error-prone PCR	NR	NR	Enhanced activity with small non-natural substrates with altered profile	[112]
Phe393	Influence the electronic nature of the heme.	His/Trp	Site-directed mutagenesis	NR	*E. coli* strain TG1	Alter the heme reduction potential and increase in the rate of first-electron transfer	[22]

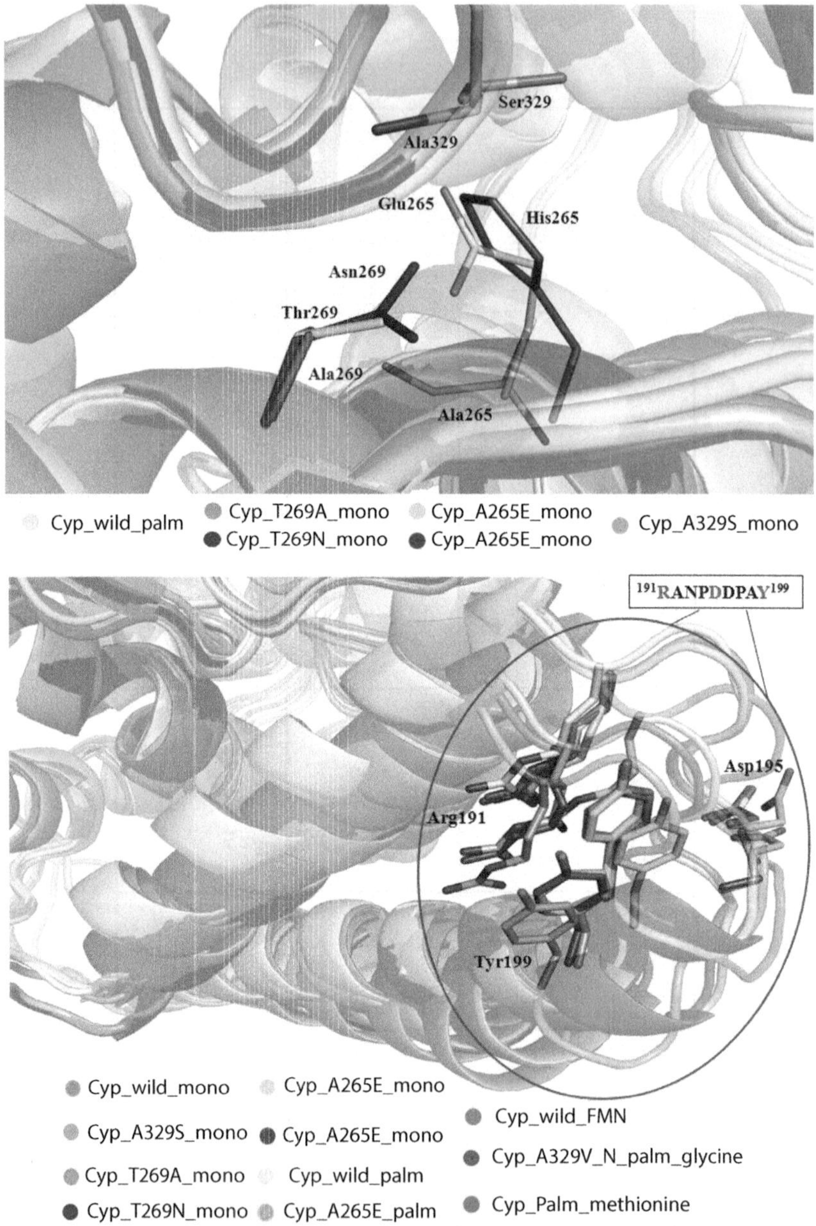

Figure 11.4 Active site analysis of CYP450 mutants with selective amino acid substitutions.

indicated that this particular tyrosine residue played a crucial role in controlling the substrate range of the P450 variant. In another report, a P450 option with F to G amino acid substitution at 87th position was reported to oxidise styrene to enantiomeric R-styrene oxide with 64% enantiomeric excess. Hydroxylation of PAHs pyrene and chrysene in different positions by actively modifying the active site thereby altering the substrate position within the catalytic domain have also been reported [52, 102]. A detailed analysis of the structures of mutant and wild-type enzyme illustrated that a single amino acid substitution at the 267th position from Glu to Arg was primarily responsible for the different catalytic properties shown by the modified enzyme towards chrysene and pyrene. The activities of another P450 BM3 towards PAH molecules naphthalene, fluorine, acenaphthene were improved by up to 4 orders of magnitude by a combination of three successive mutations A74G/F87V/L188Q. The authors concluded that the mutations largely enhanced PAH-induced heme spin-state shift and coupling efficiencies of NADPH utilization.

P450s require a cofactor to supply the single electron needed for the oxygenation of its substrates. The high cost of nicotinamide cofactors NAD(P)H, the most common source of reducing equivalents used by CYPs, has proven to be a significant barrier in technical implementation [21]. This constraint has triggered the search for both low-cost surrogate and convenient cofactor regeneration processes which have been contributory to overcome this limitation. Cofactor recycling systems have either relied on enzymatic, chemical or electrochemical means to achieve the *in-situ* recycling of $NAD(P)^+$ back to NAD(P)H. There has been some focus on the use of glucose-6-phosphate dehydrogenase or formate dehydrogenase for the regeneration of NADPH and NADH, respectively, while others have concentrated on engineering the enzyme so that it can utilise NADH rather than the more expensive NAD(P)H [108]. The idea is based on developing a cheap synthetic source with increased stability and at least similar activity to the natural cofactor. A rhodium complex has been successfully employed for the chemically mediated regeneration of both NADH and NADPH [51].

11.6 Producion Strategies of CYP450

11.6.1 Bioreactor Consideration

Being an intracellular protein, the yield and productivity of CYP450 depend on the amount of cell biomass produced during cultivation.

Hence, bioreactor setup shows several advantages over shake flask cultures regarding high-yield recombinant protein production. *E. coli* system is a well-studied host for any target protein production by recombinant DNA technology [6]. However, due to its low cell density, use of *E. coli* at industrial level is not encouraged [105]. Rather, *Bacillus* sp. are considered as better option for any industrial use due to its high cell density and permeability of membrane for the target protein expression. Besides this, the growth rate of *Bacillus* has been noticed to be much higher than that of *E. coli* on glucose. It had been found that expression of CYP450 under the influence of glycerol was better than the glucose as a carbon-source, as earlier reported in *E. coli* [67]. The specific growth rate of both wild-type and the mutant strain was more or less same having different potential for protein expression. There was hardly 5% difference in particular enzyme production in between shake flask and fermenter. However, CYPBM3 enzyme value was about 20% more higher in case of scale-up process under optimum fed-batch fermentation conditions, carried out by monitoring the impeller tip velocity, dissolved oxygen, mass-flow, and agitation speed [82]. About 15–20% increase in the production of CYPBM3 was achieved in a 50L fed-batch fermenter. This showed that the efficiency of expression of the target protein is better in the second phase of fermentation where glycerol was used as secondary carbon source [101].

11.6.2 Protein Recovery

A downstream protein recovery is designed based on its final application. Intracellular protein recovery is not cost effective and not favourable for large-scale application in bulk quantity. In such cases, use of such protein within *vivo* application as a whole cell biocatalyst may give the desired result. Crude lysate of the protein may also be tuned for desired applications. In case of complete purification, the quantitative yield of protein will be very less with the highest activity and they are used for high-value product formation through biocatalytic transformation. Intracellular protein purification constitutes up to 80% of overall production costs [95]. Various strategies and reports are available for downstream recovery of microbial CYP. However, their applications are limited to laboratory scale R&D. In most cases, understanding the protein structure and function relationship was targeted and thus CYP was purified from recombinant *E. coli.*

Depending on the expression plasmid, different purification strategies were reported (Table 11.2). Nitrilotriacetic acid (Ni-NTA) His-tag

Table 11.2 Different strategies for CYP450 purification.

Organism	Expression vector	Purification system	Molecular weight (kda)	Reference
E. coli	pET28a+	Nickel NTA	118	[78]
E. coli	pGLW11	DEAE fast flow Sepharose column	118	[16]
E. coli	pET22Hb+	DEAE 650M/ Source Q column	118	[15]
E. coli	pET20b+	DEAE 650M/ Source Q column	118	[15]

purification was largely adapted the technique for protein recovery through single-step purification. Another purification option may be of anion exchange chromatography using DEAE Hitrap columns. The crude extract was the starting material for this column, using a linear NaCl concentration gradient to elute CYP recombinant protein. By gradient elution, a step gradient was established. With the first step, *E. coli* proteins were eluted while the target protein remained on the column to avoid the interference of these two proteins during purification.

11.7 Applications

CYP450s find its limited application for its instability, low activity, narrow substrate specificity, high cofactor requirements, and electron transfer. As in case of many complex enzyme systems, which are not widely commercially available, the generation of active protein following cell transformation and gene expression remain major barrier. Close association with molecular biologists and multidisciplinary research groups play a vital role in the improved production and exploitation of such synthetically useful catalysts. In recent years, there has been an increasing recognition of CYPs for industrial synthesis of bulk chemicals, pharmaceuticals, agrochemicals, and food ingredients, due to its specific regio- and stereoselective hydroxylation characteristics [42, 43, 107]. Also, there is an increasing demand for CYP biocatalysts in the detoxification of environmental contaminants such as PAH and gene-directed enzyme prodrug therapy for cancer treatment [19].

11.7.1 Environmental Application

In this petroleum and petrochemical dominated world, our environment is facing a massive challenge regarding soil contamination by recalcitrant compounds. Many such compounds are potentially governing various sites like refineries, oil pits, storage sites. Such compounds namely PAHs, polychlorinated dibenzo-p-dioxin (PCDDs), PCDFs remain in soil due to their hydrophobic nature. To remediate such recalcitrant's, the introduction of oxygen through C-H activation to make hydrophilic is one of the primary options. Some of the major classes of industrial pollutants present in the environment are PAHs, PCDDs, polychlorinated biphenyls (PCBs), dibenzofurans, nitroaromatic compounds, pesticides, and heavy metals. Their presence creates problems because consuming products obtained from these sources could be deleterious to human health since several of these compounds (benzo(a)anthracene, benzo(a)pyrene, chrysene etc.) have been implicated in causing tumors in animals and cancer in humans. Relevant human exposure pathways include inhalation [79], skin contact [30, 103] or ingestion.

Biocatalytic hydroxylation is possible with specific enzymes like CYP450 and use of whole cell biocatalyst with engineered protein system has been conceptualized for bioremediation for a long time. However, most of the work has been performed with model compounds and projected towards bioremediation. CYP101, CYP102, CYP1A1, CYP1A2, and CYP1B1 are identified to metabolize PAHs. PCBs are metabolised by numerous CYP enzymes, and the significant factors that conclude the degree of metabolism are the extent of chlorination and position of chlorine atoms on the biphenyl nucleus. In contrast, CYP1A1 is also known to bioactivate these pollutants into genotoxic and carcinogens leading to high risk for lung cancer [10]. Hence, it is essential to have a thorough understanding of the mechanism of the various degradation processes. The present study thus notes that in-as-much-as each of these protocols in their merit are effective in depolluting PAHs from contaminated waste streams, the combined chemical and phytodegradation approach appears to be the most efficient and cost-effective environmental friendly method to decontaminate PAHs from substrates. Firstly, it is a green degradation method that is mindful of the environment, and secondly, no dead-end products are produced to initiate further contamination of the environment.

11.7.2 Medical Application

For any pharmaceutical industry, it is necessary to know how a putative drug will be processed in the human liver to prevent health risks during the lonesome and expensive trial phase or to avoid recourse expanses. Two critical steps are required during drug discovery: the first is the identification of all the metabolites produced for a given drug [7, 109], and the second is the synthesis of these compounds for toxicity tests.

CYP enzymes have a tremendous potential for the drug development and synthesis. From an industrial point of view, they are interesting targets for the production of peptide- antibiotics as well as fine chemicals. Several microorganisms have been employed historically for the synthesis of drugs using a hydroxylation reaction, the well-established commercial application of CYP in the biotransformation of steroids. Human CYP is anchored to the ER via N-terminal sequence [11, 96] and they require a redox partner, CPR to perform their function. CPR is also anchored to the ER membrane, and its part is to transfer the electrons from NADPH to the heme cofactor of the P450. In human liver, CYP plays a predominant role in phase-I drug metabolism and clearance since they turn over the vast majority of the known commercially available drugs into metabolites. The exploitation of bacterial enzyme as a biocatalyst to produce drug metabolites requires a multistep approach. This approach aimed at the optimisation of enzyme performance by protein engineering and development of platform allowing the avoidance of the NADPH cofactor and immobilization of biocatalyst to reduce costs. Thus the scientific literature on CYP has been focused on improved catalytic performance by protein engineering, substitution of the costly cofactor (NADPH), immobilization and scale-up of the process for industrial application.

CYP BM3 from *B. megaterium* can offer a reliable alternative for the synthesis of drug metabolites at industrial scale. In fact, the enzyme (CYP102A1) is a soluble and self-sufficient with diflavin-containing reductase fused to a heme-containing P450 domain in a single polypeptide chain [28, 75]. The availability of the crystal structure of P450 BM3 heme domain in the substrate-free form and complex with the palmitoleic acid offers the opportunity to identify critical residues for substrate binding and catalysis and therefore to carry out mutagenesis experiments [20, 62]. Molecular docking simulations are also a convenient tool to predict the fitness of a drug in the catalytic pocket of enzyme and to understand how the substrate should be oriented in the active site for hydroxylation [85, 97].

SDM was also used to create a panel of mutants of P450 BM3, able to metabolize probe substrates for human CYP such as 7-ethoxycoumarin and testosterone, that is broadly used to quantify the drug metabolizing activity in liver and probe the activity of P450 3A4. The P450 BM3 variants were shown to perform 3-hydroxylation and O-deethylation, producing the typical metabolites of human enzymes. The rates for the two reactions were enlarged by up to 61- and 129-fold concerning wild-type for the O-demethylation and 3-hydroxylation reactions, respectively [29].

11.8 Conclusion

The P450 enzymes are of significant interest in synthetic organic chemistry because of their impressive ability to catalyze the insertion of oxygen into non-activated C–H bonds. Very few chemical methods exist that directly hydroxylate aliphatic or aromatic C–H bonds, and most of them are not selective or of limited scope. Biocatalysts such as P450s represent a promising alternative. However, some limitations have restricted their use in synthesis and industrial applications. These include substrate specificity, the need for a complex system of cofactors, incompatibility with organic solvents, low activity and poor stability. From the very first studies on recombinant P450s, efforts were directed toward constructing fusions between P450s and redox partners in anticipation of generating more efficient enzymes. The final use of engineered P450s will require other aspects of their biology to be addressed, such as tolerance to heat, solvents, and the high substrate and product concentrations. To fully understand and exploit the power of P450s, it will be necessary to have a means by which to sample, visualize, and analyze the full range of structures accessed by these enzymes in solution, during the catalytic cycle, and in interactions with accessory proteins. Only with a truly dynamic picture of P450s, we will be able to appreciate the basis to their catalytic versatility fully.

References

1. Akyuz, M., Cabuk, H., Gas-particle partitioning and seasonal variation of polycyclic aromatic hydrocarbons in the atmosphere of Zonguldak, Turkey. *Sci. Total Environ.*, 408, 5550–5558, 2010.
2. Anzenbacher, P., Anzenbacherová, Cytochromes P450 and metabolism of xenobiotics. *E. Cell Mol. Life Sci.*, 58, 737–747, 2001.
3. Appel D., Lutz-Wahl S., *et al.*, A P450 BM-3 mutant hydroxylates alkanes, cycloalkanes, arenes and heteroarenes. *J. Biotechnol.*, 88, 167–171, 2001.
4. Ayala, M., Duhalt, R.V., Enzymatic catalysis on petroleum products. *Stud. Surf. Sci. Catal.*, 151, 67–111, 2004.
5. Bamforth, S.M., Singleton, I., Bioremediation of polycyclic aromatic hydrocarbons: current knowledge and future directions. *J. Chem. Technol. Biotechnol.*, 80, 723–736, 2005.

6. Baneyx, F., Recombinant protein expression in *Escherichia coli*. *Curr. Opin. Biotechnol.*, 10, 411–421, 1999.
7. Baranczewski, P., Stanczak, A., *et al.* Introduction to early *in vitro* identification of metabolites of new chemical entities in drug discovery and development. *Pharmacol. Rep.*, 58, 341–352, 2006.
8. Bayer, E., Hill, H.A.O., *et al.* The interaction between haem-iron and thiols. *Chem. Commun.*, 109, 1969.
9. Bayer, E., Hill, H.A.O., *et al.* Cytochrome P-450: suggestions as to the structure and mechanism of action. *Naturwissenschaften.*, 57, 69–72, 1970.
10. Billet, S., Abbas, I., *et al.* Genotoxic potential of polycyclic aromatic hydrocarbons-coated onto airborne particulate matter (PM 2.5) in human lung epithelial A549 cells. *Cancer Lett.* 270, 144–155, 2008.
11. Black, S.D., Membrane topology of the mammalian P450-cytochromes. *FASEB J.*, 6, 680–685, 1992.
12. Black, S.D., French, J.S., *et al.* Role of a hydrophobic polypeptide in the N-terminal region of NADPH-cytochrome P-450 reductase in complex formation with P-450LM. *Biochem. Biophys. Res. Commun.*, 91, 1528–1535, 1979.
13. Blattner, F.R., Plunkett, G., *et al.* The complete genome sequence of Escherichia coli K-12. *Science.*, 277, 1453, 1997.
14. Brodie, B., Axelrod, J., *et al.* Detoxication of drugs and other foreign compounds by liver microsomes. *Science.*, 121, 603–604, 1955.
15. Budde, M., Maurer, S.C., *et al.* Cloning, expression and characterization of CYP102A2, a self-sufficient P450 monooxygenase from *Bacillus subtilis*. *Appl. Microbiol. Biotechnol.*, 66(2), 180–186, 2004.
16. Carmichael, A.B., Wong, L., Protein engineering of *Bacillus megaterium* CYP102.The oxidation of polycyclic aromatic hydrocarbons. *Eur. J. Biochem.*, 268(10), 3117–3125. 2001.
17. Cerniglia, C.E., Biodegradation of polycyclic aromatic hydrocarbons. *Biodegradation.*, 3, 351–368, 1992.
18. Champion, P.M., Stallard, B.R., *et al.*, Resonance Raman detection of a Fe-S bond in cytochrome P450cam. *J. Am. Chem. Soc.*, 104, 5469–5473, 1982.
19. Chen, L., Waxman, D.J., Cytochrome P450 gene-directed enzyme prodrug therapy (GDEPT) for cancer. *Curr. Pharm. Des.*, 8(15), 1405–1416, 2002.
20. Chouhan, O.P., Bandekar, D., *et al.*, Effect of site-directed mutagenesis at the GGEEF domain of the biofilm forming GGEEF protein from *Vibrio cholerae*. *AMB Express.*, 6, 2, 2016.
21. Cirino, P.C., Arnold, F.H., A self-sufficient peroxide-driven hydroxylation biocatalyst. *Angew. Chem. Int. Ed. Engl.*, 42, 3299–3301, 2003.
22. Clark, J.P., Miles, C.S., *et al.*, The role of Thr268 and Phe393 in cytochrome P450 BM3. *J. Inor. Biochem.*, 100, 1075–1090, 2006.
23. Coelho, P.S., Brustad, E.M., *et al.*, Olefin Cyclopropanation via Carbene Transfer Catalyzed by Engineered Cytochrome P450 Enzymes. *Sci.*, 339, 307–310, 2013.

24. Cojocaru, Balali-Mood, K., *et al.*, Structure and dynamics of the membrane-bound cytochrome p450 2c9. *PLoS Comp. Boil.*, 7, e1002152, 2011.
25. Coon, M.J., Cytochrome P450: nature's most versatile biological catalyst. Ann. Rev. Pharmacol. Toxicol., 45, 1–25, 2005.
26. Cupp-Vickery, Jr., Poulos, T.L., Structure of cytochrome P450eryF involved in erythromycin biosynthesis. *Nat. Struct. Biol.*, 2, 144–153, 1995.
27. Denisov, I.G., Makris, T.M., *et al.*, Structure and chemistry of cytochromes, 450. *Chem. Rev.*, 105, 2253- 2277, 2005.
28. Di Nardo, G., Fantuzzi, A., *et al.*, Wild-type CYP102A1 as a biocatalyst: Turnover of drugs usually metabolized by human liver enzymes. *J. Biol. Inorg. Chem.*, 12, 313–323, 2007.
29. Di Nardo, G., Gilardi, G., Optimization of the bacterial Cytochrome P450 BM3 system for the production of human drug metabolites. *Int. J. Mol. Sci.*, 13, 15901–15924, 2012.
30. Dor, F., Jongeneelen, F., *et al.*, Feasibility of assessing dermal exposure to PAHs of workers on gaswork sites—the SOLEX study. *Sci. Total Environ.*, 263, 47–55, 2000.
31. Doull, J.L., Vining, Nutritional control of actinorhodin production by *Streptomyces coelicolor* A3(2): suppressive effects of nitrogen and phosphate. *Appl. Microbiol. Biotechnol.*, 32(4), 449–454, 1990.
32. Eiben, S., Kaysser, L., *et al.*, Preparative use of isolated CYP102 monooxygenases - a critical appraisal. *J, Biotechnol.*, 124, 662–669, 2006.
33. England, P.A., Harford-Cross, C.F., *et al.* The oxidation of naphthalene and pyrene by cytochrome P450cam. *FEBS Lett.*, 3, 271–274, 1998.
34. Farinas, E.T., Schwaneberg, U., *et al.*, Directed evolution of a cytochrome P450 monooxygenase for alkane oxidation. *Adv. Synth. Catal.*, 343, 601–606, 2001.
35. Garfinkel, D., Studies on pig liver microsomes. I. Enzymic and pigment composition of different microsomal fractions. *Arch. Biochem. Biophys.*, 77, 493–509, 1958.
36. Geronimo, I., Denning, C.A., *et al.*, Effect of mutation and substrate binding on the stability of cytochrome P450BM3 variants. *Biochem.*, 55, 3594–3606, 2016.
37. Girvan, H.M., Munro, H.W., Applications of microbial cytochrome p450 enzymes in biotechnology and synthetic biology. *Curr. Opin. Chem. Biol.*, 31, 136–145, 2016.
38. Girvan, H.M., Seward, H.E., *et al.*, Structural and spectroscopic characterization of P450 BM3 mutants with unprecedented P450 heme iron ligand sets. New heme ligation states influence conformational equilibria in P450 BM3. *J. Biol. Chem.*, 282, 564–572, 2007.
39. Glieder, A., Farinas, E.T., *et al.*, Laboratory evolution of a soluble, self-sufficient, highly active alkane hydroxylase. *Nat. Biotechnol.*, 20, 1135–1139, 2002.
40. Guengerich, F.P., Munro, A.W., Unusual cytochrome P450 enzymes and reactions. *J. Biol. Chem.*, 288, 17065–17073. 2013.

41. Guengerich, F.P., Mechanisms of cytochrome P450 substrate oxidation: minireview. *J. Biochem. Mol. Toxicol.*, 21, 163–168, 2007.
42. Guengerich, F.P., Cytochrome P450 enzymes in the generation of commercial products. *Nat. Rev. Drug. Discov.*, 1, 359–366, 2002.
43. Guengerich, F.P., Cytochrome P450 oxidations in the generation of reactive electrophiles: epoxidation and related reactions. *Arch. Biochem. Biophys.*, 409, 59–71, 2003.
44. Haines, D.C., Hegde, A., *et al.*, A single active-site mutation of P450BM-3 dramatically enhances substrate binding and rate of product formation. *Biochem.*, 8333–8341, 2011.
45. Hasemann, C.A., Kurumbai, R.G., *et al.*, Structure and function of cytochromes P450: a comparative analysis of three crystal structures. *Struc.*, 3, 41–62, 1995.
46. Hashimoto, Y., Yamano, T., *et al.*, An electron spin resonance study of microsomal electron transport. *J. Biol. Chem.*, 237, 3843–3844, 1962.
47. Hazra, S., Konrad, M., *et al.*, The sugar ring of the nucleoside is required for productive substrate positioning in the active site of human deoxycytidine kinase (dCK): implications for the development of dCK-activated acyclic guanine analogues. *JMC*, 53, 5792–5800, 2010.
48. Hazra, S., Szewczak, A., *et al.*, Post-translational phosphorylation of serine 74 of human deoxycytidine kinase favors the enzyme adopting the open conformation making it competent for nucleoside binding and release. *Biochem.*, 50, 2870–2880, 2011.
49. Heidari, R., Rabiee, F., *et al.*, Expression and purification of the recombinant cytochrome P450 cyp141 protein of *Mycobacterium tuberculosis* as a diagnostic tool and vaccine production. *Iran Red. Crescent Med. J.*, 17, E23191, 2015.
50. Hill, H.A.O., Röder, A., *et al.*, The chemical nature and reactivity of cytochrome P-450. *Structure and Bonding.*, 8, 123–151, 1970.
51. Hollmann, F., Witholt, B., *et al.*, $[Cp^{*}Rh(bpy)(H_2O)]^{2+}$: a versatile tool for efficient and non-enzymatic regeneration of nicotinamide and flavin coenzymes. *J. Mol. Catal. B: Enzym.*, 19–20, 167–176, 2002.
52. Huang, W.C., Cullis, P.M., *et al.*, Control of the stereo-selectivity of styrene epoxidation by cytochrome P450 BM3 using structure-based mutagenesis. *Metallomics.*, 3, 410–416, 2011.
53. Huang, W.C., Joyce, M.G., *et al.*, Cytochrome P450 BM3 mutant in complex with palmitic acid. *J. Mol. Biol.*, 373, 633–651, 2007.
54. Isin, E.M., Guengerich, F.P., Complex reactions catalyzed by cytochrome P450 enzymes. *Biochim. Biophys. Acta.*, 1770(3), 314–329, 2006.
55. Johnson, E.F., Stout, C.D., Structural diversity of human xenobiotic-metabolizing cytochrome P450 monooxygenases. Biochem. Biophys. Res. Commun., 338(1), 331–336, 2005.
56. Jung, S.T., Lauchli, R., *et al.*, Cytochrome P450: taming a wild-type enzyme. *Curr. Opin. Biotechnol.*, 22, 1–9, 2011.

57. Kim, D.H., Kim, K.H., *et al.*, Generation of human metabolites of 7-ethoxycoumarin by bacterial Cytochrome P450BM3. *Drug Metab. Dispos.*, 36, 2166–2170, 2008.
58. Klingenberg, M., Pigments of rat liver microsomes. *Arch. Biochem. Biophys.*, 75, 376–386, 1958
59. Kumar, S., Engineering Cytochrome P450 Biocatalysts for Biotechnology, Medicine, and Bioremediation. *Expert Opin. Drug Metab. Toxicol.*, 6, 115–131, 2011.
60. Lechat, P., Hummel, L., *et al.*, Genolist: an integrated environment for comparative analysis of microbial. Nucleic Acids Res., 36, 469–474, 2008.
61. Lee, D.S., Yamada, A., *et al.*, Substrate recognition and molecular mechanism of fatty acid hydroxylation by cytochrome P450 from *Bacillus subtilis* crystallographic, spectroscopic, and mutational studies. *J. Biol. Chem.*, 278, 9761–967, 2003.
62. Lenong, Li, Zhenzhan, Chang, *et al.*, Modes of heme binding and substrate access for cytochrome P450 CYP74A revealed by crystal structures of allene oxide synthase. *Proc. Natl. Acad. Sci.* U S A., 105, 13883–13888, 2008.
63. Lentz, O., Feenstra, A., *et al.*, Altering the regioselectivity of cytochrome P450 CYP102A3 of *Bacillus subtilis* by using a new versatile assay system. *Chembiochem.*, 7, 345–350, 2006.
64. Lewis, J.C., Mantovani, S.M., *et al.*, Combinatorial alanine substitution enables rapid optimization of cytochrome P450BM3 for selective hydroxylation of large substrates. *Chem. Bio. Chem.*, 11, 2502–2505, 2010.
65. Li, H.Y., Poulos, T.L., The structure of the cytochrome P450BM-3 haem domain complexed with the fatty acid substrate, palmitoleic acid. *Nat. Struct. Biol.*, 4, 140–146, 1997.
66. Li, Q.S., Ogawa, J., Schmid, R.D., *et al.*, Engineering cytochrome P450 BM-3 for oxidation of polycyclic aromatic hydrocarbons. *Appl. Env. Microbiol.*, 67, 5735–5739, 2001.
67. Liu, M.Z., Durfee, T., *et al.*, Global transcriptional programs reveal a carbon source foraging strategy by *Escherichia coli*. *J. Biol. Chem.*, 280, 15921–15927, 2005.
68. Makris, T.M., Denisov, I., *et al.*, *Activation of molecular oxygen by cytochrome P450.Cytochrome P450: structure, mechanism, and biochemistry. Ortiz de Montellano PR. 3rd ed.* New York, Kluwer Academic/Plenum.Ch. 5: 149–182, 2015.
69. Martin, J.F., Demain, A.L., Control of antibiotic synthesis. *Microbiol. Rev.*, 44, 230–231, 1980.
70. McLean, K.J., Sabri, M., *et al.*, Biodiversity of cytochrome P450 redox systems. *Biochem. Soc. Trans.*, 33, 796–801, 2005.
71. McSorley, T., Ort, S., *et al.*, Mimicking phosphorylation of Ser-74 on human deoxycytidine kinase selectively increases catalytic activity for dC and dC analogues. *FEBS Lett.*, 582, 720–724, 2008.

72. Meinhold, P., Peters, M.W., *et al.*, Direct conversion of ethane to ethanol by engineered cytochrome P450BM3. *Chem Biochem.*, 6, 1765–1768, 2005.
73. Meunier, B., De Visser, S.P., *et al.*, Mechanism of oxidation reactions catalyzed by cytochrome p450 enzymes. *Chem. Rev.*, 104, 3947–3950, 2004.
74. Mounier, N., Katlama, C., *et al.*, Drug interactions between antineoplastic and antiretroviral therapies: Implications and management for clinical practice. *Crit. Rev. Oncol. Hematol.*, 72, 10–20, 2009.
75. Munro, A.W., Daff, S., Coggins, J.R., *et al.*, Probing electron transfer in flavocytochrome P-450 BM3 and its component domains. *Eur. J. Biochem.*, 239, 403–409, 1996.
76. Munro, A.W., Girvan, H.M., *et al.*, What makes a P450 tick? *Trends Biochem. Sci.*, 38, 140–150, 2013.
77. Murataliev, M.B., Klein, M., *et al.*, Functional interactions in cytochrome P450BM3: flavin semiquinone intermediates, role of NADP(H), and mechanism of electron transfer by the flavoprotein domain. *Biochem.*, 27, 8401–8412, 1997.
78. Nebel, B.A., Scheps, D., *et al.*, Biooxidation of n-butane to 1-butanol by engineered P450 monooxygenase under increased pressure. *J. Biotechnol.*, 191, 86–92, 2014.
79. Nessel, C.S., Amoruso, M.A., *et al.*, Pulmonary bioavailability and fine particle enrichment of 2,3,7 8-tetrachlorodibenzo-p-dioxin in respirable soil particles. *Fundam. Appl. Toxicol.*, 19, 279–285, 1992.
80. Ogu, C.C., Maxa, J.L., Drug interactions due to cytochrome P450. *Proc. Bayl. Univ. Med. Cent.*, 13, 421–423, 2000.
81. Omura, T., Sato, R., The carbon monoxide-binding pigment of liver microsomes. solubilization, purification and properties. *J. Biol. Chem.*, 239, 2378–2385, 1964.
82. Oosterhuis, N.M.G., Kossen, N.W.F., Dissolved oxygen concentration profiles in a production-scale bioreactor. *Biotechnol. Bioeng.*, 26, 546–550, 1983.
83. Palrusu, M., Nagini, S., Cytochrome P450 structure, function and clinical significance: A review. *Curr. Drug Targets*, 19, 38–54, 2018.
84. Park, S.Y., Shimizu, H., *et al.*, Crystallization, preliminary diffraction and electron paramagnetic resonance studies of a single crystal of cytochrome P450nor. *FEBS Lett.*, 412, 346–350. 1996.
85. Paul, M., Hazra, M., *et al.*, Comparative molecular dynamics simulation studies for determining factors contributing to the thermostability of chemotaxis protein CheY. *JBSD.*, 32, 928–949, 2013.
86. Peters, M.W., Meinhold, P., *et al.*, Regio- and enantioselective alkane hydroxylation with engineered cytochromes P450 BM-3. *J. Am. Chem. Soc.*, 125, 13442–13450, 2003.
87. Pochapsky, T., Kazanis, S., *et al.*, Conformational plasticity and structure / function relationships in Cytochrome P450. *Antioxid. Redox. Signal.*, 13(8), 1273–1296, 2010.

88. Poulos, T.L., Johnson, E.F., Structures of cytochrome P450 enzymes, in. *Cytochrome P450: structure, mechanism, and biochemistry*, R.O. de Montellano, (Ed.), 87–114, Springer, New Jersey, US, 2015.
89. Poulos, T.L., Finzel, B.C., *et al.*, The 2.6-A crystal structure of Pseudomonas putida cytochrome P-450. *J. Biol. Chem.*, 260, 16122–16130, 1985.
90. Prasad, S., Mitra, S., Structure and mechanism of function of cytochrome P450. *Prot. Indian Natn. Sci. Acad.*, 70, 283–291, 2004.
91. Quartararo, C.E., Hazra, S., *et al.*, Structural, kinetic and chemical mechanism of isocitrate dehydrogenase-1 from *Mycobacterium tuberculosis*. *Biochem.*, 52, 1765–1775, 2013.
92. Raner, G.M., Thompson, J.I., *et al.*, Spectroscopic investigations of intermediates in the reaction of cytochrome $P450_{BM3}$–F87G with surrogate oxygen atom donors. *J. Inorg. Biochem.*, 100, 2045–2053, 2006.
93. Ravichandran, K.G., Boddupalli, S.S., *et al.*, Crystal-structure of hemoprotein domain of p450bm-3, a prototype for microsomal P450's. *Science.*, 261, 731–736, 1993.
94. Sabini, E., Hazra, S., *et al.*, Structural basis for activation of the therapeutic L-nucleoside analogs 3TC and troxacitabine by human deoxycytidine kinase. *Nucleic Acid Res.*, 35, 186–192, 2007.
95. Sadana, A., Beelaram, A.M., Efficiency and economics of bioseparation: some case studies. *Bioseparation.*, 4, 221–235, 2014.
96. Sakaguchi, M., Mihara, K., *et al.*, A short amino-terminal segment of microsomal cytochrome-P-450 functions both as an insertion signal and as a stop transfer sequence. *EMBO J.*, 6, 2425–2431, 1987.
97. Santos Rita, Hritz Jozef, *et al.*, Role of water in molecular docking simulations of Cytochrome P450 2D6. *J. Chem. Inf. Model.*, 50, 146–154, 2010.
98. Scott, E.E., Halpert, J.R., Structures of cytochrome P450 3A4. *Trends Biochem. Sci.*, 5–7, 2005.
99. Seifert, A., Vomund, S., *et al.*, Rational design of a minimal and highly enriched CYP102A1 mutant library with improved regio-, stereo- and chemoselectivity. *Chem. Bio. Chem.*, 10, 853–861, 2009.
100. Shaik, S., Kumar, D., *et al.*, Theoretical perspective on the structure and mechanism of cytochrome P450 enzymes. *Chem. Rev.*, 105, 2279–2228, 2005
101. Shukla, V.B., Parasu, V., *et al.*, Scale-up of biotransformation process in stirred tank reactor using dual impeller bioreactor. *J. Biochem. Eng.*, 8, 19–29, 2001.
102. Sideri, A., Goyal, A., *et al.*, Hydroxylation of non-substituted polycyclic aromatic hydrocarbons by cytochrome P450 BM3 engineered by directed evolution. *J. Inor. Biochem.*, 120, 1–7, 2013.
103. Skowronski, G.A., Turkall, R.M., *et al.*, Effects of soil on the dermal bioavailability of mxylene in male rats. *Environ. Res.*, 51, 182–193, 1990.
104. Spizek, J., Tichy, P., Some aspects of overproduction of secondary metabolites. *Folia Microbiol.*, 40, 43–50, 1995.
105. Stulke, J., Hillen, W., Regulation of carbon catabolism in Bacillus species. *Annu. Rev. Microbiol.*, 54, 849–880, 2000.

106. Tomb, J.F., White, O., *et al.*, The complete genome sequence of the gastric pathogen *Helicobacter pylori. Nature.*, 388, 539–547, 1997.
107. Urlacher, V.B., Lutz-Wahl, S., *et al.*, Microbial P450 enzymes in biotechnology. *Appl. Microbiol. Biotechnol.*, 64, 317–325, 2004.
108. van der Donk, W.A., Zhao, H., *Curr. Opin. Biotechnol.*, 14, 421–426, 2003.
109. Wang, W., McKinnie, S.M.K., *et al.*, Angiotensin Converting Enzyme 2 Metabolizes and Partially Inactivates Pyrapelin-13 and Apelin-17: Physiological Effects in the Cardiovascular System. *Hypertension.*, 68, 365–377, 2016.
110. Watanabe, I., Nara, F., *et al.*, Cloning, characterization and expression of the gene encoding cytochrome P-450sca_2 from Streptomyces carbophilus involved in the production of pravastatin, a specific HMG-CoA reductase inhibitor. *Gene.*, 163, 81–85, 1995.
111. Weber, E., Seifert, A., *et al.*, Screening of a minimal enriched P450 BM3 mutant library for hydroxylation of cyclic and acyclic alkanes. *Chem Commun.*, 47, 944–946, 2011.
112. Whitehouse, C.J.C., Bell, S.G., *et al.*, Evolved CYP102A1 (P450BM3) variants oxidise a range of non-natural substrates and offer new selectivity options. *Chem. Commun.*, 966–968, 2007.
113. Zanger, U.M., Schwab, M., Cytochrome P450 enzymes in drug metabolism: Regulation of gene expression, enzyme activities, and impact of genetic variation. *Pharmacol. Ther.*, 138, 103–141, 2013.
114. Zhang, L., Lu, Y., *et al.*, The retardant effect of 2-Tridecanone, mediated by Cytochrome P450, on the development of cotton bollworm, *Helicoverpa armigera. BMC Genomics.*, 17, 954–968, 2016.
115. Zhu, B.T., On the general mechanism of selective induction of cytochrome P450 enzymes by chemicals: some theoretical considerations. *Exp. Opin. Drug. Metab. Toxicol.*, 6, 483–494, 2011.
116. Zhu, G.D., Okamure, W.H., Synthesis of vitamin D (Calciferol). *Chem. Rev.*, 95, 1877–1852, 1995.

12

Production of Polyunsaturated Fatty Acids by Solid State Fermentation

Bruno Carlesso Aita, Stéfani Segato Spannemberg, Raquel Cristine Kuhn and Marcio Antonio Mazutti*

Department of Chemical Engineering, Center of Technology, Federal University of Santa Maria, Santa Maria, Brazil,

Abstract

Polyunsaturated fatty acids (PUFAs) are organic compounds with functional characteristics that regulate biological functions in humans. Currently, microorganisms producing microbial oils or single cell oils are the main alternative sources for the production of PUFAs. Submerged fermentation and solid state fermentation (SSF) are the processes used in the production of PUFAs by oleaginous fungi. Although lipid production by oleaginous microorganisms through submerged fermentation is well defined in literature and has industrial application for approximately two decades, however, it presents several disadvantages in relation to SSF. Production of PUFAs by SSF is still a relatively recent industrial process and needs to be optimized, presenting challenges such as scale up, extraction and purification of final product and monitoring of process parameters. Thus, this chapter presents information about PUFAs production by SSF, focusing on microorganisms used, substrates, bioreactors, and main parameters that influence the process.

Keywords: Polyunsaturated fatty acids, single cell oils, oleaginous microorganisms, filamentous fungi, solid state fermentation

12.1 Introduction

Polyunsaturated fatty acids (PUFAs) are organic compounds with unique structural and functional characteristics, acting as precursors of a wide

**Corresponding author*: mazutti@ufsm.br

Arindam Kuila and Vinay Sharma (eds.) Principles and Applications of Fermentation Technology, (217–238) © 2018 Scrivener Publishing LLC

variety of metabolites, which regulate critical biological functions in mammals [1]. Its deficiency can have negative consequences on the skin and kidneys, as well as affecting the nervous, immune, cardiovascular, endocrine, respiratory, and reproductive systems [2, 3].

The omega-3 (ω-3) and omega-6 (ω-6) families are fatty acids of major importance in terms of occurrence and for the biological and nutritional functions related to human health [4]. α-linolenic acid (ALA), eicosapentaenoic acid (EPA), docosahexaenoic acid (DHA), linoleic acid (LA), γ-linolenic acid (GLA), and arachidonic acid (ARA) are considered essential fatty acids [5]. For the production of essential PUFAs, several enzymes are required whose mammals are unable to synthesize, which implies the complementation of PUFAs to humans through specific diets or medications [2, 3].

PUFAs could be obtained in the animal and plant kingdom, as marine fish oils, pig liver, microorganisms and seeds of oleaginous plants as main sources [6, 7]. Currently, most of the industrial PUFAs production are obtained from fish by-products and oils and marketed in the form of capsules [8, 9]. However, the production of PUFAs from marine sources presents several problems, especially the limitation imposed by fishing quotas, the possibility of several contaminants in fish oils [7, 10], variation in the quality and quantity of the PUFAs [9], as well as the high cost of the processes used in the extraction of PUFAs from fish oil [10, 11]. Although the seeds of some oleaginous plants are good sources of PUFAs, the lack of enzymes precludes the production of long-chain PUFAs with more than 18 carbon atoms whose beneficial effects on human health have been evidenced [10]. Another problem in the use of plants to obtain PUFAs is the need of suitable land for their cultivation [8]. Therefore, the lack of adequate, safe and economically viable sources to obtain PUFAs from conventional sources (animals and plants), coupled with their current high demand, evidences the need to seek low cost and high efficiency alternative sources of PUFAs.

Currently, oil-producing microorganisms, including fungi, bacteria and algae, are the main alternative sources for the production of PUFAs [9, 12–14], although this route of production has been commercially exploited only in the last two decades [15]. It is known that the accumulation of microbial lipids occurs during the periods of metabolic stress, being able to reach 60% of their dry mass [16] Although approximately 60 microbial genera are known for efficiency in oil production, their commercial production is currently restricted to fungi (including yeasts and filamentous fungi) and algae grown heterotrophically rather than photosynthetically [9, 13, 15]. Short growing periods, continuous production under controlled conditions and greater microbial oil stability in relation to PUFAs of animal and vegetable origin, are the main advantages attributed to PUFAs production by oleaginous microorganisms [7].

Fungi have gained attention as a potential source of PUFAs in recent years, because they have high concentrations of essential fatty acids and also enable increased production of desired products through manipulation of growth conditions [8]. Furthermore, oleaginous fungi could produce PUFAs by fermentation in a low-cost medium from agricultural residues and industrial by-products [17]. Therefore, one of the current challenges regarding the use of oleaginous microorganisms is to seek biotechnological strategies that guarantee high yields and low cost in the production of the desired PUFAs.

Fermentation through filamentous and unicellular (yeast) fungi, for the production of PUFA-rich microbial biomass, is a relatively recent industrial process [18]. It is known that the oil accumulated by heterotrophic filamentous fungi is more unsaturated than that produced by yeasts, which makes them more suitable for the production of lipids rich in PUFAs of medical and dietary interest [19], such as GLA, EPA, DHA and ARA [18]. Therefore, it is important for optimizing the production of PUFAs from oleaginous filamentous fungi, through the adaptation of environmental conditions during fermentation.

In this context, Solid state fermentation (SSF) process is presented as a viable alternative for the production of PUFAs [6]. The main challenges to the development of SSF are scaling up, purification of the final product and the monitoring of environmental parameters, mainly the temperature and the humidity of the biomass, since the heat generation by the metabolic activity can lead to the degradation of the final product, whereas the excess or lack of moisture can reduce the oxygenation level of the system and the availability of nutrients to the microorganisms, respectively [20–22]. Thus, the search for low cost substrates for use in SSF and the establishment of environmental conditions necessary to optimize the production of lipids by oleaginous fungi through fermentation process are priority aspects to scientific research. This chapter discusses the studies related to the main parameters that influence the PUFAs production process by SSF.

12.2 PUFAs Production by SSF

All microorganism needs to synthesize lipids to make up its cell membrane. However, microorganisms defined as oleaginous are those that can accumulate more than 20% of their dry biomass as lipids (mainly in the form of triglycerides) [16]. Lipids produced by these microorganisms are known as single cell oils (SCO) or microbial oils, considered important sources of PUFAs and other fatty acids [13]. The term SCO was originally created to easily identify the edible fraction of lipids produced by microorganisms. However, the term is now used to denote any type of lipid containing fatty acids in a microbial cell [15].

The production and accumulation of oil in microbial cells is already well established. Although some bacteria, such as *Rhodococus opacus*, can produce triglyceride-rich oils, the commercial production of SCO is currently restricted to fungi (including yeasts and filamentous fungi) and algae grown heterotrophically rather than photo-synthetically [9, 13, 15]. Regardless of the type of microorganism, the production of SCO occurs in a similar way. The chosen microorganism is grown in a medium where the carbon source, usually glucose, is in excess, but with nitrogen (usually as NH_4^+ salts or urea) as a limiting nutrient. After an initial phase in which the growth is balanced and all the nutrients are available, the cells become depleted of nitrogen and are no longer able to multiply, since the nitrogen supply is essential for the formation of new proteins and for the biosynthesis of nucleic acids. Oleaginous microorganisms continue to assimilate glucose or other available carbon source, which is then preferably directed toward lipid biosynthesis, even stopping the growth. The accumulation of lipids after nitrogen depletion in the culture medium depends on the continuous synthesis of Acetyl-Coenzyme A (Acetyl-CoA) [15]. After their synthesis, the lipids are stored in the form of oil droplets in the cell cytosol, in order to conserve energy reserves to ensure vital functions in nutrient deficiency or stress.

Although plants are one of the main sources of essential fatty acids for human consumption, fungi have been attention as potential sources. This is because the fungi in the industrial production of essential fatty acids rather than plants reduces competition for land needed for cultivation (this only occurs with the use of a fungal growth medium that require little cultivable land), and it also isn't affected by the climatic conditions. Another advantage is that the fatty acid profiles produced by the fungi naturally have high concentrations of essential fatty acids or can be induced to produce larger amounts of the desired product by manipulating the growth conditions. This procedure is less invasive and also less time-consuming than genetic manipulation, in addition to avoiding negative marketing related to the use of Genetically Modified Organisms [8].

Oleaginous filamentous fungi can store up to 80% of their biomass as lipids, mainly in the form of GLA and ARA [23]. These microorganisms can grow using various carbon sources such as glucose, xylose, arabinose, mannose, glycerol, and agricultural and industrial residues [23]. Two processes are used for PUFAs production with fungi: submerged fermentation and SSF [3, 24]. Submerged fermentation or liquid fermentation may be defined as the fermentative process wherein the culture medium has high amount of free water. In the submerged fermentation process, the substrate consumption occurs rapidly and the final products are secreted directly into the fermenting broth [25]. The lipid production with the use of submerged fermentation of oleaginous microorganisms is well defined in the literature and has industrial application for approximately two decades[15, 26].

The SSF is defined as the fermentation process involving solid substrates in the absence (or near absence) of free water. Due to the low moisture content, SSF can only be performed by a limited number of microorganisms, mainly yeasts and fungi. It should be noted that in SSF the solid substrate acts both as a physical support and as a source of nutrients, and must have enough moisture to support microbial growth and metabolism [20–22]. SSF can be particularly advantageous for the cultivation of filamentous fungi, because it simulates the natural habitat of microorganisms, since the filamentous fungi develop through hyphae that penetrate the solid substrate, absorbing water and nutrients, favoring productivity of enzymes or other products of interest. An important advantage of SSF is the possibility of using agro-industrial residues (e.g., sugarcane bagasse, rice bran, soybean meal, among others) as substrates that serve as sources of carbon and energy for microbial growth, also contributing to an appropriate destination of this waste. In addition, the SSF process has several biotechnological advantages, such as higher fermentation capacity, greater stability of the final product, reduction of substrate inhibition, lower energy consumption, reduction in the amount of effluent generated and lower operating cost. Thus, SSF has received more attention in recent decades [20, 27, 28].

12.3 Microorganisms Used for PUFAs Production by SSF

According to Ratledge [29], many microorganisms are capable of producing high amounts of fatty acids, mainly PUFAs. In the case of oils obtained from plants and animals, there is the disadvantage of always mixing a large variety of saturated and PUFAs, making it difficult to obtain the desired product. Most of the studies published on the production of PUFAs by SSF involve the application of filamentous fungi, with emphasis on the genus *Mortierella*. The *Mortierella* fungal genus of the Mortierellaceae family belongs to the order Mucorales and to the phylum Zygomycota. According to its morphology, the genus can still be divided into two subgenera known as *Mortierella* (*Mortierella alpina*, *Mortierella hyaline*, *Mortierella elongate*, etc.) and *Micromucor* (*Mortierella isabellina*, *Mortierella ramanniana*, *Mortierella vinacea*, etc.) [30, 31]. According to Dyal and Narine [8], the main difference between these two subgenera is the way in biosynthesis of fatty acids. Studies performed by Amano *et al.* [32] demonstrated that the fungi belonging to the subgenus *Mortierella* contained C20 PUFAs (PUFAs with 20 carbon atoms, e.g., ARA, EPA, etc.) [7, 8] while those of the subgenus Micromucor are the best sources of C18 PUFAs (e.g., ALA, GLA, etc.). A screening of fungi of the genus Mucorales led to the selection of the microorganisms *Thamnidium elegans*, *Cunninghamella echinulata*, *Cunninghamella elegans*, *Mucor mucedo* and *Mortierella isabellina* as producers of GLA and *Mortierella alpina* as producers of Dihomo γ-Linoleic Acid (DGLA), AA, and EPA [33, 34]. Moreover, studies have shown that the oils produced by the *Mortierella alpina* fungus are completely safe for human consumption and that this result can be extended to other oleaginous species of the genus [8, 35]. Biotechnological processes for the enrichment of cereals with GLA, DGLA, AA, and EPA using SSF processes with oleaginous fungi (*Thamnidium* sp., *Cunninghamella* sp., *Mucor* sp., *Mortierella* sp.) were reported and these products were successfully tested as foods or feed additives [36].

Studies have examined the production of PUFA by *Mortierella* fungi in submerged culture, finding that many species of the genus produce a high amount of PUFAs [37–39]. However, the submerged fermentation for the production of microbial oils consumes large amounts of energy, produce a high amount of effluents and sometimes have a low fermentation capacity. According to Peng and Chen [40] for an economically competitive process, the production of microbial oils should preferably be performed by SSF. Considering the economic viability of the process, the production of PUFAs is predominantly affected by the cost of the substrate, being necessary to select a raw material of low cost and that allows the appropriate microbial development. Thus, agro-industrial residues such as straw and bran of rice and wheat, cassava husk, fruit residues, sugarcane bagasse and corn cob can be considered as suitable substrates for SSF with fungi [7].

12.4 Main Process Parameters

Recent research in SSF has focused on maximizing PUFAs yields by optimizing fermentation process conditions [40–45]. The main parameters that influence the efficiency in SSF process for PUFAs production are initial moisture, pH, nitrogen source, C/N ratio, temperature, incubation time, and aeration level [7, 8]. Table 12.1 presents some results on the production of PUFAs by SSF, as well as the main operational parameters used.

12.4.1 Moisture Content of the Substrate

In SSF the microorganism growth and product formation occur on the surface of solid substrate particles. According to Pandey *et al.* [46], optimization of moisture content present in the substrate is essential to maintain physical-chemical characteristics of the solid and to guarantee the productivity of the process. Similarly, Khosravi-Darani and Zoghi [47] indicate that the initial moisture concentration in the substrate have importance for the SSF process. Water guarantees the intracellular mass transport, for the evaporative cooling of substrate layers, and probably, the most important, as a solvent for nutrients. Balanced water availability can prevent the growth of undesirable microorganisms in non-sterile fermentations, especially in combination with extreme pH, and thus reduce sterilization costs [36].

Zhang and Hu [42] tested four levels of initial moisture in soybean SSF using *M. isabellina*. According to the authors, the initial moisture content of the substrate affected the growth of microbial cells and the accumulation of lipids. At all tested levels, there was an increase in moisture over the 8 weeks of fermentation, ranging from 50.0%, 66.7%, 75.0%, and 78.6% to 76.2%, 78.5%, 83.6% and 88.1%, respectively. The authors also report that the total lipid production increased with the increase of the humidity of the process, and the highest lipid production (38.9 $mg.g^{-1}$ soybean hull) was obtained with the initial humidity of 78.6%.

In the study of Jang and Yang [6], it was verified that optimal initial moisture for rice bran SSF was 75% for the total production of PUFAs (mainly LA and ARA), while a value of 65% caused higher production of ALA and EPA. According to the authors, for the production of ω-6 the initial moisture content is 70 to 75% and for ω-3 of 60 to 65%. Similarly, moisture levels in the range of 60 to 75% are considered optimal for the growth and production of GLA by Mucorales fungi in SSF processes in grains [33, 51]. According to the authors, this is due to a better assimilation of the cereal starch by the fungi when the substrate is more moistened. Low moisture contents can reduce the solubility and availability of nutrients and increase the surface tension of the water layer, reducing the growth of fungi. On the other hand, high moisture levels reduce substrate porosity, impair oxygen diffusion, increase risk of bacterial contamination, and

Table 12.1 PUFA production by various microorganisms and substrates in SSF.

Microorganism	Substrate	Lipid yield (% m/m)	PUFA yield (mg / g de dried substrate)						Best fermentation conditions	Reference
			LA	GLA	ALA	DGLA	ARA	EPA		
M. alpina CCF 185	Dehulled millet	16.03	11.4	3.45	-	2.97	38.00	-	Incubation temperature: 28°C Incubation time: 21 days	[48]
	Barley	15.44	18.6	2.86	-	3.51	32.9	-		
Pythium ultimum MUCL 16164	Coconut oil+spent malt grain	16.7	-	-	-	-	2.2	1.8	Incubation temperature: 21°C Incubation time: 9 days Best substratecomposition: 28.5% pearled barley, 5.75% spent malt grains, 5.75% linseed oil, and 60% nutrient solution	[49]
	Peanut oil+spent malt grain	15.4	-	-	-	-	2.1	2.7		
	Sunflower oil + spent malt grain	16.1	-	-	-	-	2.0	2.6		
	Linseed oil + spent malt grain	16.7	-	-	-	-	2.0	2.9		
	Linseed oil + barley + spent malt grain	24.4	-	-	-	-	2.7	3.6		

M. alpina ATCC 32222	Peanut meal residue	-	29.32	ND	4.62	-	8.06	5.88	Incubation temperature: 20 °C Incubation time: 10 days Initial moisture: 65% Initial pH= 6.5	[50]
	Rice bran	-	48.72	ND	8.7	-	33.35	12.77		
	Wheat bran	-	1.18	ND	ND	-	0.15	0.11		
	Sweet potato residue	-	1.14	ND	ND	-	0.11	0.07		
T. elegans CCF 1456	Pearl barley	15.6	-	7.97	-	-	-	-	Incubation temperature: 28 °C Incubation time: 7 days Initial moisture: 75%	[51]
C. echinulata CCF 103		14.1	-	6.66	-	-	-	-		
C. elegans CCF 1318		15.8	-	9.26	-	-	-	-		
M. isabelina CCF14		12.1	-	5.27	-	-	-	-		
M. vinacea MUCL 15067		12.7	-	4.74	-	-	-	-		
M. ramanniana MUCL 8691		11.3	-	5.32	-	-	-	-		
M. circinelloides CCF 127		9.8	-	6.59	-	-	-	-		
M. circinelloides MUCL 15438	Pearl barley	10.6	-	6.54	-	-	-	-		[51]

(Continued)

Table 12.1 Cont.

Microorganism	Substrate	Lipid yield (% m/m)	PUFA yield (mg / g de dried substrate)						Best fermentation conditions	Reference
			LA	GLA	ALA	DGLA	ARA	EPA		
Rhizopus stolonifer CCF 445		8.8	-	4.37	-	-	-	-		[51]
Thamnidium elegans CCF 1465	Spelt wheat flakes and spent malt grain at a ratio of 3:1	7.2	-	7.23	-	-	-	-	Incubation temperature: 24 °C Incubation time: 4 days Initial moisture: 67%	[33]
M. alpina ATCC 32222	Rice bran	-	117	-	5.0	-	6.0	12.0	Incubation temperature of 20°C for 5 daysand then 12°C for 7 days Initial moisture: 57% Initial pH = 6 - 7	[6]
M. isabellina ATHUM 2935	Pear pomace	12.0	-	2.9	-	-	-	-	Incubation temperature: 28°C Incubation time: 10 days Initial pH: 6.5	[52]

M. alpina CBS 754.68	Sunflower oilcake	-	-	-	-	-	1.14	-	Incubation temperature of 20°C for 6 days and then 12°C for 6 days	[53]
	Soybean oilcake	-	-	-	-	-	0.53	-		
	Colza oilcake	-	-	-	-	-	0.38	-		
	Olive oilcake	-	-	-	-	-	0.06	-		
M. alpina ATCC 16266	Oat bran	8.7 (PUFA only)	-	-	-	-	87.0	-	Incubation temperature of 20°C for 21 days (on day 15 the best results for PUFA profile were obtained)	[54]
M. circinelloides CCF 2617	Rye bran/spent malt grains (3:1)	-	-	24.2	-	-	-	-	Incubation temperature: 25°C Incubation time: 6 days Suplement of 30% of sunflower oil on the substrate	[55]

PUFA: polyunsaturated fatty acids;–: not reported; ND: not detectable; LA: linoleic acid; GLA: γ-linolenic acid; ALA: α-linolenic acid; DGLA: Dihomo-γ-linolenic acid; ARA: arachidonic acid; EPA: eicosapentaenoic acid.

restrict gas exchange, significantly influencing fungal growth and PUFA production [7, 36, 56].

12.4.2 Temperature

Incubation temperature is a critical factor that affects production of microbial oils by SSF, as it controls the microbial growth rate, the lipid synthesis and changes the composition of the fatty acids produced [7]. In general, elevated temperatures increase microbial growth, while low temperatures increased the production of PUFAs and their degree of unsaturation [8, 12, 57]. An example of this is the study of SSF with *Mortierella alpina* performed by Jang *et al.* [50], which obtained an increase in PUFA production of 12% reducing the growth temperature from 20 to 12°C on the fifth day of cultivation. In a similar study, Jang and Yang [6], when optimizing PUFA production in a solid-state column reactor packed with rice bran, identified that reducing the temperature (20 to 12 °C) of cultivation in the fifth day increased yield in production of EPA around 24%. Michinaka *et al.* [58] performed the growth of *M. circinelloides* at different temperatures and found that Δ6 desaturase enzyme activity was twice higher at 15 °C in relation to the activity produced at 28 °C. Thus, the authors concluded that the culture grown at a lower temperature had a higher amount of unsaturation in the lipids produced.

According to Armenta and Valentine [57], a suitable strategy for PUFA production is to first perform the growth at elevated temperatures (25 °C) to maximize cell growth during the exponential phase and then switch to a lower temperature (i.e., 10 to 15 °C) in order to increase PUFA production. However, total yields of biomass and lipids obtained at lower temperatures need to be taken into account when selecting the ideal temperature for the production of PUFAs. In the case of very low biomass and/or lipid yields, the selection of low temperatures may not be suitable for the process [8].

12.4.3 Substrate

Selection of a suitable substrate is critical in SSF in order to assure total lipid yield as well as the desired PUFAs. In SSF processes, agro-industrial grains or residues are the substrates usually used, such as rice, barley, rice bran, wheat and soybean, spent malt grain, pear pomace, dehulled millet, soybean hull, rye meal, sprout wheat, oat flakes, crushed corn, among others [7, 36].

Substrate conversion to PUFAs is significantly improved when an internal support is present on the substrate. The internal support provides better

O_2 diffusion and contributes to microbial respiration due to increased particle space, which also helps to remove the heat generated during fermentation [59]. Substrates that do not have an inner support are more likely to lead to particle agglomeration which negatively interferes with microbial respiration and process efficiency. However, an excess of carrier material can greatly reduce substrate surface available for microbial growth [36].

Conti *et al.* [51] evaluated the Mucorales for the production of GLA by SSF and found that *Cunninghamella elegans* strain produced 14.2 mg of GLA per gram of a mixture of barley, spent malt grains, and peanut oil. According to the authors, spent malt grains were incorporated into the substrate mixture, because these particles serve as an inert carrier, preventing agglomeration of cereal grains and being an absorbent material, it was able to retain moisture and serve as a reserve of water for microorganisms.

Substrate particles size also affects SSF efficiency. Small particles provide a larger surface area and nutrient availability, but particles of very small size cause agglomeration of the substrate, inhibiting oxygen transfer [7]. However, the large particles, the contact surface is limited, although they provide a suitable porous solid medium, facilitating the diffusion of oxygen and favoring the development of microorganisms [46, 51]. The results of Ghobadi *et al.* [53] demonstrated that substrate particles (residual cake from the extraction of sunflower, soybean, canola, and olive oil) in the range of 1 to 1.4 mm were more suitable than 0.2 to 0.6 mm for the production of ARA by SSF with *Mortierella alpina*.

12.4.4 Carbon to Nitrogen (C/N) Ratio

The C/N ratio of the solid substrate in the SSF is one of the main variables influencing the microbial growth and lipid production. A very high or low C/N ratio alters the metabolic pathway between microbial growth and PUFA production, impairing the fermentation efficiency [60].

When evaluating the addition of nitrogen sources (yeast extract and nitrates) in different solid substrates (rice and wheat bran, peanut flour, and sweet potato residues) in SSF, Jang *et al.* [50] verified that the production of PUFAs depends on the amount of nitrogen supplied, with an addition of 2.3 to 5% of nitrogen being the most adequate value. According to Asadi *et al.* [7] for oleaginous microorganisms growth, nitrogen supplementation may provide additional nutrients for microbial growth and increase the production of PUFAs in the next stage (stationary phase). However, it is important that there is a nitrogen limitation and the presence of a large amount of carbon in stationary phase to guarantee the production of PUFAs. It is also worth noting that organic nitrogen sources are more suitable for fungal growth and

PUFA accumulation than inorganic sources [41]. According to Certik and Adamechova [36], the use of cereals as a substrate, such as rice bran, wheat bran, oat flakes, peeled or pearled barley provide an adequate source of nutrients for fungi growth and production of lipids, as they have a high carbon content in starch and adequate levels of organic nitrogen, resulting in a C/N ratio generally of 20 to 60.

12.4.5 pH

The pH is an important parameter to be evaluated in SSF processes, because when microbial cells are exposed to fermentative media with a pH outside their optimum range, they are forced to use energy to maintain intracellular pH, mainly through proton pumps (e.g., K^+ and Na^+). This energy use for pH maintenance ends up negatively interfering with cell growth and process efficiency [57, 61]. According to Dyal and Narine [8], the main effect of pH on SSF is on fungus growth and not on PUFA biosynthesis.

For the production of PUFAs by SSF with microorganisms of the genus Mortierella, the appropriate pH is slightly acidic to neutral [7]. In the study of Jang *et al.* [50], the optimal initial pH range of rice bran for SSF with *Mortierella alpina* is 6 to 7, with values above 8 and below 5 reducing the yield. Li *et al.* [62] reported that production of PUFAs and total lipids reached their maximum values at a pH of 6.0 and decreased sharply when pH went from 7.0 to 8.0 in the fermentation of fungus *M. recurvus.*

12.4.6 Incubation Time

Certik *et al.* [55] analyzed the GLA production in SSF of a mixture of rye meal/spent malt grain (3:1) with the fungus *Mucor circinelloides.* According to the authors, the GLA level gradually increased to almost 21%, reaching maximum production (3.4 $mg.g^{-1}$ of fermented cereals) after 120 hours of fermentation. Zhang and Hu [42] studied SSF of soybean peel with *Mortierella isabellina* over a period of 8 weeks and found that cell growth and lipid accumulation increased over the first 4 weeks and then remained stable for the remainder of the period. However, the authors did not analyze lipid profile variation produced during the fermentation period.

In the study of Fakas *et al.* [52], the oil content of a fermented mass of pear pomace reached its maximum value of 12% (m/m) after 212 hours of inoculation. After 280 hours of cultivation, the oil content began to reduce, reaching 9% (m/m) in 330 hours. However, Jang *et al.* [50] and Ghobadi *et al.* [53] reported that the production of PUFA by SSF remained stable during an incubation period of 288 hours (12 days). This difference may have occurred due to the fungus strain, the substrate composition and the form of culture used by the authors.

12.5 Bioreactors

The SSF process with the use of fungi still needs to be improved for the consolidation of its application in industrial scale. The major challenges encountered are process scale up, extraction and purification of the final product and the monitoring of biomass [21, 22]. The scale up of SSF processes is one of main problems, there are difficulties such as: maintaining an adequate transfer of heat and mass throughout the solid layer; monitor and control process parameters; and mixing the solid layer without damaging the microorganisms and their carrier medium. Another problem is the generation of heat by the metabolic activity of microorganisms, which can lead to the degradation of the final product. Moisture can also interfere with the process, since when excess can reduce the porosity of the substrate and prevent the entry of oxygen and when deficits can reduce the availability of nutrients for microorganisms [20]. In order to avoid or minimize these problems, different types of bioreactors have been developed, depending on the type of aeration and agitation system. One such type is tray bioreactors where SSF occurs in shallow trays where humid air is blown through the substrate to ensure sufficient water in the medium as well as oxygen and heat transfer. Another configuration is that of rotating drums, where the biomass is continuously mixed in order to provide substrate and oxygen to the microorganism, besides maintaining a homogeneous temperature and humidity during the fermentation period. It is worth mentioning that few of these bioreactors were used in real scale, making it necessary to further develop these processes [22].

Conti *et al.* [51] investigated the feasibility of increasing the scale of PUFA production by SSF with the fungus *Cunninghamella elegans* CCF 1318. The authors cultivated the microorganism in rotary bottles and plastic bags (through which a stream of moist air was passed), simulating rotating drums and tray bioreactors, respectively. Cultivation in the rotating bottles produced a slightly lower lipid yield than that obtained in static culture in Erlenmeyer flasks, while the cultures in plastic bags showed a superior lipid yield. According to the authors, the efficiency was higher in the culture in plastic bags due to the increase in oxygen availability through forced aeration, associated with a lower thickness of the substrate. Jang and Yang [6] demonstrated that the decrease in oxygen concentration in the lower layer of a column reactor used for SSF is a limiting factor for microbial growth and PUFA production, with optimum O_2 concentration being about 20%, which increased by up to 26% the production of PUFAs.

12.6 Extraction of Microbial Oil

After fermentation, it is necessary to carry out the extraction and purification of PUFAs from the microbial biomass. For this, efficient and reliable methods and solvents that do not promote the degradation of the final product are needed [63]. For a method to be efficient in extracting PUFAs and lipids in general, it is necessary that the solvent used be able to completely penetrate the cell mass and have a polarity similar to the compounds to be extracted [64]. In addition, if extracted PUFAs are used in pharmacological, medical or food applications, solvents must possess acceptable characteristics and parameters in terms of toxicity, safety and cost [65]. Currently, there are several methods available for extracting lipids from biological materials, most of them using organic solvents, generally in mixtures containing chloroform (apolar) and methanol (polar), as in the case of Bligh and Dyer [66] and Folch *et al.* [67], which produce large quantities of toxic waste and are difficult to handle/operate [68]. Automated extraction equipment, such as Soxhlet or Goldfisch, are successfully used in lipid extraction, but its use requires long extraction times and generally demands the use of organic solvents such as hexane [68]. These conventional methods of lipid extraction are relatively efficient and already understood technology, but have high energy expenditure. In this context, Jin *et al.* [26] and Chuck *et al.* [69] mention that the high costs associated with extracting and recovering PUFAs is one of the major factors limiting the large-scale production of microbial oils. Thus, increasing interest in the extraction of high value-added lipids has led to the search for more efficient, fast, easy to use, solvent-free and less toxic extraction methods [68, 70]. Among these methods, extraction with supercritical fluids and pressurized fluids (i.e. supercritical carbon dioxide and ethanol; pressurized butane and propane) can be highlighted [71].

12.7 Concluding Remarks

The use of genetic and metabolic engineering in order to produce microbial strains with a greater efficiency in the production of PUFAs and the discovery of new strains and substrates of low cost capable of being used in the process of production of PUFAs by SSF are necessary to feasible the production of microbial oil. In general, the results obtained with SSF confirm the potential of this process for the production of PUFAs on a larger scale in order to meet the market demands mainly in the food and pharmaceutical area. However, there is still a need to optimize this biological process in order to increase the efficiency and quality of the final product and reduce the production costs, thus requiring further studies in this area.

References

1. Patterson, E., *et al.*, Health Implications of High Dietary Omega-6 Polyunsaturated Fatty Acids. *J. Nutr. Metab.*, 2012.
2. Certik, M., Shimizu, S., Biosynthesis and regulation of microbial polyunsaturated fatty acid production. *J. Biosci. Bioeng.* v. 87, n.1, p.1–14, 1999.
3. Zhu, H., Utilization of Rice Bran by *Pythium Irregulare* for Lipid Production. 79 p. Tese - Faculty of the Louisiana State University and Agricultural and Mechanical College. Department of Biological and Agricultural Engineering, 79, 2002.
4. Xiao, L., Mjøs, S.A., Haugsgjerd, B.O., Efficiencies of three common lipid extraction methods evaluated by calculating mass balances of the fatty acids. *J. Food Compos. Anal.*, v. 25, p. 198–207, 2012.
5. FAO-WHO., Fats and fatty acids in human nutrition. Report of an Expert Consultation. *FAO Food and Nutrition Paper*, v. 91,.
6. Jang, H., Yang, S., Polyunsaturated fatty acids production with a solid-state column reactor. *Biores. Technol.*, v. 99, p. 6181–6189, 2008.
7. Asadi, S.Z., *et al.*, Evaluation of the effect of process variables on the fatty acid profile of single cell oil produced by *Mortierella* using solid-state fermentation. *Crit. Rev. Biotechnol.*, v. 35, n.1, p. 94–102, 2013.
8. Dyal, S.D., Narine, S.S., Implications for the use of *Mortierella* fungi in the industrial production of essential fatty acids. *Food Res. Int.*, v. 38, p. 445–467, 2005.
9. Cantrell, K.B., Walker, T.H., Influence of temperature on growth and peak oil biosynthesis in a carbon-limited medium by *Pythium irregulare*. *J. Amer. Oil Chem. Soc. September*, v. 86, p. 791–797, 2009.
10. Ratledge, C., *et al.*, Down-stream processing, extraction, and purification of single cell oil, in: *Single cell oils: microbial and algal oils*, Zvi Cohen, Colin Ratledge, 2nd ed. (Eds.), Elsevier, p. 179–197, 2010.
11. Yen, H., *et al.*, Supercritical fluid extraction of valuable compounds from microalgal biomass. *Biores. Technol.*, v. 184, p. 291–296, 2015.
12. Peng, C., *et al.* A temperature-shift strategy for efficient arachidonic acid fermentation by *Mortierella alpina* in batch culture. *Biochem. Eng. J.*, v.53, p. 92–96, 2010.
13. Dey, P., Banerjee, J., Maiti, M.K., Comparative lipid profiling of two endophytic fungal isolates – *Colletotrichum* sp. and *Alternaria* sp. having potential utilities as biodiesel feedstock. *Biores. Technol.*, v. 102, p. 5815–5823, 2011.

14. Carvalho, A.K.F., *et al.*, Biosynthesis, characterization and enzymatic trans¬esterification of single cell oil of Mucor circinelloides – A sustainable pathway for biofuel production. Bioresource Technol., v.181, p. 47–53, 2015.
15. Ratledge, C., Microbial oils: an introductory overview of current status and future prospects. *Oilseeds & fats Crops and Lipids - OCL 2013*, v. 20, n.6, D602, 2013.
16. Donot, F., Fontana, A., Baccou, J.C., Strub, C., Schorr-Galindo, S., Single cell oils (SCOs) from oleaginous yeasts and moulds: Production and genetics. Biomass Bioener., v.68, 135–150, 2014.
17. Matsakas, L., Sterioti, A., Rova,U., Christakopoulos, P., Use of dried sweet sorghum for the efficient production of lipids by the yeast *Lipomyces starkeyi* CBS 1807. *Ind Crops Prod.,* v.62, 367–372, 2014.
18. Catchpole, O.J., *et al.*, Extraction of lipids from fermentation biomass using near-critical dimethylether. *J. Supercrit. Fluids*, v.53, p. 34–41, 2010.
19. Papanikolaou, S., Aggelis, G., Lipids of oleaginous yeasts. Part I: Biochemistry of single cell oil production. *Eur. J. Lipid Sci. Technol.*, v. 113, p. 1031–1051, 2011.
20. Pandey, A., Solid-state fermentation. *Biochem. Eng. J.*, v. 13, p. 81–84, 2003.
21. Couto, S.R., Sanromán, M.A., Application of solid-state fermentation to food industry—A review. *J. Food Eng.*, v. 76, p. 291–302, 2006.
22. Singhania, R.R., *et al.*, Recent advances in solid-state fermentation. *Biochem. Eng. J.*, v. 44, p. 13–18, 2009.
23. Subramaniam, R., *et al.*, Microbial lipids from renewable resources: production and characterization. *J. Ind. Microbiol. Biotechnol.*, v. 37, p. 1271–1287, 2010.
24. Fakas, S. *et al.*, Compositional shifts in lipid fractions during lipid turnover in *Cunninghamella echinulata. Enzyme Microbial. Technol.*, v. 40, p. 1321–1327, 2007.
25. Subramaniyam, R., Vimala, R., Solid state and submerged fermentation for the production of bioactive substances: a comparative study. *Int. J. Sci. Nat.*, v. 3, n. 3, p. 480–486, 2012.
26. Jin, M., *et al.*, Microbial lipid-based lignocellulosic biorefinery: feasibility and challenges. *Tren. Biotechnol.*, v. 33, n. 1, p. 43–54, 2015.
27. Farinas, C.S., Developments in solid-state fermentation for the production of biomass-degrading enzymes for the bioenergy sector. *Renew. Sustain. Ener. Rev.*, v.52, p. 179–188, 2015.
28. Beheraa, S.S., Ray, R.C., Solid state fermentation for production of microbial cellulases: Recent advances and improvement strategies. *Int. J. Biol. Macromol.*, v. 86, p. 656–669, 2016.
29. Ratledge, C., Fatty acid biosynthesis in microorganisms being used for Single Cell Oil production. *Biochimie.*, v. 86, p. 807–815, 2004.
30. Gams, W., A key to the species of *Mortierella. Personia*, v. 9, p. 381–391, 1977.
31. Batrakov, S.G., Konova, I.V., Sheichenko, V.I., *et al.*, Unusual fatty acid composition of cerebrosides from the filamentous soil fungus Mortierella alpina. *Chem. Phys. Lipids*, v.117, p.45–51, 2002.

32. Amano, N., *et al.*, Chemotaxonomic significance of fatty acid composition in the genus *Mortierella* (*Zygomycetes, Mortierellaceae*). *Mycotaxon*, v. 44, p. 257–265, 1992.
33. Certik, M., Slavikova, L., Masrnova, S., Sajbidor, J., Enhancement of nutri¬tional value of cereals with g-linolenic acid by fungal solid state fermentations. *Food Technol. Biotechnol.* v.44, p.75–82, 2006.
34. Certik, M., Biotransformation of oils to value-added compounds, in: Biocatalysis and Bioenergy, C.T. Hou and J.-F. Shaw (Eds.), Wiley, New Jersey, USA, pp. 571–585, 2008.
35. Lewis, K.D., *et al.*, Toxicological evaluation of arachidonic acid (ARA)-rich oil and docosahexaenoic acid (DHA)-rich oil. *Food Chem. Toxicol.*, v. 96, p. 133– 144, 2016.
36. Certik, M., Adamechova, Z., Cereal-based bioproducts containing polyun¬saturated fatty acids. *Lipid Technol.*, v.21, p.250–253, 2009.
37. Lai, M.H., Slugen, D., Sajbidor, J., Production of dihomogamma- linolenic acid by *Mortierella alpina*. *Biologia Bratislava*, v. 53, p. 801–806, 1998.
38. Yu, L.J., *et al.*, Improved arachidonic acids production from the fungus *Mortierella alpina* by glutamate supplementation. *Biores. Technol.*, v. 88, p. 265–268, 2003.
39. Jang, H.D., Lin, Y.Y., Yang, S.S. Effect of culture media and conditions on polyunsaturated fatty acids production by *Mortierella alpina*. *Biores. Technol.*, v. 96, p. 1633–1644, 2005.
40. Peng, X., Chen, H., Single cell oil production in solid-state fermentation by *Microsphaeropsis* sp. from steam-exploded wheat straw mixed with wheat bran. *Biores Technol.*, v. 99, p. 3885–3889, 2008.
41. Dong, M., Walker, T.H., Addition of polyunsaturated fatty acids to canola oil by fungal conversion. *Enzyme and Microbial Technology*, v. 42, p.514–520, 2008.
42. Zhang, J., Hu, B., Solid-State Fermentation of *Mortierella isabellina* for Lipid Production from Soybean Hull. *Appl. Biochem. Biotechnol.*, v.166,p.1034–1046, 2012.
43. Cheirsilp, B., Kitcha, S., Solid state fermentation by cellulolytic oleaginous fungi for direct conversion of lignocellulosic biomass into lipids: Fed-batch and repeated-batch fermentations. *Ind. Crops Prod.*, v.66, p.73–80, 2015.
44. Dulf, F.V., Vodnar, D.C., Socaciu, C., Effects of solid-state fermentation with two filamentous fungi on the total phenolic contents, flavonoids, antioxidant activities and lipid fractions of plum fruit (*Prunus domestica* L.) by-products. *Food Chem.* v.209, p.27–36, 2016.
45. Yang, S., Zhang, H. Enhanced polyunsaturated fatty acids production in *Mortierella alpina* by SSF and the enrichment in chicken breasts. *Food Nutr. Res.*, v.60,p. 30842, 2016.
46. Pandey, A., *et al.*, Solid state fermentation for the production of industrial enzymes. *Curr. Sci.*, v. 77, p. 149–162, 1999.

47. Khosravi-Darani, K., Zoghi, A., Comparison of pretreatment strategies of sugarcane bagasse: experimental design for citric acid production. *Biores. Technol.*, v. 99, p. 6986–6993, 2008.
48. Stred'anská, S., Slugeti, D., Stred'ansky, M., Grego, M., Arachidonic acid production by*Mortierella alpina* grown on solid substrates. *World J. Microbiol. Biotechnol.*, 9, 51 l-513, 1993.
49. Stredansky, M., Conti, E., Salaris, A., Production of polyunsaturated fatty acids by *Pythium ultimum* in solid-state cultivation. *Enzyme Microbial Technol.,* 26, 304–307, 2000.
50. Jang, H.D., Lin, Y.Y., Yang, S.S., Polyunsaturated fatty acid production with *Mortierella alpina* by solid substrate fermentation. *Bot. Bull. Acad. Sinica*, v.41, p. 41–48, 2000.
51. Conti, E. *et al.*, γ-Linolenic acid production by solid-state fermentation of *Mucorales* strains on cereals. *Biores. Technol.*, v. 76, p. 283–286, 2001.
52. Fakas, S. *et al.*, Fatty acid composition in lipid fractions lengthwise the mycelium of *Mortierella isabellina* and lipid production by solid state fermentation. *Biores. Technol.*, v. 100, p. 6118–6120, 2009.
53. Ghobadi, Z., Hamidi-Esfahani, Z., Azizi, M.H., Determination of effective variables on arachidonic acid production by *Mortierella alpina* CBS 754.68 in solid-state fermentation using Plackett-Burman screening design. *World Acad. Sci. Eng. Technol.*, v. 81, p. 678–680, 2011.
54. Fischer. L., Stressler, T., Eisele, T., Rost, J., Haunschild, E., Kuhn, A., Production of Polyunsaturated Fatty Acids by *Mortierella alpina* Using Submerse and SolidState Fermentation. *Chemie Ingenieur Technik.*, 85 No. 3, 318–322, 2013.
55. Certík, M., Adamechová, Z., Guothová, L., Simultaneous enrichment of cereals with polyunsaturated fatty acids and pigments by fungal solid state fermentations. *J. Biotechnol.*, 168, 130–134., 2013
56. Zandrazil, F., Brunert, H., Investigation of physical parameters important for solid-state fermentation of straw by white rot fungi. *Appl. Microbiol. Biotechnol.*, v. 11, p. 183–188, 1981.
57. Armenta, R.E., Valentine, M.C., Single-cell oils as a source of omega-3 fatty acids: An overview of recent advances. *J. Am. Oil Chem. Soc.*, 90, 167–182, 2013.
58. Michinika, Y., Aki, T., Shimachi-Nakajima, T., Kawamoto, S., Shigeta, S.O., Suzuki & Ono., Differential response to low temperature of two D6 desaturase from *M. cercinelloides*. *Appl. Microbiol. Biotechnol.*, 62, 362–368, 2003.
59. Certik, M., Zuzana Adamechova, and Lucia Slavikova., Biotechnological enrichment of cereals with polyunsaturated fatty acids, in: *Biocatalysis and biomolecular engineering*, T. Ching Hou and Jei-Fu Shaw (Eds.), John Wiley & Sons, Inc., Hoboken, New Jersey, 2010.
60. Sajbidor, J., Dobronova, S., Certik, M., Arachidonic acid production by *Mortierella* sp. S-17: influence of C/N ratio. *Biotechnol. Lett.*, v. 12, n. 6, p. 455- 456, 1990.

61. Jiang, Y., Chen, F., Effects of medium glucose concentration and pH on docosahexaenoic acid content of heterotrophic *Crypthecodinium cohnii*. *J. Am. Oil Chem. Soc.*, 35, 1205–1209, 2000.
62. Li. N., Deng, Z.N., Qin, Y.L., Chen, C.L., Liang, Z.Q., Production of Polyunsaturated fatty acids by *Mucor recurvus* sp. with Sugarcane Molasses as the carbon source. *Food Technol. Biotechnol.*, 46, 73–79, 2008.
63. Arjuna, A., Somal, P., Effect of extraction methods on lipid and fatty acid composition by *Mortierella Ramanniana*. *Int. J. Sci. Res. Pub.*, v.3, n.3, 2013.
64. Bonturi, N., *et al.*, Single cell oil producing yeasts *Lipomyces starkeyi* and *Rhodosporidium toruloides*: Selection of extraction strategies and biodiesel property prediction. *Energies*, v. 8, p. 5040–5052, 2015.
65. Somashekar, D., *et al.*, Efficacy of extraction methods for lipid and fatty acid composition from fungal cultures. *World J. Microbiol. Biotechnol.*, v. 17, n.3, p. 317–320, 2001.
66. Bligh, E.G., Dyer, W.J., A rapid method of total lipid extraction and purification. *Can. J. Biochem. Physiol.*, v. 37, p. 911–917, 1959.
67. Folch, J., Lees, M., Sloane-Stanley, G.H., A simple method for the isolation and purification of total lipids from animal tissues. *J. Biol. Chem.*, v. 226, p. 497–509, 1957.
68. Sahena, F., *et al.*, Application of supercritical CO2 in lipid extraction - A review. *J. Food Eng.*, v. 95, p. 240–253, 2009.
69. Chuck, C.J. *et al.*, Simultaneous microwave extraction and synthesis of fatty acid methyl ester from the oleaginous yeast *Rhodotorula glutinis*. *Energy*, v. 69, p. 446–454, 2014.
70. Nguyen, P.L.T., Go, A.W., Huynh, L.H., Ju, Y., A study on the mechanism of subcritical water treatment to maximize extractable cellular lipids. *Biomass Bioenergy*, v.59, p.532 - 539, 2013.
71. Cescut, J., *et al.*, Optimizing pressurized liquid extraction of microbial lipids using the response surface method. *J. Chromatography A*, v. 1218, p. 373–379, 2011.

13

Solid State Fermentation – A Stimulating Process for Valorization of Lignocellulosic Feedstocks to Biofuel

Arpan Das[1]* and Priyanka Ghosh[2]

[1]*Department of Microbiology, Maulana Azad College, Kolkata, West Bengal, India*
[2]*Food Technology and Biochemical Engineering Department, Jadavpur University, Kolkata, West Bengal, India*

Abstract

Ethanol, unlike gasoline, is an oxygenated fuel that contains 35% oxygen, which reduces particulate and NOx emissions from combustion. It can be made synthetically or by bioconversion of agro-wastes through microbial fermentation. Lignocelluloses are often major components of different waste streams from various industries, forestry, agriculture, and municipalities. These wastes can be valorized through its simultaneous saccharification and fermentation for production of ethanol and other biofuels. Primary benefits of microbial hydrolysis together with the fermentation are the reduced end product inhibition of the enzymatic hydrolysis and the reduced investment costs. But the major constrain of the processes are the need to find suitable agro wastes, microorganisms and favorable conditions for both the enzymatic hydrolysis and the fermentation and the difficulty to recycle the enzymes. In this chapter, a brief overview of recent strategies of solid state fermentation using lignocellulosic feedstock for bioethanol production will be discussed.

Keywords: Lignocellulose; pre-treatment; saccharification; fermentation; consolidated bioprocessing

**Corresponding author*: arpan_das85@yahoo.co.in

Arindam Kuila and Vinay Sharma (eds.) Principles and Applications of Fermentation Technology, (239–262) © 2018 Scrivener Publishing LLC

13.1 Introduction

The progressive depletion in global energetic resources based on non-renewable fuel and energy consumption is rising day by day. Moreover, it is widely known that fossil fuel combustion is the primary cause of global warming and environmental pollution [1, 2]. In order to get a stable energy alternative that will meet world demand and at the same time while moderating climate change, it is necessary to develop renewable clean fuels. Bioethanol is an attractive alternative fuel since it can be blended with gasoline and used as clean alcohol in engines with higher octane number and heat of vaporization [3, 4]. Bioethanol production has increased rapidly because many countries targeted toward reducing oil imports, boosting rural economies along with improving the quality of air [5]. During the last decade, the production of ethanol from corn based biomasses received more attention in the United States (U.S.) and worldwide. Although corn-based and sugar based-ethanol are promising substitutes to gasoline production mainly in the transportation sector, they are not sufficient to replace a considerable portion of the one trillion gallons of fossil fuel presently consumed worldwide each year. Furthermore, the ethical concerns about the use of food as fuel raw materials have encouraged research efforts to be more focused on the potential of inedible feedstock alternatives. Millions of tons of agricultural residues are abundantly available since no economically viable technologies are available for their conversion. Open-field burning of lignocellulosic waste is a globally common practice as it represents a cheap, fast and practical means of preparing the field for the next crop. Air emissions from the burning process of wood, crop residues and other lignocellulosic biomass are not only a threat to public health but also wasting our natural resources. The burning process reduces the local air quality, creating a variety of health concerns from the discharge of carcinogenic oxides (NOx, SO2, and COx) into the atmosphere. So the conversion of lignocellulosic biomass into biofuels is an important eco-friendly and sustainable choice for the production and exploitation of alternative energy sources. Thus, due to partial substitution of gasoline, the development of biofuels, especially bio-ethanol, has gained strategic importance [4]. Lignocellulosic biomass derived ethanol is often termed as "second generation" as the "first generation" ethanol is derived from sugar cane, corn, wheat, and other starchy feedstocks [5]. Studies suggest that the net energy return on second generation ethanol is much higher than ethanol derived from corn [6, 7]. In addition, second generation ethanol has much higher potential for greenhouse gas (GHG) emissions reduction than first generation ethanol [8]. However, due to recalcitrant nature of cellulosic biomass, the current processing cost of second generation ethanol is still high. The reasons for high processing costs of cellulosic biomass to biofuels are several including inherent recalcitrant nature of cellulosic biomass than corn, energy and chemical intensive pre-treatment, inefficient and expensive enzymes resulting in low conversion at high solids loadings required for commercial application, incomplete conversion of all sugars to fuels and chemicals, and distillation [9]. Figure 13.1 shows a schematic diagram indicating the process of second generation bioethanol production from lignocellulosic substrates. In this way, the design and implementation technologies go from a simple way of sugar conversion by fermentation process to the multi-stage conversion of biomass into ethanol [10]. Although production of bio-ethanol has been significantly enhanced by new technologies, there are still challenges for further improvement and investigation. A number of reviews have been published on fuel ethanol

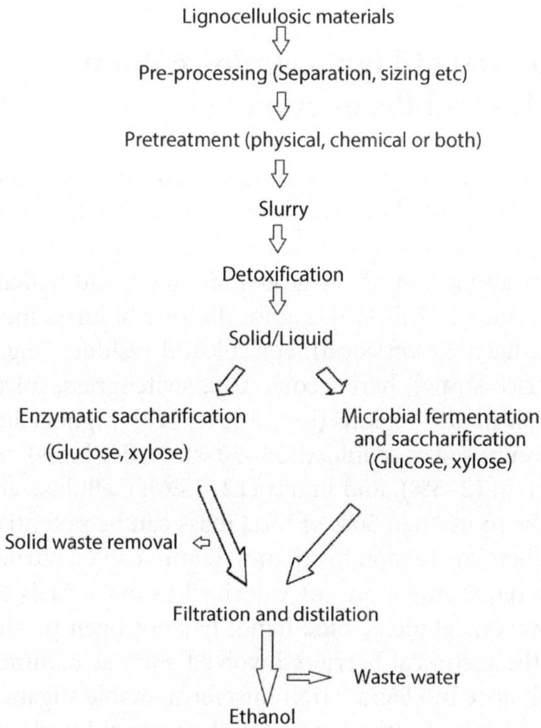

Figure 13.1 Schematic diagram indicating the process of second generation bioethanol production from lignocellulosic biomasses

production, specifically from lignocellulosic biomass. Therefore, the objective of the current chapter is to present an overview of fuel ethanol production from lignocellulosic biomass.

Among renewable fuels, ethanol due to its long history, use, and inherent characteristics, such as low toxicity to microbes and environment, low boiling point, high octane number, and comparable energy content, is considered to be a primary fuel candidate for near/long term applications [9]. Although ethanol's energy content is roughly 2/3rd of gasoline and butanol, it has higher research octane number (RON; 107) than butanol and gasoline [11]. Research shows that ethanol can be used up to 85% (v/v) in vehicles without major modifications [12]. In the U.S., bioethanol is primarily produced from corn starch feedstocks while in Brazil biofuel is mainly produced from sugarcane juice and molasses. Together, these countries account for 89% of the current global bioethanol production [13]. Moreover, ethanol can also serve as a precursor for several other chemicals and intermediates that are currently derived from non-renewable resources [14, 15].

13.2 Potential of Lignocellulosic Biomass for Biofuel Production

Agricultural residues are the widespread lignocellulosic biomass source and represent the most abundant biomass on earth. It was estimated that the annual production of these materials were to be 1010 Mt. globally, accounting for about half of the global biomass yield which corresponds to an energy value of 47 EJ [16]. Lignocellulosic biomass includes forestry residues (e.g., hard & softwood), agricultural residues (e.g., corn stover, wheat straw, rice straw), herbaceous (e.g., switchgrass, miscanthus), and plants that grow in arid regions (e.g., Agave) [17]. Lignocellulosic biomass is primarily composed of cellulose (35–50 wt. %, dry basis), hemicelluloses (15–30%), pectin (2–5%), and lignin (12–35%). Cellulose and hemicelluloses that make more than 50% of total mass can be potentially converted to sugars for their conversion to ethanol. Lignin can be burned to meet the plants energy requirement and/or valorized to make fuels and chemicals [18, 19]. However, cellulosic bioethanol has not been produced on large scale due to the technical barriers involved such as commercial feasibility of lignocellulose biodegradation into fermentable sugars. Wheat straw and rice straw are by far the most abundant agricultural wastes globally. Globally about 200 billion tonnes of plant biomass are produced annually and more than 90% of the total production of plant biomass is classified as lignocellulosic waste (LCW) [20]. These waste materials are often available at very low cost and as a cheap substrate for commercial biofuel production. Different types of lignocellulosic residues could be used for biofuels production such as straws, crop residues, wood pellets, wood chips and agro-waste [21]. Because of the low price, availability throughout the year and wide distribution geographically, LCW is considered not only the most feasible option for biofuel production but also for fossil fuel replacement since these raw materials do not compete with food crops and have the significant potential of bioethanol productivity compared to edible resources. For instance the bioethanol production from wheat straw (inedible) is expected to be 290 L/1000 kg of dry biomass compared to edible bioethanol generated from wheat 340 L/1000 kg of dry biomass [22]. LCW has the potential capability to produce about 419 GL of bioethanol annually [23]. An earlier estimate reported that the potential ethanol that could be derived from corn stover alone in the US was 15 billion litres per year. Lignocellulosic biomass composition plays a very crucial role in the performance and efficiency of both pre-treatment and biodegradation stages. Table 13.1 presents the compositions of several suitable lignocellulosic biomasses used for bioethanol production. Production of lignocellulosic bioethanol delivers several advantages over gasoline: like utilization of abundant and inexpensive sources of renewable resources; reduction in emission of GHG and toxic substances; economic benefits for rural community as well as pertaining in national energy security [27].

13.3 Structure of Lignocellulose

The main components of lignocellulosic biomass are cellulose (30–35%), hemicellulose (25–30%), lignin (10–20%) lignin and other extractable components like protein, lipids, and water [28, 29]. Cellulosic and hemicellulosic polymers constitute approximately 70% of the entire biomass and are connected to the lignin component through a variety of covalent bonds that give the lignocellulosic biomass significant robustness and resistance to (bio-)chemical or physical treatment [30]. The effective utilization of all the three components would play a significant role in economic viability of the cellulose to ethanol process.

13.3.1 Cellulose

Cellulose $(C_6H_{10}O_5)_x$, the main constituent of lignocellulosic biomass, is a linear polysaccharide that consists several thousand of D-glucose linked

Table 13.1 Compositions of different lignocellulosic biomass (% dry basis).

Raw materials	Hemicelluloses	Cellulose	Lignin	Others (i.e., ash)	References
Hardwood	25–40	45–47	20–25	0.80	[13, 24–26]
Wheat straw	24–35.5	32.9–50	8.9–17.3	-	
Softwood	25–29	40–45	30–60	0.50	
Corn stalks	16.8–35	35–39.6	7–18.4		
Waste papers from chemical pulps	12–20	50–70	6–10	2	
Newspaper	25–40	40–55	18–30	5–8	
Rice straw	19–24	36.2–47	9.9–24	-	
Switch grass	30–35	40–45	12	4-5	
Barley straw	33.8	21.9	13.8	-	
Rye straw	21.5	30.9	22.1	-	
Corn cobs	31.9	33.7	6.1	-	
Corn stover	25.5	38.3	17.4	-	

by β-(1,4)-glycosidic bonds to each other. The cellulose strains are associated together to make cellulose fibrils. These fibers are linked by a number of intra- and intermolecular hydrogen bonds and are attached to each other by hemicelluloses and amorphous polymers of different sugars as well as other polymers such as pectin and covered by lignin. The cellulose microfibrils which are present in the hemicellulose-lignin matrix are often associated in the form of bundles or macrofibrils. The molecules of individual microfibrils in crystalline cellulose are packed so tightly that not only enzymes but even small molecules like water cannot enter the complex framework. Some parts of the microfibrils have a less ordered, non-crystalline structure referred to as amorphous region [31]. High molecular weight and ordered tertiary structure make natural cellulose insoluble in water and most organic solvents. The crystalline regions of cellulose are more resistant to biodegradation than the amorphous parts. Cellulose with low degree of polymerization (DP) will be more susceptible to cellulolytic enzymes. The isolation and derivatization/dissolution of cellulose are crucial steps in determining cellulose DP [30].

13.3.2 Hemicellulose

Hemicelluloses, located in secondary cell walls, have a vague and changeable structure of branched heteropolymers containing pentoses (β-Dxylose, α-L-arabinose), hexoses (β-D-mannose, β-D-glucose, α-D galactose) and/or uronic acids (α-D-glucuronic, α-D-4-Omethylgalacturonic and a-D-galacturonic acids). They are relatively easy to hydrolyze because of their amorphous and branched structure (with short lateral chain) as well as their lower molecular weight [30]. In order to increase the digestibility of cellulose, large amounts of hemicelluloses must be removed as they cover cellulose fibrils limiting their availability for the enzymatic hydrolysis. Xylan is the primary component of hemicellulose and its composition varies in each feedstock. For this reason, hemicellulose stands in need of wide variety of enzymes to be completely hydrolyzed into free monomers [32]. Hemicelluloses are relatively sensitive to operation condition. Therefore, parameters such as temperature and retention time must be controlled to avoid the formation of unwanted products such as furfurals and hydroxyl methyl furfurals which later inhibit the fermentation process

13.3.3 Lignin

Lignin is a complex hydrophobic polymer coupled via covalent bonds to xylans rendering massiveness and stability to the plant cell wall. It is composed of three major phenolic components, namely p-coumaryl alcohol, coniferyl alcohol, and sinapyl alcohol [33]. Lignin is synthesized by polymerization of these components and their ratio varies between different plants, wood tissues, and cell wall layers. It is a dominant constituent of wood (30–60% for softwoods and 30–55% for hardwoods), while agricultural residues and grasses contain 3–15% and 10–30% respectively [34]. Contrarily, crop residues like corn stover, rice and wheat straws contain particularly hemicellulose.

13.4 Biomass Recalcitrance

The major impediments toward development of an economically viable technology for biodegradation of cellulose are the association with lignin and hemicellulose, crystallinity, DP and surface area which provides an inherent resistance of cellulosic biomass to pathogens, enzymes/microbes, and/or chemicals. During the biocatalytic valorization of lignocellulosic substrate, a residual fraction survives the attack. This fraction absorbs a significant amount of the original enzyme and restricts the use of these enzymes on added, fresh substrate. Most potential cellulosic substrates for bioconversion are heavily lignified. Thus, most of the cellulose in nature is unsuitable for bioconversion unless effective and economically viable procedures (pre-treatments) are developed to remove or modify lignin. Plants are being engineered to make them less resistant to break down, consequently requiring less harsher pre-treatments and low enzyme loadings for high product yields [35]. Besides lignin, changes in hemicelluloses, pectins, and other components in terms of backbone composition, chain length, branching, and content have also shown promise for reduction in plants recalcitrance [36, 37].

13.5 Pre-Treatment of Lignocellulosic Biomass

Cellulose is coated or sheathed by hemicellulose which acts as a blocking seal limiting the access of cellulase and hemicellulase enzymes to the cellulose–hemicellulose complex. In addition, the polysaccharide biopolymers (cellulose–hemicellulose complexes) are encapsulated with lignin which forms a physical barrier that increases resistance to chemical and microbial attack and hinders polysaccharide polymer hydrolysis into reducing sugars [38]. Therefore, a pre-treatment step is required which can disrupt the recalcitrant materials and enable the cellulose to undergo hydrolysis with higher efficiency and lower energy consumption. Among all the steps, it is one of the most costly steps accounting for 33% of the total production cost. The main challenge in the bioethanol production lies in pre-treatment step as it is very expensive and time consuming. In the process of pre-treatment, the complex structure of lignocellulosic biomass is distorted so that its cellulose component can be free for enzyme action. The enzymatic action hydrolyses cellulose into sugars that are further fermented. After pre-treatment procedure the cellulose crystallinity also reduced and the porosity of the raw substrate improved, increasing the sugars formation and improving enzymatic hydrolysis, avoiding formation of any kind of inhibitor that create problem in the hydrolysis or fermentation step. The main purpose of this step is to make cellulose more accessible to enzymatic hydrolysis. There are many ways by which this step is completed like physical pre-treatment, biological pre-treatment, chemical pre-treatment and solvent pre-treatment. There are many biological, chemical, physio-chemical, and physical processes used singly or in combination for the pre-treatment of lignocellulosic biomass. The pre-treatment process required for each feedstock was chosen according to its characteristics. Zhu and Pan [39] reported that agricultural biomass treatment differs from woody biomass because of its physical properties and chemical composition. Unlike agricultural biomass, woody biomass requires high content of energy to reach size reduction for further enzymatic saccharification. Toxic compounds have also to be considered for evaluating the pre-treatment cost. Different substances may act as inhibitors of microorganisms that are used in the ethanol fermentation. These inhibitors include phenolic compounds, furans (furfurals and 5- hydroxymethylfurfural (HMF)), aliphatic acids and inorganics compounds (iron, chromium or nickel). Several alternative measures can be taken to avoid problems caused by inhibitors [40]. Detoxification process is an important step which can affect the pre-treatment performance. General feedstock versatility and toxic inhibitors produced have to be considered on the pre-treatment efficiency in order to reach optimal conditions.

13.5.1 Chemical Pre-Treatment

This is the most studied method and it involves the use of variety of chemicals like acids, alkalis, organic solvents, peroxides, inorganic solvents like hydrochloric acid, sulphuric acid. Mostly dilute-acid pre-treatment method is used; it is finished in two steps: depolymerisation of hemicelluloses at 140 °C for at least 15 min avoiding the formation of carboxylic acids or furan. In second step, the treatment with dilute acidis done at 190 °C for10 min helping better cellulose interaction with enzyme during enzymatic hydrolysis [41]. Low temperature, usually 121 °C is often used for dilute-acid pre-treatment that avoids breaking of sugars into HMF and furfural [42].But dilute-acid pre-treatment has disadvantage as the acid used causes corrosion of the instruments; to avoid this expensive coatings or use of stainless steel that is acid resistant is recommended. Chemical pretreatment is also done with the help of alkalis like NaOH, lime, aqueous ammonia that resultin dilute base addition, increase incrystallinity, increase in DP and internal surface area and degradation of lignin. Alkali pre-treatment method is more reliable asit decreases the degradation of sugars that are to be hydrolysed and requires lower temperature and pressure compared to the dilute-acid pre-treatment.

13.5.2 Physical Pre-Treatment

Its main aim is to reduce the cellulose crystallinity, degradation of lignin and hemicelluloses without affecting cellulose. Physical pre-treatment process includes steam treatment, grinding, milling, chipping etc. Radiations such as microwaves are also being used for this purpose as they can easily penetrate the biomass surface and simultaneously heat the surface. This results in an easy access of enzyme to the surface of cellulose during hydrolysis, decrease in the crystallinity and increase in the DP. These processes demand lot of power supply because of which the cost increases. Physio-chemical pre-treatment: It is one of the oldest, most studied processes. Ammonia fibre steam explosion is the favourite option always, as it helps in converting hemicelluloses and lignin into soluble oligomers using auto hydrolysis reaction by steam at very high pressure followed by depressurizing to disrupt the structure of lignocellulosic biomass [43]. The main factors that are considered during steam explosion include temperature, holding time, chip size, and moisture content. This process is recommended for agricultural waste and hardwood but it is not good for softwood.

13.5.3 Biological Pre-Treatment

While all the other methods have high energy demands and need proper equipments, biological pre-treatment methods need much less energy as compared to physical/chemical methods and are eco-friendly. Mainly white rot fungi are used in this process as they help in the proper degradation of lignin [36]. Brown rot fungi are also used as they help in degrading lignin by attacking it with the help of enzymatic laccase and peroxides. Laccase is a copper containing compound oxidase enzyme, which can easily remove lignin, increasing the interaction of cellulase enzyme with cellulose [43].

13.5.4 Inhibitors Released During Pre-Treatment

In certain types of pre-treatments, such as steam explosion or acid pretreatments inevitably generate higher amounts of degradation compounds (e.g., acetic and formic acid, furfural, 5-HMF, and phenolic compounds) that are potentially inhibitory to yeast and enzymes [44]. Different approaches have been studied in order to produce lower concentrations of inhibitors during pre-treatment, such as avoiding the use of chemicals [43], reducing high solids loading, optimizing the temperature and retention time, and optimizing the chip size of the raw material before pre-treatment [45]. Alkali pre-treatment, wet oxidation, liquid hot water, and ammonia fiber explosion have been identified as methods that release low concentrations of inhibitors. Apart from the pre-treatment itself, the types and concentrations of degradation compounds also depend on the lignocellulosic feedstock. Therefore, each feedstock requires specific pre-treatment methods and conditions. Reactors should also be optimized for different biomass feedstocks in order to limit the amount of undesirable products and energy costs.

13.6 Hydrolysis

During this reaction, cellulose and hemicellulose are hydrolysed into simplistic and soluble compounds available for further conversion (fermentation) to ethanol. The cellulose that is left behind after the pre-treatment step is converted into glucose, while majority of hemicellulose is converted in to xylose.

$$(C5H8O4)n + nH2O \rightarrow nC5H10O5 \text{ (for pentose)} \quad (1)$$

$$(C6H10O5)n + nH2O \rightarrow nC6H12O6 \text{ (for hexose)} \quad (2)$$

Saccharification can be done by acids or enzymes among which enzymes are mostly preferred due to their low cost of processing, requirement of mild operating conditions, high sugar yield, and lack of corrosion problem. Eggeman and Elander [46] have demonstrated that Trichoderma reesei is a very efficient fungus to produce industrial grade cellulolytic enzymes. Cellulases are specific enzymes for celluloses and they constitute a mixture of enzymes that help in the hydrolysis of cellulose to glucose. The main 3 three enzymes present in cellulases are: I. Endo glucanase which creates free chain ends by attacking regions of low crystallinity in celluloses. II. Exo glucanase/Cellobiohydrolase which removes cellobiose units from the free chain ends. III. β-glucosidase which produces glucose by hydrolyzing cellobiose. The factors affecting the rate of enzymatic hydrolysis are mainly the activity of cellulase, pH, temperature and concentration of the substrate. Cellulases for hydrolysis of celluloses can be produced by a variety of organisms like fungi and bacteria; these microorganisms can be either anaerobic or aerobic, thermophillic or mesophillic. Bacteria from families of *Bacillus, Clostridium, Cellulomonas, Thermomonospora, Erwinia, Ruminococcus,*

Bacteriods, Acetovibrio, Streptomyces are being widely used in industrial purposes [23, 47, 48]. Among fungi *Aspergillus, Trichoderma* and *Penicillium* families are known as potential producer of cellulases [49, 50]. Enzymatic pre-treatment can be attained in simultaneous through biological saccharification and fermentation process in order to produce ethanol from woody biomass [51]. In this process the concentration of saccharides is kept low and cellulose inhibition is deterred. In a separate hydrolysis and fermentation (SHF) process cellulases (hydrolytic enzymes) are inhibited by glucose and cellobiose (saccharide products) resulting in a slower process and a lower yield of fermentable sugars.

13.7 Limitations of Enzymatic Hydrolysis

Regardless of the process configuration, enzymatic hydrolysis of cellulose and hemicellulose is an important part of most biofuel production from lignocellulosic raw materials. For many years, the cost of lignocellulolytic enzymes has been one of the main constraints for commercialization. Although the cost of enzymes has been significantly reduced over the past decade, it is still considerable and importantly contributes to the overall cost of the bioconversion process. In addition to the persistent efforts to reduce the enzymes costs, enzyme technology has also focused on reduction of enzyme loadings used for the enzymatic breakdown of biomass sugar polymers. This could be done either by improving properties such as activity, thermal stability, and pH stability or by optimizing the amounts and proportions of the different enzyme activities in commercial preparations [21]. Low enzymatic hydrolysis yields and rates have a great impact on the overall cellulose-to-ethanol conversion efficiency and thus have been identified as bottlenecks of bioethanol production processes during the past decade. Due to the inefficiency of the enzymatic preparations used, one-third of the total sugar product from enzymatic hydrolysis has been identified as oligomers or polymers that cannot be used by *S. cerevisiae* [52]. Van Dyk and Pletschke [53] highlighted the fact that a large number of enzymes like endoglucanase, exoglucanase, β-glucosidase, xylanase etc are required to bioconvert lignocellulose's carbohydrates into monomeric sugars effectively. Furthermore, the hydrolysis activity of multiple enzyme combinations working cooperatively together can be higher than adding individual enzymes, with the hydrolysis yield depending on the specific characteristics of the enzymes involved, their ratios and the characteristics of the substrate. A glimpse of enzymatic saccharification of different agro-wastes is shown in Table 13.2.

Table 13.2 Enzymatic saccharification of different pretreated feed stocks using cellulolytic enzymes.

Enzyme/Organism employed	Lignocellulosic biomass	Result	References
Commercial cellulase (20 IU/ml), β-glucosidase (In-house, 10 IU/ml) and xylanase (In-house, 5000 IU/ml) by *Aspergillus* sp.	Sodium hydroxide pre-treated rice straw	The cocktail containing the three enzymes resulted a maximum recovery of 574.8 mg/g of total reducing sugars	[54]
Commercial enzyme Accellerase 1500 (26 U/g)	0.75% H2SO4 at 100 °C for 2 h treated wheat Straw	Recovery of 45.6 g/L reducing sugars	[55]
Commercial cellulase (Celluclast 1.5 L, 15 FPU/g) supplemented with β-glucosidase (Novozym 188, 15 IU/g)	Sugarcane bagasse pre-treated with dilute acid and organosolv	29.1 g glucose/100 g sugarcane bagasse	[56]
NS22146 enzymes (Novozymes) 1.67%, 3.33% and 6.66% (genzyme/g glucan×100)	Sulphuric acid pre-treated empty fruit bunches	81.4% xylan and 74.8% of glucan	[57]
Cellulase from *Aspergillus fumigatus* ABK9 (30 FPU/g substrate),	Dilute sulfuric acid pretreated Water hyacinth	The maximum sugar yield (425.6 mg/g)	[58]
Cellic® CTec2 and Cellic® HTec2 from novozyme	10% sodium hydroxide treated empty fruit bunch at temperature 150°C during 30 minutes.	4.74% of ethanol in 72 hours fermentation	[59]

13.8 Fermentation

Fermentation of lignocellulosic hydrolysates involves the conversion of sugars to ethanol, mainly performed by bacteria or yeast. The most commonly employed technologies in fermentation include SHF, simultaneous saccharification and fermentation (SSF), and consolidated bio processing (CBP).

13.8.1 Separate Hydrolysis and Fermentation (SHF)

Separate hydrolysis and fermentation is the traditional method for bioethanol production. In SHF, the hydrolysis is completed in one reactor and fermented in other reactor. This transfer that is being done after hydrolysis increases the cost of reaction and consumes a lot of time. Typically *S. cerevisiae* is a prevalent microorganism used for this purpose, because it can grow and vigorously ferment in media containing as much as 40% (2.2 M) of glucose with the production of high amount of ethanol [60]. But the major weakness *of S. cerevisiae* is its incapability to ferment other than hexose sugars and thus interest for versatile-acting microorganisms are increasing. Industrial utilization of lignocelluloses for efficient bio-ethanol production is delayed due to lack of ideal microorganisms, which can efficiently ferment both pentose and hexose sugars available from the hemicellulose fraction and sustain under inhibitory conditions [21]. However, some attempts have been made recently to overcome this problem. A recombinant *Escherichia coli* strain has been reported which is able to ferment both xylose and glucose produced from wheat straw hydrolysate and achieved 0.47 g ethanol g^{-1} available sugars, but with low volumetric productivities, due to the low tolerance to ethanol [61]. Jahnavi *et al.* [29] reported that some microorganisms like *Pichia stipites* and *Candida tropicalis* are also able to ferment pentose mainly xylose into ethanol. Suitability of some microbial strains for ethanol production is shown in Table 13.3.

13.8.2 Simultaneous Saccharification and Fermentation (SSF)

Simultaneous saccharification and fermentation (SSF) processes, firstly described by Takagi *et al.* [69], combine enzymatic hydrolysis of cellulose with simultaneous fermentation of the obtained sugars to ethanol and are one of the most promising process option for bioethanol production from lignocellulosic materials. SSF is more preferred when compared with SHF because of low cost, reduced risk of contamination, less sugar degradation to its inhibitory compounds and can be done in a single vessel.

Table 13.3 Advantages and limitations of organisms used in lignocellulosic-based bioethanol fermentation.

Microorganism	Characteristics	Advantages	Limitation	References
Saccharomyces cerevisiae	Facultative anaerobic yeast	Widely adapted to ethanol fermentation. High alcohol yield (90%) with high tolerance to ethanol (up to 10% v/v). Amenability to genetic modifications.	Unable to ferment xylose and arabinose Sugars. Unable to survive high temperature of enzyme hydrolysis.	[62–68]
Zymomonas mobilis	Ethanologenic Gram negative bacteria	High ethanol productivity (five-fold more than S.cerevisiae volumetric productivity). High ethanol tolerance (upto 14% v/v). Amenability to genetic modifications. Does not require additional oxygen.	Not able to ferment xylose .Low tolerance to inhibitors. Neutral pH range.	
Candida shehatae	Micro-aerophilic yeast	Xylose fermenter.	Low tolerance to ethanol and low yield of ethanol. Unable to ferment xylose at low pH and require micro-aerophilic conditions.	

(*Continued*)

Table 13.3 Cont.

Microorganism	Characteristics	Advantages	Limitation	References
Pichia stiplis	Facultative anaerobic yeast	Good xylose fermenter with high ethanol yield (82%). Able to ferment most of cellulosic-material sugars including glucose, galactose and cellobiose. Possess cellulose enzymes favorable to SSF process.	Intolerant to high concentration of ethanol above 40 g/l. Unable to ferment xylose at low pH. Sensitive to chemical inhibitors and requires micro-aerophilic conditions to reach peak performance. Re-assimilates formed ethanol.	
Pachysolen tannophilus	Aerobic fungus	Ferment xylose.	Low yield of ethanol and does not ferment xylose at low pH. Require micro-aerophilic conditions.	
Kluveromyces narxianus	Thermophilic yeast	Able to grow at high temperature above 52°C, hence reduces chance of contamination. Ferment abroad spectrum of sugars. Amenability to genetic modifications.	Excess of sugars affect its alcohol yield. Low ethanol tolerance, fermentation of xylose is poor and leads mainly to the formation of xylitol.	

Wyman *et al.* [70] showed that using SSF process could increase yields and concentration of ethanol with less capital investment. Enzyme hydrolysis is generally carried out at 45–50 °C, therefore to carry out fermentation simultaneously with hydrolysis there should be a yeast strain that is equally capable of tolerating that temperature. Dahnum *et al.*, [59] also considered SSF method as a better process than SHF due to rapidly ethanol production and the highest concentration of produced ethanol. In such cases use of thermotolerant yeast strains would be more suitable and generally preferred. Saini *et al.* [71] reported that some of the thermotolerant yeast strains like *Kluyveromyces marxianus, Saccharomyces uvarum, Candida brassicae, Candida lusitaniae* are usually employed for SSF. A thermotolerant yeast strain *K. marxianus* DBTIOC-35 was isolated by which SSF was carried out at 42 and 45 °C using wheat straw as substrate. Maximum ethanol concentration of 29.0 and 16.1g L^{-1} were achieved, corresponding to the ethanol yields of 73% and 40.5% at 42 and 45 °C respectively. SSF was performed using *K. marxianus* and commercial cellulase on soybean cake and corn cobs. This produced maximum ethanol of 5.68g L^{-1} on corncob and 2.14g L^{-1} on soybean cake after 48 h of incubation (Meng *et al.*, 2010). Dahnum *et al.* (2015) conducted SSF by using NaOH pretreated empty fruit bunch and was digested with Cellic® CTec2 and Cellic® HTec2. When steam exploded, duckweed was subjected to SSF using an enzyme blend of Cellic CTec 2 (0.87 FPU g^{-1} substrate) together with Novozyme 188 (2 U g^{-1} substrate), has resulted in the ethanol production of 80% at a substrate concentration of 1%. The ethanol yield has reduced with an increase in substrate concentration. Upon fermentation of this mixture using S. cerevisiae have resulted in ethanol concentration of 97% at 50 °C in 24 h. Chu *et al.* [72] performed a three stage SSF using *S. cerevisiae* DQ1, a thermotolerant strain with high quantum of substrate (corn stover) loading at a rate of 30% that finally resulted in the ethanol yield of and 65.6%. Ruiz *et al.* [73]also performed SSF process for bioethanol production from hydrothermal pretreated wheat straw using a thermotolerant strain of *Saccharomyces cerevisiae* CA11. Results showed that the maximum ethanol concentration (14.84 g L^{-1}) were obtained at 45 °C, 3% substrate and 30 FPU of enzyme loading, corresponding to an ethanol yield of 82.4%, demonstrating a low enzyme inhibition and a good yeast performance during SSF process.

13.8.3 Consolidated Bioprocessing

Consolidated bioprocessing (CBP) is the most promising and potential strategy which includes enzyme production, saccharification, and fermentation into a single reactor, for effective production of ethanol from lignocellulosic materials. In this process, single or combined consortia of microorganisms are commonly used to ferment cellulose directly to ethanol [74]. The application of CBP requires no resources investment for purchasing enzyme or its production. But, CBP is not an efficient process because of poor ethanol yields and long fermentation periods (3–12 days) [75]. So, there is a requirement of highly engineered microbial strain that is capable of hydrolyzing biomass with enzymes produced on its own and producing high ethanol titer within a short period. CBP becomes feasible when an engineered CBP microorganism or microbial consortium could be developed [76]. CBP is expanding its recognition as a promising leap forward to produce bioethanol with low cost, yet its feasibility extraordinarily relies upon ethanol yield, fermentation period and whether a suitable microorganism can be found in nature or built by engineering strategies in the laboratory [77]. Genetic engineering approaches in this area like integration of genes from a cellulase producing strain into the ethanol producing strains like *S. cerevisiae* was found to be inappropriate since transfer of very high number of these genes might influence the execution of cell, their co-expression at the transcriptional level is often lop-sided causing ER-stress to the host cell [77]. Although the fungi like *T. reesei* are found to be efficient producers of cellulase, they are not broadly being proposed as possibility for CBP applications because of low ethanol yields obtained, as well as the slow fermentation rates. However, few fungi like Mucor have the ability to be fermenting lignocellulosic components to ethanol [78]. *Fusarium oxysporum* is the best contemplated filamentous organism for CBP applications with cellulolytic and hemicellulolytic properties to the enhancement of its CBP execution through Genetic engineering methodologies [79, 80]. The ascomycete Paecilomyces variotii (ATHUM8891) was evaluated as a candidate species in CBP applications. The fungus is capable of fermenting glucose and xylose toethanol, closer to the maximum theoretical yields, edifying an unusually powerful pentose metabolic pathway and the fungus possesses the necessary enzyme factory for the exploitation of lignocellulosic biomass, as it can grow and produce ethanol on commonagro-industrial derivatives. One of the major disadvantages of CBP is that the saccharification and fermentation, usually carriedout at ambient temperatures but the hydrolysis by cellulases is usually higher at higher temperature. This drawback could be avoided if thermophilic microbes could be utilized as a host for consolidated bioprocessing since it allows saccharification and fermentation to be carried out at higher temperature.

13.9 Concluding Remarks

In the next decades, biomass will be the most meaningful renewable energy source as an alternative to fossil fuels. Agro-residues biomass has been proposed to be one of the main renewable resources for cost-effectively attractive bioethanol production. Bioethanol production will be probably the most successful biofuel because it has plenty of usable forms (heat, power and electricity or vehicle fuel). Though technological advances and research efforts are still progressing, multiple configurations of systems and techniques are developed in order to design efficient, sustainable and economically feasible bioethanol production technologies and confront issues concerning the feedstocks and operations costs. The processes of pre-treatment, enzymatic hydrolysis, fermentation and distillation are the four major obstacles in bio-ethanol production and are required to overcome by efficient technology. Extensive research has been carried out in order to increase fermentable carbohydrate recovery, decrease inhibitors produced from sugar degradation during pre-treatment process, diminish utilization of chemical materials and energy input, produce valuable by-products and decrease cost of bioethanol process. Beside effective pre-treatment process, recent advances in functional genomics, metagenomics, genetic and metabolic engineering imply that the future of economic bioethanol production form biomass will strongly depend on achievements in artificially designed plants, containing high levels of cellulose while capable of producing hydrolases. In the last, it may be supposed that to solve the technology bottlenecks of the conversion process, novel science and efficient technology are to be applied, so that bioethanol production from agro-residues may be effectively developed and optimized in the near future.

References

1. Pandey, A., Srivastava, N., Sinha, P., Optimization of hydrogen production by *Rhodobacter sphroides* NMBL-01. *Biomass Bioenergy.*, 37, 251–256, 2012.
2. Kurian, J. K., Nair, G. R., Hussain, A., Vijaya Raghavan, G. S., Feedstocks, logistics and pretreatment processes for sustainable lignocellulosic biorefineries: a comprehensive review, *Renew. Sust. Energ. Rev.*, 25, 205–219, 2013.
3. Demirbaş, A., Combustion characteristics of different biomass fuels. *Prog. Energy Combus. Sci.*, 30(2), 219–230, 2004.
4. Chovau, S., Degrauwe, S., Bruggen, B. V., Critical analysis for techno-economic estimate for the production cost of lignocellulosic bioethanol production. *Renew. Sust. Energ. Rev.*, 26, 307–321, 2013.

5. Jordan, D. B., Bowman, M. J., Braker, J. D., Dien, B. S., Hector, R. E.. Lee, C. C., Mertens, J. A., Wagschal, K., Plant cell walls to ethanol. *Biochem. J.*, 442(2), 241–252, 2012.
6. Lynd, L., Greene, N., Dale, B., Laser, M., Lashof, D., Wang, M., Wyman, C., Energy returns on ethanol production. *Science,* New York, N.Y. 312(5781), 1746–1748, 2006.
7. Schmer, M. R., Vogel, K. P., Mitchell, R. B., Perrin, R. K., Net energy of cellulosic ethanol from switchgrass. *Proc. Natl. Acad. Sci.,* 105(2), 464–469, 2008.
8. Hsu, D.D., Inman, D., Heath, G. A., Wolfrum, E. J., Mann, M. K., Aden, A., Life cycle environmental impacts of selected U.S. ethanol production and use pathways in 2022. *Environ. Sci. Technol.,* 44(13), 5289–5297, 2010.
9. Lynd, L. R., Laser, M. S., Bransby, D., Dale, B. E., Davison, B., Hamilton, R., Himmel, M., Keller, M., McMillan, J. D., Sheehan, J., Wyman, C. E., How biotech can transform biofuels. *Nat. Biotechnol.,* 26(2), 169–172, 2008.
10. Feria, M., Rivera, J. A., Ruiz, R., Grandal, E., Domínguez, J. C. G., Pérez, A., López, F., Energetic characterization of lignocellulosic biomass from southwest Spain. *Int. J. Green Ener.,* 8(6), 631–642, 2011.
11. Lynd, L. R., Overview and evaluation of fuel ethanol from cellulosic biomass: Technology, economics, the environment, and policy. *Annu. Rev. Energy. Env.,* 21, 403–465, 1996.
12. Balat, M., Balat, H., Öz, C., Progress in bioethanol processing. *Prog. Energy Combust. Sci.,* 34(5), 551–573, 2008.
13. Limayem, A., Ricke, S. C., Lignocellulosic biomass for bioethanol production: Current perspectives, potential issues and future prospects. *Prog. Energy Combust. Sci.,* 38(4), 449–467, 2012.
14. Angelici, C., Weckhuysen, B. M., Bruijnincx, P. C. A., Chemocatalytic Conversion of ethanol into Butadiene and Other Bulk Chemicals. *Chem. Sus. Chem.,* 6(9), 1595–1614, 2013.
15. Sun, J., Wang, Y., Recent Advances in Catalytic Conversion of Ethanol to Chemicals. *ACS Catal.,* 4(4), 1078–1090, 2014.
16. Gabrielle, B., Gagnaire, N., Life-cycle assessment of straw use in bio-ethanol production: A case study based on biophysical modelling. *Biomass Bioenergy.,* 32(5), 431–441, 2008.
17. Somerville, C., Youngs, H., Taylor, C., Davis, S. C., Long, S. P., Feedstocks for Lignocellulosic Biofuels. *Science.,* 329(5993), 790–792, 2010.
18. Ragauskas, A. J., *et al.,* Lignin Valorization: Improving Lignin Processing in the Biorefinery. *Science.,* 344(6185), 1246843, 2014.
19. Wyman, C. E., Ragauskas, A. J., Lignin Bioproducts to Enable Biofuels. *Biofuels Bioprod. Biorefin.,* 9(5), 447–449, 2015.
20. Michelin, M., Maria de Lourdes T., Application of lignocelulosic residues in the production of cellulase and hemicellulases from fungi. *Fungal Enzymes.* USA: CRC Press Taylor & Francis Group., 31–64, 2013.
21. Chiaramonti, D., Rizzo A. M., Prussi, M., Tedeschi, S., Zimbardi, F., Braccio, G., Viola, E., Pardelli, P. T., 2nd generation lignocellulosic bioethanol: is

torrefaction a possible approach to biomass pretreatment? *Biomass. Conv. Bioref.*, 1(1), 9–15, 2011
22. Conde-Mejía, C., Jimé nez-Gutié rrez, A., El-Halwagi, M., A comparison of pretreatment methods for bioethanol production from lignocellulosic materials. *Process Saf. Environ. Prot.*, 90(3), 189–202, 2012.
23. Sarkar, N., Ghosh S. K., Bannerjee S., Aikat, K., Bioethanol production from agricultural wastes: an overview. *Renew. Energy.*, 37(1), 19–27, 2012.
24. Swart, J. A. A., Jiang, J., Ho, P., Risk perceptions and GM crops: the case of China. *Tailoring Biotechnol. Soc. Sci. Technol.*, 3(3), 11–28, 2008.
25. Harel, A., Noritech seaweed biotechnology Inc. Algae World Conference. Rotterdam, NL., 2009.
26. Mood, S. H., *et al.*, Lignocellulosic biomass to bioethanol, a comprehensive review with a focus on pretreatment. *Renew. Sust. Energ. Rev.*, 27, 77–93, 2013.
27. Lin, Y., Tanaka, S., Ethanol fermentation from biomass resources: current state and prospects. *Appl. Microbiol. Biotechnol.*, 69(6), 627–642, 2006.
28. Badger, P.C., In: Jannick J., Whipsekey A., editors. Trends in new crops and new uses. *Alexandria, VA: ASHS Press.* 17–21, 2000.
29. Girio, F.M., Fonseca, C., Carvalheiro F., Duarte, L.C., Marques, S., Bogel-Lukasic, R., Hemicelluloses for fuel ethanol: A review. *Bioresour. Technol.*, 101(13), 4775–4800, 2010.
30. Kumar, R., Singh, S., Singh, O. V., Bioconversion of lignocellulosic biomass: biochemical and molecular perspectives. *J. Ind. Microbiol. Biotechnol.*, 35(5), 377–391, 2008.
31. Michelin, M., Ruiz, H. A., Silva D. P., Ruzene, D. S., Teixeira, J. A., Polizeli, M. L. T. M., Cellulose from lignocellulosic waste. *Polysaccharides.*, 475–511, 2011.
32. Kim, S., Dale, B. E., Global potential bioethanol production from wasted crops and crops residues. *Biomass. Bioenergy.*, 26(4), 361–375, 2005.
33. Grabber, J. H., Ralph, J., Hatfield, R. D., Model studies of ferulateconiferyl alcohol cross-product formation in primary maize walls: Implications for lignification in grasses. *J. Agri. Food Chem.*, 50(21), 6008–6016, 2012.
34. Demirbas, A., Bioethanol from cellulosic materials: A renewable motor fuel from biomass. *Energy Source.*, 27(4), 327–337, 2005.
35. Shuai, L., Questell-Santiago, Y. M., Luterbacher, J. S., A mild biomass pretreatment using [gamma]-valerolactone for concentrated sugar production. *Green Chem.*, 18(4), 937–943, 2016.
36. Doblin, M. S., Johnson, K. L., Humphries, J., Newbigin, E. J., Bacic, A., Are designer plant cell walls a realistic aspiration or will the plasticity of the plant's metabolism win out? *Curr. Opin. Biotechnol.*, 26, 108–114, 2014.
37. Wu, M., Yan, Z. Y., Zhang, X. M., Xu, F., Sun, R. C., Integration of mild acid hydrolysis in γ-valerolactone/water system for enhancement of enzymatic saccharification from cotton stalk. *Bioresour. Technol.*, 200, 23–28, 2016.
38. Khanna, M., Chen, X., Huang, H., Önal, H., Supply of cellulosic biofuel feedstocks and regional production pattern. *Am. J. Agr. Econ.*, 93(2), 473–480, 2011.

39. Zhu, J. Y., Pan, H. J., Woody biomass pretreatment for cellulosic ethanol production: Technology and energy consumption evaluation. *Bioresour. Technol.*, 101(13), 4992–5002, 2010.
40. Jönsson, L. J., Alriksson, B., Nilvebrant, N. O., Bioconversion of lignocellulose: Inhibitors and detoxificatiom. *Biotechnol. Biofuels.*, 6(1), 16, 2013.
41. Schell, D. J., Farmer, J., Newman, M., McMillan, J. D., Dilute sulphuric acid pretreatment of corn stover in pilot scale reactor. Investigation of yields, kinetics and enzymatic digestibility of solids. *Appl. Biochem. Biotechnol.*, 105(108), 69–85, 2003.
42. Saha, B. C., Cotta, M. A., Enzymatic saccharification and feremeritation of alkaline peroxide pretreated rice hulls to ethanol. *Enzyme Microb. Technol.*, 41(4), 528–532, 2007.
43. Koppram, R., Toma´ s-Pejo, ´E., Xiros, C., Olsson, L., Lignocellulosic ethanol production at high-gravity: challenges and perspectives. *Trends in Biotechnol.*, 32(1), 46–53, 2014.
44. Liu, Z. L., Slininger, P. J., Dien, B. S., Berhow, M. A., Kurtzman, C. P., Gorsich, S. W., Adaptive response of yeasts to furfural and 5-hydroxymethylfurfural and new chemical evidence for HMF conversion to 2,5-bis-hydroxymethylfuran. *J. Ind. Microbiol. Biotechnol.*, 31(8), 345–352, 2004.
45. Watanabe, I., *et al.*, Ethanol production by repeated-batch simultaneous saccharification and fermentation (SSF) of alkali-treated rice straw using immobilized *Saccharomyces cerevisiae* cells. *Bioresour. Technol.*, 123, 695–698, 2012.
46. Eggeman, T., Elander, R. T., Process and economic analysis of pretreatments technologies. *Bioresour. Technol.*, 96(18), 2019–2025, 2005.
47. Sun, Y., Cheng, J., Hydrolysis of lignocellulosic material for ethanol production: a review. *Bioresour. Technol.*, 83(1), 1–11, 2002.
48. Rabinovich, M. L., Melnik, M. S., Boloboba, A. V., Microbial cellulases (review). *Appl. Biochem. Microbiol.*, 38(4), 305–321, 2002.
49. Gusakov, A. V., Alternatives to Trichoderma reesei in biofuel production. *Trends Biotechnol.*, 29(9), 419–425, 2011.
50. Das, A., Paul, T., Halder, S. K., Jana, A., Maity, C., Das Mohapatra, P. K., Pati, B. R., Mondal, K. C., Production of cellulolytic enzymes by *Aspergillus fumigatus* ABK9 in wheat bran-rice straw mixed substrate and use of cocktail enzymes for deinking of waste office paper pulp. *Bioresour. Technol.*, 128, 290–296, 2013.
51. Banerjee, S., Mudliar, S., Sen, R., Giri, L., Satpute, D., Chakrabarti, T., Pandey, R. A., Commercializing lignocellulosic bioethanol: technology bottlenecks. *Biofuels Bioprod. Biorefining.*, 4(1), 77–93, 2003.
52. Lau, M. W., Dale, B. E., Cellulosic ethanol production from AFEX-treated corn stover using *Saccharomyces cerevisiae* 424A(LNHST). *Proc. Natl. Acad. Sci. U.S.A.*, 106(5), 1368–1373, 2009.
53. Van, D. J., Pletschke, B., A review of lignocellulose bioconversion using enzymatic hydrolysis and synergistic cooperation between enzymes – factors

affecting enzymes, conversion and synergy. *Biotechnol. Adv.*, 30(6), 1458–1480, 2012.

54. Leya, T., Binod, P., Ashok, P., Hydrolysis of pretreated rice straw by an enzyme cocktail comprising acidic xylanase from *Aspergillus* sp. for bioethanol production. *Renew. Energ.*, 98, 9–15, 2016.
55. Govumoni, S. P., Koti, S., Srilekha, Y. K., Venkateshwar, S., Rao, L. V., Evaluation of pretreatment methods for enzymatic saccharification of wheat straw for bioethanol production. *Carbohyd.Polym.*, 91(2), 646–650, 2013.
56. Mesa, L., González, E., Cara, C., González, M., Castro, E., Mussatto, S. T., The effect of organosolv pretreatment variables on enzymatic hydrolysis of sugarcane bagasse. *Chem. Eng. J.*, 168(3), 1157–1162, 2011.
57. Ryan, J., Bouzaa, Z. G., Evan, J. H., Screening conditions for acid pretreatment and enzymatic hydrolysis of empty fruit bunches. *Ind. Crop. Prod.*, 84, 67–71, 2016.
58. Das, A., Ghosh, P., Paul, T., Ghosh, U., Pati, B. R., Mondal, K. C., Production of bioethanol as useful biofuel through the bioconversion of water hyacinth (*Eichhornia crassipes*). *3 Biotech.*, 6(1), 70, 2016.
59. Dahnum, D., Sri, O. T., Eka, T., Muhammad, N., Haznan, A., Comparison of SHF and SSF processes using enzyme and dry yeast for optimization of bioethanol production from empty fruit bunch. *Energy Procedia.*, 68, 107–16, 2015.
60. Saito, H., Posas, F., Response to hyperosmotic stress. *Genetics.*, 192(2), 289–318, 2012.
61. Saha, B. C., *et al.*, Ethanol production from wheat straw by recombinant *Escherichia coli* strain FBR5 at high solid loading. *Bioresour. Technol.*, 102(23), 10892–10897, 2011.
62. Kumar, S., Singh, S. P., Mishra, I. M., Adhikari, D. K., Recent advances in production of bioethanol from lignocellulosic biomass. *Chem. Eng. Technol.*, 32(4), 517–526, 2009.
63. Zaldivar, J., Nielsen, J., Olsson, L., Fuel ethanol production from lignocellulose: a challenge for metabolic engineering and process integration. *Appl. Microbiol. Biotechnol.*, 56(1–2), 17–34, 2001.
64. Nigam, J. N., Ethanol production from wheat straw hemicellulose hydrolysate by Pichia stipitis. *J. Biotechnol.*, 87(1), 17–27, 2001.
65. Jorgensen, H., Effect of nutrients on fermentation of pretreated wheat straw at very high dry matter content by Saccharomyces cerevisiae. *Appl. Biochem. Biotechnol.*, 153(1–3), 44–57, 2009.
66. Jeffries, T. W., *et al.*, Genome sequence of the lignocelluloses-bioconverting and xylose fermenting yeast *Pichia. stipilis. Nat. Biotechnol.*, 25(3), 319–326, 2007.
67. Talebnia, F., Karakashev, D., AngelidikaI. Production of bioethanol from wheat straw. An overview on pretreatment, hydrolysis and fermentation. *Biores. Technol.*, 101, 4744–4753, 2010.

68. Gamage, J., Howard, L., Zisheng, Z., Bioethanol production from lignocellulosic biomass. *J. Biobased Mater. Bioenerg.*, 4(1), 3–11, 2010.
69. Takagi, M., Abe, S., Suzuki, S., Emerth, G., Yata, N. A., A method for production of ethanol directly from cellulose using cellulase and yeast. In: Ghose TK, editor. *Proceedings of bioconversion symposium, Delhi*, 551–571, 1977.
70. Wyman, C. E., *et al.*, CAFI, Consortium for Applied Fundamentals and Innovation. Comparative Sugar Recovery and Fermentation Data Following Pretreatment of Poplar Wood by Leading Technologies. *Biotechnol. Prog.*, 25(2), 333–339, 2009.
71. Saini, A., Neeraj, K. A., Anuja, S., Anita, Y., Prospects for irradiation in cellulosic ethanol production. *Biotechnol. Res. Int.*, 13, 2015.
72. Chu, D., Jian, Z., Jie, B., Simultaneous saccharification and ethanol fermentation of corn stover at high temperature and high solids loading by a thermotolerant strain *Saccharomyces cerevisiae* DQ1. *Bioenerg. Res.*, 5, 1020–6, 2012.
73. Ruiz, H. A., Silva, D. P., Ruzene, D. S., Lima, L. F., Vicente, A. A., Teixeira, J. A., Bioethanol production from hydrothermal pretreated wheat straw by a flocculating *Saccharomyces cerevisiae* strain – Effect of process conditions. *Fuel.*, 95, 528–536, 2012.
74. Hasunuma, T., Kondo, A., Consolidated bioprocessing and simultaneous saccharification and fermentation of lignocellulose to ethanol with thermotolerant yeast strains. *Proc. Biochem.*, 47(9), 1287–1294, 2012.
75. Szczodrak, J., Fiedurek, J., Technology for conversion of lignocellulosic biomass to ethanol. *Biomass Bioenergy.*, 10(5–6), 367–375, 1996.
76. Parisutham, V., Tae, H. K., Sung, K. L., Feasibilities of consolidated bioprocessing microbes: from pretreatment to biofuel production. *Bioresour. Technol.*, 161, 431–440, 2014.
77. Xu, Q.; Singh, A.; and Himmel, M. E.; Perspectives and new directions for the production of bioethanol using consolidated bioprocessing of lignocellulose. *Curr. Opin. Biotechnol.*, 20(3), 364–371, 2009.
78. Sharifia, M., Karimi, K., Taherzadeh, M. J., Production of ethanol by filamentous and yeast-like forms of Mucor indicus from fructose, glucose, sucrose, and molasses. *J. Ind. Microbiol. Biotechnol.*, 35(11), 1253–9, 2008.
79. Xiros, C., Christakopoulos, P., Enhanced ethanol production from brewer's spent grain by a *Fusarium oxysporum* consolidated system. *Biotechnol. Biofuels.*, 2(1), 4, 2009.
80. Anasontzis, G., *et al.*, Homologous overexpression of xylanase in *Fusarium oxysporum* increases ethanol productivity during consolidated bioprocessing (CBP) of lignocellulosics. *J. Biotechnol.*, 152(1–2), 16–23, 2011.

14

Oleaginous Yeasts: Lignocellulosic Biomass Derived Single Cell Oil as Biofuel Feedstock

Neha Bansal[1,2], Mahesh B Khot[1,2], Arijit Jana[1,2], Abhilek K Nautiyal[1,2], Tripti Sharma[1,2], Diptarka Dasgupta[1,2], Swati Mohapatra[3], Sanoj Kumar Yadav[3], Saugata Hazra[3,4,*] and Debashish Ghosh[1,2,*]

[1]*Biotechnology Conversion Area, Bio Fuels Division; CSIR-Indian Institute of Petroleum, Mohkampur, Dehradun, Uttarakhand India*
[2]*Academy of Scientific and Innovative Research (AcSIR)*
[3]*Department of Biotechnology & [4]Center of Nanotechnology, Indian Institute of Technology, Roorkee, Uttarakhand, India*

Abstract

Increasing demand for biofuels and oleochemicals is largely supported by plant-based oils. Among biofuels, the transesterification of plant oil yields corresponding mono-alkyl esters of long-chain fatty acids (e.g. methyl esters, FAME) as biodiesel. It is currently being produced on commercial scale from edible vegetable oils but their use as biodiesel feedstock is receiving criticism due to low sustainability, conflict with food for the utilization of arable land, with high water and fertilizer requirements resulting in high oil price. Moreover, the feedstock price is the principal governing factor for the economic viability of biodiesel market and accounts for 70–95% of the total biodiesel production cost. Hence, non-edible cost-effective feedstocks are now being studied to produce biodiesel in a sustainable and economical way to avoid previous drawbacks. The use of nonedible plant oils is also not feasible over use of land. Microbial lipids referred to as single cell oils (SCOs) accumulated by oleaginous yeasts have emerged as a potential complementary feedstock for biodiesel production. The applications of yeast lipids as an input for biodiesel would be competitive with other commodity type oils only by the use of

**Corresponding author*: saugata.iitk@gmail.com; dghosh@iip.res.in

Arindam Kuila and Vinay Sharma (eds.) Principles and Applications of Fermentation Technology, (263–306)

low-cost and renewable substrates such as lignocellulosic biomass as carbon and energy source for oleaginous yeasts. This chapter describes the bioprocess of SCO production from oleaginous yeast with focus on lignocellulosic pretreatment; genetic engineering and downstream processing.

Keywords: Oleaginous yeast; single cell oil; lignocellulosic biomass; biodiesel; genetic modification

14.1 Introduction

Chemistry has made a vast array of materials, medicines, fertilizers, and fuels available at low cost; however, it is based almost entirely on processes that convert petroleum derived resources into these highly valued products. During 20th century, cheap and abundant petroleum was readily available; however, it has become shockingly clear that these feedstocks will be priced very differently in the 21st century, as the declining supplies can no longer keep pace with an ever growing demand [113]. Furthermore, the dramatic climate change brought about by the increase in greenhouse gas (GHG) emissions due to uncontrolled fossil fuel combustion may result into extreme weather events, food crisis and human conflicts while threatening the world economy [63]. This necessitates the lookout for renewable sources for energy and fuels.

In this scenario, liquid biofuels such as bioethanol and biodiesel have arisen as an attractive alternative to the fossil fuels. Bioethanol produced from sugars derived from corn, sugar cane and biodiesel derived from edible oils of oleaginous plants such as soybean, canola, palm are the most used "first generation biofuels" worldwide [114]. Due to concerns over use of edible feedstocks for production, the first generation biofuels are being criticized globally giving rise to Food or Fuel issue [45]. The bioethanol obtained from non–food crops and biodiesel originating from non-edible oils are the so called "Second Generation Biofuels" as these are more sustainable because their production does not compete with food, have a favourable energetic balance and lead to greater reduction in GHG emission than in their first generation counterparts [57].

In case of second generation bioethanol, the residues from agro-forestry and wood industry are the source of raw material including those obtained from forest and agricultural harvest, sawdust, tipping, burdocks and smaller fragments from pulping and sawing [41]. However, bioethanol do require changes in the distribution chain and it increases the vapour pressure when used blended with gasoline at low concentration (i.e. 5%). These restrictions are limiting the incorporation of bioethanol blended with gasoline in new markets because of which drop-in biofuel are gaining popularity [185].

Biodiesel is an environmentally friendly alternative fuel to petrodiesel consisting of mono-alkyl esters of long-chain fatty acids (mostly fatty acid methyl esters, FAME) [144]. Today biodiesel can be used without any major modifications for the diesel engines in vehicles and it is compatible with the current fuel infrastructure. Biodiesel is currently being produced on commercial scale from vegetable oils such as rapeseed, soybean, sunflower, palm etc [114]. A variety of alternative non-food feedstocks are being developed for biodiesel such as animal fats, waste cooking oils, non-edible plant oils (e.g. Jatropha) [113]. However, competition of land for growth of non-edible plants may not be feasible and hence use of non-edible plant oils for biodiesel production still is not a viable proposition. Therefore, to meet the demand of the biodiesel industry, alternative sources of oil rich biomass have to be explored and developed.

14.2 Oleaginous Yeasts: A Brief Account

Microbial lipids accumulated by oleaginous microalgae, yeasts and molds referred to as single cell oils (SCOs), have emerged in the last decade as a potential complementary feedstock for biodiesel production [82, 92, 114]; Meng *et al.*, 2009; Kosa & Ragauskas, 2011). Oleaginous yeasts may provide unique platforms for the sustainable production of biodiesel due to some advantages over their plant and algal counterparts [1] such as-

- No competition with food production.
- Ability to grow in conventional microbial bioreactors.
- Rapid growth rates.
- Growth is unaffected by space, light or climatic variations.
- Ability to utilize sugars derived from lignocellulosic biomass (LCB).
- Easy to scale up.
- Amenable to metabolic engineering approaches, which can be utilized to enrich specific desired fatty acids within the oils.

Using oleaginous yeasts in integrated biorefineries has many advantages. Oleaginous yeast species are robust, can tolerate high sugar loading, and resist pretreatment inhibitors, are viable for multiple generations, versatile being able to utilize a range of different carbon and nitrogen sources.

Furthermore, the oleaginous yeast *Yarrowia lipolytica* has been regarded as "safe to use" industrial microorganism; thus, potential exists for other oleaginous yeast species to be regarded as safe for industrial manufacturing purposes [46]. Existing infrastructure from first-generation biorefineries could be used for oleaginous yeast cultivation, and fermentation vessels would need to be modified or replaced to include agitation and air sparging systems.

The oleaginous yeast genera include, but are not limited to, *Yarrowia*, *Candida*, *Rhodotorula*, *Rhodosporidium*, *Cryptococcus*, *Trichosporon*, and *Lipomyces*. Oleaginous yeasts can produce SCO heterotrophically from a variety of low-cost feedstocks such as agricultural residues, food waste streams, and industrial co-products. The maximum calculated theoretical yield of lipid from yeast for sugar consumed is for glucose - 320 g/ kg and xylose 340 g/ kg, and the practical yield are for glucose 200 to 220 g/ kg for xylose 230 g/ kg [121].

Since the accumulation of lipids by oleaginous yeasts is species and growth substrate specific, not all isolates can be used as a feedstock for biodiesel production. Therefore, careful selection of the strains, growth substrates and characterization of lipid composition need to be performed to ascertain their suitability for biodiesel production [71].

SCO as a commodity-type oil has been hindered by competition from oilseed crops, where costs range from approximately \$1.5–3/kg less. Although the use of SCO as a feedstock for biodiesel has received interest in recent years, high manufacturing costs prevent the stand-alone production of biodiesel from SCO [136]. A recent techno-economic analysis by [83] determined that the cost of yeast SCO would be \$3.4/kg, excluding feedstock costs or \$5.5/kg including the cost of glucose as a feedstock. As more emphasis is placed on advancement of integrated biorefineries, the possibility of incorporating a yeast SCO biochemical platform for renewable fuels, chemicals, power, and products may become a reality [128].

The following sections describe the yeast SCO biochemistry, challenges and approach towards its genetic modification to maximize lipid yield, its cultivation, recovery and finally conversion into fuel entity. Throughout the chapter, yeast lipid has been highlighted as an alternative feedstock for biofuel production.

Before that, a brief account of LCB and its deconstruction strategies have been highlighted in the next section, considering various aspects of utilizing its fermentable sugars to produce yeast lipid as biofuel feedstock.

14.3 Lignocellulosic Biomass and its Deconstruction

Lignocelluloses refers to vegetal dry matter (biomass), so called LCB, may be classified primarily into forest residues such as wood [6]; agricultural residues such as rice straw [194], wheat straw [152], cane bagasse [147], corn cob [5], corn stover [165]; industrial residues such as pulp and paper processing waste, and energy crops such as switchgrass [107]. It is composed of carbohydrate polymers (cellulose, hemicellulose), and an aromatic polymer (lignin). Biomass is increasingly recognized as a valuable commodity, since it is an alternative to petroleum for the production of biofuels and chemicals.

LCB is a carbohydrate polymer that contains different sugar monomers (six and five carbon sugars) and they are tightly bound to lignin in a 4:3:3 [108] ratios wherein the individual composition varies significantly (Table 14.1) based on climatic conditions, agricultural practices, and other environmental considerations [32]. The remainder contains a small amount of pectin, nitrogenous materials, ash and other extractives [65, 87, 117]. This polymer has evolved to resist degradation and to confer hydrolytic stability and structural robustness to the cell walls of the plants. This robustness or "recalcitrance" is attributable to the cross-linking between the polysaccharides (cellulose and hemicellulose) and the lignin via ester and ether linkages (USDEOS, 2006). Ester linkages arise between oxidized sugars, the uronic acids, and the phenols and phenylpropanols functionalities of the lignin. To extract the fermentable sugars, one must first disconnect the celluloses from the lignin, and then use acid or enzymatic methods to hydrolyze the newly freed celluloses to break them down into simple monosaccharides which are called cellobiose units. Lignin is reported to constitute of some alcoholic substances such as: *p*-coumaryl alcohol, coniferyl alcohol, snapyl alcohol etc. Another challenge to biomass fermentation is the high percentage of pentoses in the hemicellulose, such as xylose, or wood sugar. Other than xylose there are arabinose, galactose, mannose, rhamnose etc. also in hemicelluloses. These different types of pentose and hexose sugar moieties are linked to form galactomannan and gucuronoarabinoxylan like units in hemicelluloses. Unlike hexoses such as glucose, pentoses are difficult to ferment (Figure 14.1).

One of the greatest problems that the world is facing today is that of environmental pollution, increasing with every passing year and causing grave and irreparable damage to the earth [66]. Different international unions worldwide are conscious about the impact of environment started to increase their initiatives in identifying environment friendly materials which are not synthetic and primarily based on natural resources. In this

Table 14.1 Composition of various LCB raw materials.

Biomass	Category	Hemicellulose	Cellulose	Lignin	Silica	Extractives	Reference
		%	%	%	%	%	
Eucalyptus	Forest residue	11–19	45–55	25.9–31.3	0.1–0.25	1.96–2.8	[35]
Aspen		14.6–19	49–52.4	19–26.7	0.07	0.95	[80]
Birch wood		23–26	40–45	18–23		0.3	[38]
Poplar		16–23	42–49	21–29	0.02	0.6–2.7	[146]
Corn cob	Agricultural residue	35–45	38–45	11.9–15	0.13–1.1	4.8–7.3	[68, 126]
Corn stover		25.3–26	32.7–38	15.3–19	3.9–8	9.7–11.23	[154, 194]
Rice Straw		19–27	32–47	5–24	11.2–13	19–23	[118, 173]
Sweet sorghum		24–27	39–45	21–23	0.8	0.8–1.40	[4]
Sugarcane Bagasse (SCB)		25.9	35–42	14–20	2–3	1.9–3.5	[74, 154]
Wheat Straw		20–25	33–40	15–20	3–4.35	7–9.51	[7, 102]
Switchgrass	Energy crop	10–40	30–50	5–20	1.1–3.2	0.4–4.5	[58, 102]
Miscanthus		17–20	39–44	18–22	0.55–2.4	3.2–6	[96, 180]

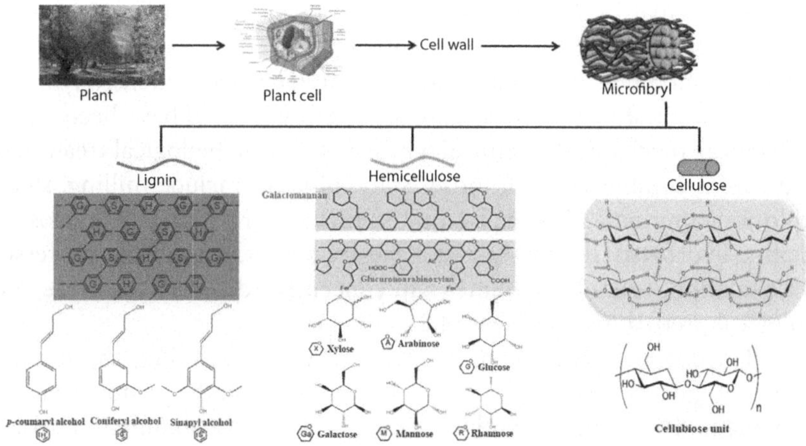

Figure 14.1 The main components and structure of LCB from the plant source.

senario, alternative straeties are urgently needed so that sustianable polymers could be developed from renewable natural sources. This would significantly reduces our present dependence on petrifacted resources and balacing between the consumption and producion rate of carbon dioxide. It could be easy to predict that biomass and biomass derived materials would be one of the most potential and highly promising alternative [64, 190]. Biomasses including agro masses like straw, baggasse etc. are synthesized by utilizing atmospheric carbon dioxide, natural water and sunlight through the process of photosynthesis.

LCB is considered as one of the most demanding feedstock for energy production because it is a renewable and widely-distributed energy resource, and can be developed sustainably in the future with net zero carbon emission [131, 191]. Worldwide many scientists have believed that LCB derived fuel chemicals (second generation cellulosic fuels) are the sustainable energy source to satisfy both energy replacement. [99, 159, 162]. Furthermore, it has positive environmental properties, such as low sulfur and nitrogen content (relative to coal) which allows LCB usage more eco-friendly than the fossil fuels having higher sulfur and carbon contents. With this expectation, over the past few decades, biorefining of lignocellulose feedstock is attracting worldwide interest. This lignocellulose feedstock can be utilized to produce biofuels, biomolecules and biomaterials [3, 13, 160].

To derive chemicals from LCB based on sugar platform [8, 56], deconstruction of the feedstock by the pre-treatment methods is essential before it can be effectively utilized. These processes alter the bonding characteristics of the individual LCB components by significantly modifying the supramolecular structure of the polymeric matrix [25].

Various methods of biomass fractionation have been extensively investigated in the last decade by researchers worldwide and have been classified such as mechanical, chemical, physicochemical, biological treatments or different combinations of those. Some of them include milling, steam explosion, ammonia fiber explosion (AFEX), supercritical CO_2 explosion, alkaline hydrolysis, liquid hot-water pre-treatment, organosolv processes, wet oxidation, ozonolysis, dilute- and concentrated acids hydrolyzes, and biological pretreatments (Table 14.2).

LCB hydrolysis generates a mixture of sugars including the hexoses such as D-glucose, D-galactose, and D-mannose, and the pentoses such as D-xylose and L-arabinose, and uronic acids. The most abundant hexose is glucose while xylose is typically the principal pentose. Arabinose fraction found in LCB hydrolysates can be significant depending on the materials and the process, e.g., a dilute acid hydrolysate of sugar cane bagasse used for fermentation which contain 75.7 g/L xylose, 13.5 g/L arabinose (with mannose) and 13.2 g/L glucose [100]. On the other hand, the sulfite cooking of spruce resulted in the liquor containing 34.3% arabinose, 25.5% xylose and 4.4% glucose [178].

SCO production from oleaginous yeasts has been demonstrated using biomass hydrolysates as the feedstock (Table 14.3) but is deficient of high lipid coefficient and excessively high costs as well [61, 62, 183].

Y. lipolytica Po1g cultivation was reported on sugarcane bagasse hydrolysate that gave lipid titer of 6.68 g/L corresponding to 58.5% lipid content [171]. Tsigie and coworkers [170] have also demonstrated that *Y. lipolytica* could produce up to 48.02% lipid on defatted, detoxified rice bran hydrolysate. Sorghum bagasse hydrolyzate, another lignocellulosic residue, was utilized by *Cryptococcus curvatus* to produce 0.19 g/g neutral lipid/gram of sugars (glucose and xylose) [94]. In another study, *C. curvatus* was found to accumulate 33.5% (w/w) lipid when grown on wheat straw pretreated with dilute sulfuric acid [183]. Corn cob hydrolyzate has been used for lipid production by *Trichosporon cutaneous* [26] and *Lipomyces starkeyi* [60] with lipid contents of 36% and 47%, respectively. The lipid coefficient was 17.4% (w/w) in case of *T. cutaneum* and 20.9% in case of *L. starkeyi*. Corn cob hydrolyzed by cellulase obtained from *Trichoderma reesei* was used for lipid production using *Trichosporon dermatitis* which accumulated up to 40.1% lipid content resulting in 9.8 g/L lipid concentration corresponding to 16.7% lipid content on dry weight basis (w/w) [59]. Corn cob residue which is the cellulose-rich residue left over after xylan extraction from corn cobs for xylitol production has also been used for lipid production [40]. It

Table 14.2 Biomass hydrolysis strategies.

Pretreatment technique		Mechanistic action	Recovery of fermentable sugars	Inhibitor formation	Energy requirement	Flexibility to biomass feedstock	Reference
Physical	Chipping, milling, and grinding	Chipping, grinding to reduce cellulose crystallinity	-	-	Low	Yes	[161]
Physico-chemical	Mild acid and steam hydrolysis	Selective hydrolysis of hemicellulose	High	Low/High (depends on process conditions)	Low	Yes	[115, 156]
	AFEX	Primarily removes lignin and minor removal of hemicellulose Partial decrystallization of cellulose	Low	Low	Low	No	[9, 85]
	CO_2 explosion	Removal of hemicellulose and Decrystallization of cellulose	Low	Low	High	No	[24]
	Ozonolysis	Removal of lignin	Low	No	Low	-	[14]

(*Continued*)

Table 14.2 Cont.

Pretreatment technique		Mechanistic action	Recovery of fermentable sugars	Inhibitor formation	Energy requirement	Flexibility to biomass feedstock	Reference
Chemical	Liquid hot water	Solubilizes hemicellulose and lignin; Minor cellulose hydrolysis	High xylose recovery	Low	High	Yes	[75, 125]
	Concentrated acid	Hydrolysis of hemicellulose and cellulosic fractions	High	High	Low	Yes	[179]
	Organosolv	Extraction of lignin employing organic solvent or mixture of solvents in combination with water, Removal of hemicellulose	Low/High Depending on operational scheme	Solvent used may be inhibitory to saccharification, growth and fermentation	High	Yes	[72, 105]

Chemical	Alkaline process	Removal of lignin, Decrease in the DOP of cellulose	High	Low	Low	No	
	Wet oxidation	High degree of solubilization of hemicellulose and lignin	High	Low	High	Yes	[11, 12]
	Ionic liquids	Hydrolysis of cellulose	Low	Solvent may be inhibitory to saccharifying cellulases and affect fermentation	Low	-	[19]
Biological	Microbial/ Enzymatic hydrolysis	Delignification and saccharification	High	Low	Low	Yes	[127, 175]

Table 14.3 Biomass and lipid production by various oleaginous yeasts on different lignocellulosic wastes.

Yeast species	Substrate	Time	x	L	$Y_{P/x}$	Reference
		h	g/L	g/L	g/g	
Trichosporon fermentans	Rice straw hydrolysate	192	28.6	11.5	0.40	[61]
Trichosporon cutaneum	Corn stover hydrolysate	168	10.2	3.11	0.30	[62]
Yarrowia lipolytica	Sugarcane bagasse hydrolysate	96	11.4	6.68	0.59	[171]
Cryptococcus curvatus	Wheat straw hydrolysate	168	17.2	5.8	0.33	[183]
Rhodotorula graminis	Corn stover hydrolysate	60–80	48	16.3	0.34	[39]
Yarrowia lipolytica	Rice bran hydrolysate	-	10.75	5.16	0.48	[170]
Trichosporon cutaneum	Corn cob hydrolysate	120	38.4	12.3	0.32	[40]
Rhodosporidium kratochvilovae	*Cassia fistula* biomass	144	8.9	4.73	0.53	[123]
Lipomyces kononenkoae	Switchgrass hydrolysate	-	47.7	28.1	0.59	[157]
Rhodotorula mucilaginosa	Sugarcane bagasse hydrolysate	-	15.3	-	0.097	[73]

- = Not mentioned; x = Biomass; L = Lipid concentration / titre; $Y_{P/x}$ = Lipid per unit biomass was hydrolyzed enzymatically by a commercial enzyme preparation and used to grow *T. cutaneum* ACCC 20271 yielding a lipid content of 32.1%. The similar substrate after de-lignification was also used for lipid production by *Cryptococcus curvatus* ATCC 96219 with a lipid content of 44.36% [97]. In contrast, *Rhodotorula graminis* was grown on undetoxified corn stover hydrolysate and produced 34% lipid content with lipid productivity of 0.21 g/L/h [39]. *Rhodotorula mucilaginosa* TYJ15a was also shown to accumulate 52.9% (w/w) of lipid by fed-batch cultivation on cassava starch hydrolysate [91].

Not all oleaginous yeasts can utilize all of the lignocellulosic sugars efficiently. Particularly, the pentose sugars (e.g., xylose) making up the hemicellulosic fraction are not readily metabolized by some yeasts or so with weak efficiency. *Y. lipolytica* model organism for lipid accumulation is unable to naturally grow on xylose as the sole carbon source. Consequently, the xylose sugars act as a barrier for efficient conversion of whole biomass to lipids by oleaginous yeasts making it crucial factor to achieve favorable process economics. The production of yeast SCO from hemicellulosic hydrolysates is being explored recently with reports on *R. mucilaginosa, Rhodosporidium toruloides* with sugarcane bagasse as source of xylose rich stream [10, 21, 73]. However, most yeasts ferment glucose before consuming pentoses because of the glucose repression. This preferential utilization of glucose results in delayed fermentation time and lower productivity.

Therefore, co-fermentation of lignocellulosic sugars could achieve cost-effective conversion of LCB to microbial lipid. Significant effort has focused on identifying yeast species which can simultaneously consume the multiple sugars and efficiently use them for lipid accumulation. [54] reported production of similar biomass yields, lipid contents and fatty acids on glucose plus xylose mixed in varying proportions by *Candida curvata* D in a single stage chemostat culture. *Lipomyces starkeyi* gave 61.5% (w/w) lipid content through co-fermentation of glucose and xylose (2: 1 wt/wt) in an optimized medium [189]. *Trichosporon cutaneum* assimilated both the glucose and xylose, and simultaneously accumulated intracellular lipid up to 59 wt% with a lipid coefficient up to 0.17 g/g sugar. This values of assimilation has been recorded upon the cultivation on a 2:1 glucose/xylose mixture in a 3-liter stirred-tank bioreactor. Simultaneous utilization of glucose and xylose by the yeast was also seen during corn-stover hydrolysate fermentation with a lipid content and coefficient of 39.2% and 0.15 g/g sugar, respectively [58]. Yu and team [184] explored the co-utilization of glucose, xylose, and cellobiose by *C. curvatus* wherein the consumption of both xylose and cellobiose was repressed by glucose, while xylose and cellobiose could be simultaneously consumed at similar rates. Recently, notable simultaneous utilization of a mixture of the sugars - glucose, xylose and arabinose was shown by *Pseudozyma hubeiensis* but it took ten days [167].

To integrate biomass hydrolysis and lipid production processes and enhance the overall efficiency, [42] evaluated three different processes, namely, separated hydrolysis and enhanced lipid production (SHELP), simultaneous saccharification and lipid production (SSLP) and simultaneous saccharification and enhanced lipid production (SSELP). The SSELP process integrates the biomass hydrolysis step and an enhanced lipid accumulation step, to effectively convert lignocellulosic materials into lipids. Specifically, cells prepared in a nutrient-rich medium were inoculated at high dosage for lipid production in biomass suspension in the presence of hydrolytic enzymes without auxiliary nutrients [42]. In a recent study, ionic liquid-pretreated corn stover was converted into lipids by the oleaginous yeast *Cryptococcus curvatus* according to the SSELP process, and the lipid coefficient for the regenerated corn stover was 112 mg/g in the presence of adequate amounts of cellulase, xylanase, and β -glucosidase. More recently, it was found that *C. curvatus* had the unique feature of converting oligomeric sugars of biomass origin into lipids in the absence of exogenous cellulolytic enzymes and that oligomeric sugars were transported into *C. curvatus* cells. The oligomeric sugars are hydrolyzed by cytoplasmic enzymes. Special emphases were put on reducing enzyme loading as well as enabling enzyme recovery. Attempts were also made to use recycled enzymes from the SSELP process [43].

14.4 Biochemistry of Lipid Biosynthesis

Yeasts can accumulate lipids by two pathways depending on the carbon source. The *ex novo* pathway occurs in presence of hydrophobic substrates like fatty acids, triglycerides, and alkanes. In *ex novo* pathway, yeasts like *Yarrowia, Candida,* and *Torulopsis* utilize hydrophobic substrates (HS) by active transport and assimilation occurs in an unchanged or modified form at different rates for substrate fatty acids [119]. In the case of the *ex novo* pathway, lipid accumulation occurs along with cell growth is independent of extracellular nitrogen concentration [2]. When fats are the sole source of carbon and energy the principle enzymes (ATP-citrate lyase (ACL) and FAS) of the *de novo* pathway are not expressed, and metabolism of lipids does not proceed via acetyl-CoA [121]. Oleaginous yeasts modify the fatty acid composition of the hydrophobic substrate on which they are grown resulting in significantly altered fatty acid profile of the stored intracellular lipid [120].

The *de novo* pathway occurs when cultivation is carried out on sugar-based media with acetyl-CoA as the main intermediate in cellular metabolism [121]. Yeasts are grown in the presence of a high carbon concentration and nitrogen-limiting conditions, wherein, they utilize sugars like glucose, xylose and different wastes like molasses, crude glycerol, whey, etc. to accumulate intracellular lipid [137]. Under nitrogen-limiting conditions, citrate is excreted from the mitochondrial matrix into the cytoplasm where it is broken down to acetyl-CoA and oxaloacetic acid by the enzyme ACL [27]. The presence of ACL is an important marker to determine the oleagenicity of a given microorganism with the absence of the enzyme leading to an accumulation of citric acid rather than lipid [28, 121]. The exhaustion of phosphorous from the medium resulted in similar biochemical events as during nitrogen depletion. There was a decrease in mitochondrial isocitrate dehydrogenase activity due to the depletion of the allosteric activator adenosine monophosphate (AMP). Phosphate exhaustion induced AMP breakdown to release inorganic phosphate for other cellular processes. However, the cell number, as well as lipid free biomass, continued to increase until the carbon source exhausted [98].

Fatty acid synthesis (FAS) is a step by step addition of acetyl-CoA units (mostly as malonyl-CoA) finally resulting elongated fatty acid chain, for example, palmitate (C16 saturated). The biosynthesis of fatty acids, therefore, requires the constant supply of acetyl-CoA as initial biosynthetic unit, malonyl-CoA as the elongation unit and reducing equivalents in the form of NADPH [176].

The fatty acids produced by the FAS complex are then esterified in the glycerol backbone to form Triacyl Glycerols (TAG s) [27, 49]. The whole process is known as the storage lipid pathway, and its end products form the neutral lipid fraction of the cell, packed inside the LB [18].

SCO or neutral storage lipid accumulation is the biochemical process of storing carbon (often from sugars) as TAG. The theoretical yield of TAG synthesized from sugars is dependent on FA chain length and the degree of unsaturation. According to [134], the synthesis of one molecule (mol) of trioleoylglycerol ($C_{57}H_{104}0_6$) requires 16 mol of glucose or xylose. To characterize the production of lipid or oil, researchers often refer to the yield as the mass of oil produced per mass of dry cell weight ($Y_{P/x}$). Because oil is produced intracellularly, this "yield" is essentially the oil content of the dry yeast cell. Although this can be used to describe a microorganism's oil-accumulating ability, it does not reflect the chemical conversion yield derived from the stoichiometric balance described above. The yield of product per unit of substrate ($Y_{P/S}$), also known as the lipid coefficient, is a more suitable representation. One mole of glucose, when metabolized exclusively via glycolysis, generates two moles pyruvate; thus, it can be calculated that approximately15 mole glucose are needed to synthesize 1 mole triacylglycerol; i.e., 100 g glucose will provide maximally 32 g lipid, assuming that glucose is not used for the synthesis of any other product-which, of course, it is.

This calculation is valid if all the sugar that is taken up by the cell is utilized in lipid biosynthesis. The maximum theoretical yield of SCO from xylose is 0.34 g/g [98]. Similarly, glycerol which it is considered as the best substrate for many oil-producing organisms due to its availability as main byproducts of biodiesel has a maximum theoretical yield of 0.30 g/g. However, inside the cell, it does not hold true as the carbon flux is diverted towards other metabolites as well. Under the best growth condition (i.e., in a chemostat) the highest yields of lipid that have been obtained are 22 g/100 g glucose used [31, 134, 182].

14.5 Genetic Modification for Enhancing Lipid Yield

Genetic exploitation of microbial lipid pathway is being pursued to improve lipid production with an ultimate aim of increasing biofuel production. Numerous works have been reported in this direction with yeasts, fungi, algae as host organisms for modifications. Recently [155] reviewed the bottlenecks of yeast metabolic engineering to produce fatty acid derived biofuels. They have identified five factors related to yeast metabolic engineering, namely (1) FAB precursors' supply limitation, (2) cofactor supply limitation, (3) tight regulation of FAS pathway, (4) toxicity of fatty acid and its derivatives, and (5) lack of genetic information for oleaginous yeasts. However, lipid overproduction and qualitative improvement could be broadly achieved through two ways as discussed below.

14.5.1 Over-Expression of Key Metabolic Genes

Since the enzyme, acetyl CoA carboxylase encoded by the gene ACC catalyzes this first committed step of fatty acid biosynthesis. The **ACC1** enzyme, therefore, seems to be directly responsible for providing the malonyl-CoA for cytoplasmic FAS. The over expression of ACC1 from fungi, *Mucor rouxii* in non-oleaginous yeast, *Hansenula polymorpha* led to a 40% increase in total fatty acid content [88, 138]. On the other hand,

[163] achieved two folds increase in the lipid content in *Y. lipolytica* by over expression of endogenous ACC1. This two fold increase could be resulted in tight regulation of ACC, an assumption further strengthened by the observation that accumulation of acyl-CoA in acyl-CoA synthetase mutant decreased the ACC activity by eight folds in *Y. lipolytica* [15, 70]. A multiple gene approach involving co-expression of ACC with other genes such as FAS1 and FAS2 in the model yeast *Saccharomyces cerevisiae* resulted in a 17% increase the lipid accumulation [18]. The replacement of native promoters of the fatty acid biosynthesis genes with strong constitutive promoters, such as TEF1 promoter also increased the lipid production in *S. cerevisiae.* [18]. Studies have also been carried out to know the synergistic effect of ACC and **DGA** on biosynthesis of fatty acid. Co-expression of ACC1 and DGA1 in *Y. lipolytica* led to a five-fold increase in the overall lipid yield compare to non-engineered yeast [53, 163]. The enhanced lipid accumulation can be attributed to a better balance between the fatty acid and TAG synthesis pathways

ACL is a key enzyme for lipid accumulation in mammals and oleaginous yeasts and fungi that catalyzes the conversion of citrate to acetyl-CoA and Oxaloacetate [47, 89]. This enzyme is possibly catalyzes the rate-limiting step for lipid biosynthesis as the specific activity of ACL enzyme seems to correlate with the specific rate of lipid synthesis [22, 124]. A 1.7-fold increase in the productivities of fatty acids and 1.9-fold increase of TAG relative to the parental strain was observed in the ACL-enhanced expression in *Aspergillus oryzae* [130, 164].

DGAT catalyzes the last step of TAG formation to form TAG from DAG and fatty acyl CoA [52, 181]. The substrate of DGAT, DAG, could be allocated to either phospholipid biosynthesis or TAG formation. Overexpression of DGAT would commit more DAG to TAG formation rather than phospholipid formation suggesting that the reaction catalyzed by DGAT is an important rate-limiting step in lipid biosynthesis. Recently, DGA has been emphasized for its effect on growth and lipid synthesis in yeast [48, 50, 163]. Yeast transformed with the Arabidopsis DGAT led to a 3–9-fold increase in TAGs accumulation [23, 51]. In *Y. lipolytica*, three acyltransferases (DGA1/DGA2/PDAT) are involved in the final step of TAG biosynthesis, in conversion of diacylglycerol (DAG) into TAG. Transcriptomics analysis of these acyltransferases revealed that only DGA2 is expressed differently during lipid accumulation phase [111, 141].

In another strategy involving the GUT2 gene which catalyzes the dihydroxyacetone phosphate (DHAP) formation from G3P was deleted to boost G3P availability in *Y. lipolytica*, leading to a 3-fold increase in lipid accumulation compared to the wild-type strain [18, 142]. It was also understood from the studies that the overexpression of the Glycerol-3-phosphate dehydrogenase1 (GPD1) gene, catalyzing the conversion of DHAP to G3P, results in a fourfold increase in lipid accumulation. Above mentioned modifications combined with deletions of the acyl-CoA oxidase (POX) genes, nullify the effect of β-oxidation, resulting in recombinant yeast strain with a potential of accumulating more than eighty percent of its mass as lipids [139].

ME have been reported to be widely distributed in range of eukaryotic organisms, from fungi to mammals and participate in diverse metabolic pathways such as photosynthesis, energy metabolism and lipogenesis [44, 55, 110]. They are localized in range of subcellular locations, including the cytosol, mitochondria, and chloroplast. There are three types of malic enzymes have been reported according to the coenzyme specificity and ability to catalyze the decarboxylation of oxaloacetate, [174]; (1) *L-Malate:NAD$^+$ oxidoreductase* (*oxaloacetate decarboxylating; EC 1.1.1.38,* uses NAD+ as coenzyme); (2) *L-Malate: NAD+ oxidoreductase (decarboxylating; EC 1.1.1.39,* uses NADP+ as a coenzyme in some cases, but prefers NAD+) and (3) *L-Malate: NADP+-oxidoreductase* (*oxaloacetate decarboxylating; EC 1.1.1.40,* NADP+- dependent) catalyzes the oxidative decarboxylation of L-malate to pyruvate with the concomitant reduction of NADP$^+$ in presence of divalent cations (Mg^{2+} or Mn^{2+}) as cofactors. The NADP$^+$ dependent cytosolic malic enzyme (ME; EC 1.1.1.4 0) has been proposed to be the key supplier of NADPH for lipid biosynthesis in oleaginous yeasts. It is suggested that most of the oleaginous microorganisms are believed to have a concerted lipogenic metabolon complex between the ME, ACL and FAS to create a pathway for the metabolites (NADPH and acetyl-CoA) toward the FAS active sites.

Over-expression of this enzyme resulted in increase in lipid accumulation in some yeasts while in others it was found to be absent and did not increase lipid content to significant extents. Amongst one of the earlier studies a 2.5-fold increase in lipid accumulation in the oleaginous fungus, *M. circinelloides* was observed on over-expression of its malic enzyme when placed under the control of GPD1 promoter [188]. Similarly, *R. glutinis* over-expression of malic enzyme from *M. circinelloides* increased the lipid content from 19% of the biomass to 39% [93]. Thus, this has proved to be a good strategy of enhancing malic enzyme activity and concomitantly increasing lipid yield.

Some other strains were identified in which lipid production took place but without occurrence of NADP$^+$ dependent malic enzyme contradicting its role as sole provider of NADPH [166, 186]. This included strain of *Yarrowia lipolytica* which contained only one ME gene which is the mitochondrial form [33, 186]. The homologous over-expression of this ME (YALI0E18634g) did not result in any increase in the lipid content of wild type *Y. lipolytica*. This was attributed to ME's low affinity for NADP+ to provide NADPH and secondly its location in the mitochondria. It was therefore implied that ME might not play an important role in lipid production in *Y. Lipolytica* [16].

The above observation concluded that although ME plays critical part in most of species, it couldn't supply the required reducing prowess [135]. The other observation is that the NADPH provided by the Malate Dehydrogenase-Malic Enzyme cycle might not be essential for lipid accumulation, but it might be significant to maintain high lipid productivity [116]. Overall malic enzyme (ME) does play a role in the regulation of fatty acid biosynthesis. This is further supported from the stoichiometry of FAS which predicts that if PPP is the only provider of NADPH in presence of ME, predicted yield of TAG from glucose falls to 27.6% w/w from 31.6% [135, 140].

Investigation of the regulation of ME on lipid production in oleaginous microorganisms is expected to have good application prospect as it will influence the lipid production abilities of the strains. Although malic enzyme has been found not to be playing any role in the FAS of some yeasts yet it plays an important part in the lipid accumulation process of most other oleaginous microorganisms. If it could become possible to modulate the activity of malic enzyme in any way either by making biochemical changes or through genetic engineering, this would offer an economic advantage to the overall process of yeast lipid production. Malic enzyme activities can be influenced by optimization of culture conditions like oxygen concentration, carbon, and nitrogen source. On another hand, genetic engineering approaches for improved lipid productions have demonstrated ME as one of the most hopeful targets for gene transformation [140, 153]. ME over-expression or co-expression with other gene targets can increase lipid accumulation in transgenic strains. The aspect of genetic modification should be explored to greater extents because of its great effectiveness.

14.5.2 Blocking Competing Pathways

Decreasing lipid catabolism can be a complementary strategy to increase lipid accumulation for which the genes directly involved in β-oxidation of fatty acids can be a target for inactivation. Several reports have shown that knocking out genes involved in b-oxidation in *S. cerevisiae* not only led to increased amounts of intracellular free fatty acids but also results in extracellular fatty acid secretion in some instances [106, 150]. *Y. lipolytica* contains six AOXs, encoded by the POX1 to POX6 genes, which catalyze the limiting step of peroxisomal β-oxidation can accumulate lipids to levels exceeding 50% of cell dry weight [16]. A modification of the POX genotype is useful in preventing lipid degradation and therefore leads indirectly to an increase in lipid accumulation [17]. Elimination of peroxisomal oxidation pathway and engineering of glycerol metabolism showed 40–70% increase in lipid content by ex-novo lipid accumulation in *Y. lipolytica* [163]. Besides carbon flux provision of reducing equivalents is another strategy to enhance biofuel production. Over-expression of genes involved in this process has led to substantial increase in lipid production in some yeast. The Malic enzyme gene is considered the provider of NADPH in most oleaginous organisms and plays an important role in lipid accumulation.

14.5.3 Challenges in Genetic Engineering of Yeast

Although the biosynthetic biochemical pathways of oleaginous yeast are not very different from that of low oil containing yeasts, such as *S. cerevisiae*, only few of the high oil-producing yeast like *Y. lipolytica* are the model organisms for genetic engineering studies. Among the oleaginous fungus members of *Mortierella, Mucor* are the most extensively studied microbes. However, several other oleaginous yeasts such as *C. curvatus, L. starkeyi, R. toruloides, R. glutinis* are yet to be explored. Tools for genetic modification of these organisms need to developed and more genome information will be required to enable them for increased lipid yields. Moreover, although several genetic engineering processes to improve biodiesel production have yielded satisfactory results at laboratory scale level, commercialization requires scaling up of these processes to reach very high yields and productivities without losing performance, which is the greatest challenge in commercialization.

14.6 Fermentative Cultivation, Recovery of Yeast Lipids as SCO and Production of Biofuel

The first commercially viable SCO process was the production of an oil-rich in γ-linolenic acid (GLA) using oleaginous mold *Mucor circinelloides* [30]. Thus, microbial lipids have only been produced commercially for high-value unsaturated fatty acids so far; process development for bulk production is still in infancy.

Unlike filamentous fungi, yeasts can produce limited amounts of PUFA and high contents of stearic acid (18:0) can be observed by in SCOs of some strains. This fact was utilized for the large-scale (pilot) production of a cocoa butter equivalent (CBE) fat using wild-type strain of yeast, *Candida curvata* (now *Cryptococcus curvatus*). The process was scaled to 250 m^3 and a production process was demonstrated using oleaginous yeast that produced palm oil equivalent using lactose as feedstock originating from the cheese creamery processes in New Zealand [30]. Recently a successful pilot-scale process for yeast oil based biodiesel production was reported by [158] using yeast *R. toruloides* DEBB 5533. It involved fed-batch fermentation at 1000 L working volume and a low-cost medium composed by sugarcane juice and urea.

Submerged fermentation has been conducted for de novo lipid accumulation in shake-flask, batch-bioreactor, fed-batch bioreactor, and continuous culture modes using a large variety of substrates as carbon sources e.g. analytical/industrial sugars, cheese whey, molasses, waste glycerol, organic acids, lignocellulosic hydrolysates, municipal waste water, food waste hydrolysate [121]. The lipid yield and energy input was estimated by [103] for large-scale production of SCO for biodiesel from agricultural waste using lipid-accumulating yeast. The study evaluated both submerged and solid-state fermentation processes. Submerged fermentation a well-controlled cultivation system is suitable for yeasts while solid-state fermentation is the culture of usually a filamentous fungus, on a (wet) solid matrix without free water. A detailed techno-economic process evaluation the production of microbial oil or biodiesel was provided in case of the glucose-based media fermentation by the yeast strain *R. toruloides* [84]. The study emphasized the importance of the feedstock used and the fermentation stage where higher productivities and final yeast oil concentrations should be achieved.

The intra-cellular lipid synthesis and its storage via de novo lipid accumulation is a non-growth associated process, and thus SCO is a secondary fermentation product. A two-stage fermentation process is often employed to induce lipid accumulation. The first stage is targeted at biomass formation by promoting cellular growth while, a nutrient stress-response is induced during the second stage, mostly nitrogen limitation, causing metabolic shift to storage lipid synthesis and accumulation. The lipid accumulation process has been considered a combination of two different mechanisms: The first mechanism is a balanced cell growth phase, in which lipid accumulation is proportional to non-lipid biomass generation. The second one involves lipid biosynthesis independent from the production of non-lipid cell mass and was performed through the formation of a rate-controlling lipid intermediate between sugar and storage lipid describing, the nitrogen-limited phase (unbalanced growth phase) [121]. However, a very limited number of modeling studies of growth and lipid accumulation have been conducted so far on oleaginous yeasts using both hydrophilic and hydrophobic substrates.

The yeast lipid content is known to be dependent on some factors namely the strain and the species, growth phase, type and concentration of carbon source, nitrogen type and level, C/N ratio, pH temperature, phosphorus level, growth factors, trace metals. This information resource has been utilized for fermentation process parameter optimization aimed at increasing cell mass and lipid production from some oleaginous yeast species. Some representative studies are summarized in Table 14.4.

Oleaginous yeasts are known to accumulate different amounts of microbial oil with varying fatty acid profiles, depending on the substrate or growing conditions. Leiva-Candia *et al.* [90] evaluates the most relevant aspects regarding yeast oil production using agro-industrial waste as culture media and the potential of yeast oil as feedstock for biodiesel production.

The statistical optimization strategies have been used successfully for improving the SCO production in oleaginous yeasts using response surface methodology (RSM) [10, 59, 143]. RSM not only identifies the variables affecting a process but also specifies the levels of the variables that maximize product formation. It can also infer on the interactions that exist between the variables that affect the outcome. The mathematical process model generated by RSM may be used for prediction purposes, and its analysis could suggest ways for obtaining even higher production [109].

Recovery of biomass, extraction of lipids and fatty acid profiling are the major steps of whole downstream processing of oleaginous yeast-based bioprocess. Yeast lipids are intracellular and protected by rigid cell wall with other membrane systems. Yeast cell is known to possess a quite complex set of lipid molecular species (around 150) identified to date and includes membrane phospholipids, storage triacylglycerols, and sphingolipids. A total lipid analysis from yeast cells is a difficult task due to major differences in chemical nature and abundance, starting with cell breakage and extraction, and various qualitative and quantitative analysis levels [81]. There has been lack of a single method of choice that results in 100% lipid recovery. An ideal yeast lipid extraction method should allow

Table 14.4 Medium and process optimization for SCO production by oleaginous yeasts.

Yeast strain	Optimized conditions	x	L	Lipid content	Reference
		g/L	g/L	wt%	
***Rhodotorula glutinis* TISTR 5159**	Fed batch-3L glycerol - 9.5%; C/N- 85; Tween20 - 1.5 g/L pH- 6.0; aeration - 2 vvm	13.77	8.36	60.7	[143]
***Lipomyces starkeyi* AS 2.1560**	Batch - Shake flask glucose - 48.9 g/L; xylose- 24.4 g/L; yeast extract – 7.9 g/L; $FeSO_4$- 4 mg/L	20.5	12.6	61.0	[189]
***Lipomyces starkeyi* CBS 1807**	Batch - Shake flask C/N- 190; yeast extract	12	5.81	47.3	[101]
Cryptococcus curvatus	Fed batch -26 L removal of malt extract; addition of deoiled yeast lysate	50.4	-	45	[169]
Trichosporon fermentans	Batch- Shake flask C/N 165; Total sugar conc. 123.5 g/L; Inoculum 11% pH 7.6; Fermentation time 9 days	-	15.8	-	[59]

(*Continued*)

Table 14.4 Cont.

Yeast strain	Optimized conditions	x	L	Lipid content	Reference
		g/L	g/L	wt%	
***Pichia kudriavzevii* MTCC 5493**	Fed batch - 26 L Corn Steep Liquor - 20 g/L; Deoiled yeast autolysate -5 g/L; crude glycerol - 45 g/L; C/N- 5.14 to 200	32.1	-	16.6–23	[145]
***Rhodosporidium toruloides* A29**	Fed Batch 30L yeast extract- 2.5 g/L; $NaNO_3$ - 2.75 g/L; $MgSO_4$ -0.5 g/L; Glucose - 75 g/L	23.36	12.5	53.5	[148]

= Not mentioned; x = Biomass; L = Lipid concentration / titre.

comprehensive and quantitative determination of all the diverse molecular lipid species. Multiple extraction protocols have been developed over the years in the literature, and a great deal has been reported on laboratory scale extraction of lipids from yeasts. Jacob [67] provided the state of the art of the yeast-lipid extraction methodologies conducted in the laboratory, as well as pilot-plant operation. A given solvent system may not be able to extract all lipid classes present in yeast cell with the same efficacy, and thus, alternative procedures may need to be considered depending on the type of lipid under analysis. In other words, the extraction of total lipids or a specific component requires specific procedures. As yeast lipids are polar and non-polar, a suitable solvent system is essential for effective lipid extraction.

Biomass conditioning, solvent extraction, and lipid extract washing are the three general steps of yeast lipid recovery. Conditioning of the yeast biomass involves the treatment of the cells to make it accessible to efficient solvent contact and extraction by affecting cell wall permeability. These include treatment by enzyme, chemical (acid, alkali, and detergent) or any physical or mechanical means (ball milling, pressure extrusion, sonication). These conditioning methods result in breakage of the rigid cell wall and thus favor more accessibility to solvents for extraction without affecting the native state of the lipids.

The liquid-liquid extraction of lipids is carried out by one of the three methods widely cited in the literature [20, 37, 177]. Chloroform/methanol (2:1, v/v) is used as the extracting solvent in all three methods. The laboratory protocols based on glass bead disruption in the presence of organic solvents have been developed to quantitatively extract lipids from yeast cells [76, 151] and have been applied to oleaginous yeasts [71] for SCO recovery. Lipid extracts tend to trap water soluble non-lipid material, and such impurities are removed by washing with various aqueous salt solutions such as 0.88% KCl, 0.0034% $MgCl_2$.

Fatty acid composition of the lipid extract is typically performed by gas-liquid chromatography after transmethylation of lipid fatty acids to their respective volatile methyl esters. One of the advantage of using Yeast as a production system lies on the fact that, polyunsaturated fatty acids are absent in yeast lipids, which are susceptible to (per) oxidation. So, in case of yeast during the isolation procedure, no specific precautions need to be taken.

The conversion of yeast oil to biodiesel is accomplished through transesterification reaction. It is conducted in the presence of a suitable acid/base catalyst and alcohol (e.g. methanol/ethanol) yielding corresponding alkyl esters (e.g. FAME/ FAEE) along with glycerol as a side product. The glycerol produced alongside has other industrial applications. Acid catalysts are preferred when SCO is with high free fatty acid content. Lipase mediated biocatalytic transesterification is efficient and eco-friendly but yet commercially non-viable.

The methods investigated for yeast oil transesterification are based on either direct transesterification (without extraction of SCO from the yeast biomass) or conventional two-step extraction-transesterification where SCO is extracted and then transesterified. The techno-economic evaluation of biodiesel production from biomass of oleaginous yeast *R. toruloides* via two transesterification processes was recently investigated [84].

Earlier, [95] demonstrated the production of FAME with acceptable CN (56.4–63.5) and reported the SCO to FAME conversion yields higher than 90% (w/w) via direct acid-catalyzed transesterification of three oleaginous microorganisms including yeasts *Lipomyces starkeyi* and *R. toruloides*. Zhu *et al.* [193] reported the conversion of *Trichosporon fermentans* derived oil into FAME via base catalysis after removal of free fatty acids at a conversion yield of 92% (w/w). Recently, [73] reported direct acid catalyzed transesterification for estimating the total lipid content of *R. mucilaginosa* IIPL32 and to determine its biofuel potential. The commercialization of yeast SCO transesterification for biodiesel production is dependent on the development of fermentation processes with high values of lipid yield, productivity, and lipid titer.

Other than trans-esterification, selective catalytic deoxygenation of SCO can lead to production of renewable hydrocarbon while catalytic cracking and isomerization can further yield hydrocarbon of desired fuel range, i.e., gasoline, ATF or diesel.

14.7 Characterization of Yeast SCO: Implications towards Biodiesel Properties

The biodiesel properties have been included as specifications in different biodiesel standards (ASTM D6751; EN14214). Those include a range of physical properties such as density, kinematic viscosity, cetane number (CN), distillation range, flash point, pour and cloud point and chemical properties such as acid number, iodine value, ester content, copper corrosion, sulfur, and phosphorus content, total sulfur, oxidative stability.

Biodiesel composition reflects the fatty acid profile of the feedstock used and being a mixture of different fatty esters; each ester component contribute to the properties of the resulting fuel [77, 79].

Few key fuel properties are known to be directly dependent on the structure of its component fatty esters defined by chain length and degree of saturation such as ignition quality (CN), oxidative stability, kinematic viscosity, cold flow (cloud and pour point).

As composition of SCO varies with the oleaginous yeast strain used, with different proportions of saturated and unsaturated fatty acids, the resulting biodiesel would have different fuel properties depending on the feedstock yeast oil. For example, a higher content of saturated FAME results in high CN, better oxidative stability, but poor cold flow properties. On the other hand, a higher total polyunsaturated content result in better cold flow properties but may lower oxidative stability of the fuel.

Moser and Vaughn [112], demonstrated the potential of fatty acid composition as an important tool for screening and selection of biodiesel feedstock. Monounsaturated fat content is desirable, and a prerequisite for good biodiesel quality as such fatty acids impart better characteristics concerning ignition quality (CN), fuel stability, and flow properties. Therefore, methyl palmitoleate (16:1) or oleate (18:1) is warranted being liquid at room temperature and are desirable for good CN without any adverse effect on cold-flow properties.

An ideal biodiesel is made mainly of monounsaturated with balanced levels of saturated and polyunsaturated methyl esters [69, 112, 133] Moser & Vaughn, 2012; Kakkad *et al.*, 2015). Table 14.5 summarizes the fatty acids and their influence on fuel properties followed by specifications of physicochemical biodiesel fuel properties and their significance.

14.8 Concluding Remarks

SCOs of oleaginous yeasts, though known since 1940, have not seen the commercial process development for biodiesel production because of bottlenecks associated with productivity on lignocellulosic, oil recovery and its conversion into FAMEs. Biodiesel production from oleaginous yeast starting with LCB should be developed as a multiproduct biorefinery allowing complete use of starting material with minimal waste generation. The coupled production of value-added products and biodiesel may lead to its economic success. Authors propose a thorough investigation of product profiling through life cycle analysis by data generation from integrated pilot plant with multiproduct biorefinery concept to achieve real-time analysis.

Table 14.5 Characterization of microbial oil and their implication in fuel properties.

Fuel properties	Unit	Biofuel standard		Influence in fuel properties	Remarks	References
		ASTM D6751	EN 14214			
FATTY ACID CHAIN LENGTH AND SATURATION						
Saturated Fatty Acid content (SFA)	Wt%	ns	ns	High SFA decreases the possibility of auto-oxidation and increases the shelf life of the fuel	Quantitative fatty acid profile of the microbial oil determines the qualitative property of the microbial oil as feed for biodiesel	[29]
Monounsaturated Fatty Acid content (MUFA)	Wt%	ns	ns	High MUFA content enhances the fuel flow properties and balances the effect of SFA and PUFA		
PolyUnsaturated Fatty Acid content (PUFA)	Wt%	ns	ns	Mitigates the fuel stability and influence cold filter plugging point (CFPP)		
Linolenic Acid content (C18:3)	Wt%	ns	12 max	Undesirable in biofuel and an optimum ratio of SFA and UFA (MUFA & PUFA) is required for better CFPP, density, viscosity and oxidative stability		
Degree of Unsaturation (DU)	-	ns	ns	Low DU facilitates oxidation stability and facilitates long time storage	-	[29]
Long Chain Saturated Factor (LCSF)	-	ns	ns	A higher LCSF has positive effect on the cetane number (CN)		

PHYSICO-CHEMICAL PARAMETERS						
Acid Value/Acid number	mg KOH /g	0.8 max	0.5 max	An important characteristic to be evaluated to ensure the suitability of SCO for use as fuel because at high temperature the free fatty acids present in oil tend to react with the metallic engine parts increasing wear.	high free fatty acid value is problematic in transesterification using alkaline catalyst as it leads to soap formation and the separation of products i.e biodiesel from glycerine becomes difficult resulting in a low yield of FAMEs	[36]
Saponification value	-	ns	ns	a measure of the average molecular weight of the triacylglycerols in a sample.	-	[29]
Iodine Value (IV)	g /100 g	ns	≤ 120	Determines degree of fatty acid unsaturation and high IV results gum formation during heating of fuel.		
CN	-	47–65	51	Combustion behavior of the fuel and readiness towards the auto ignition. Higher cetane numbers have also been associated with reduced engine roughness, lower starting temperatures for engines, cold start combustion and noise of the engine.	To calculate exact cetane number high amount of oil is required (min 1 L) and mostly CN of microbial oil is reported based on empirical formulae as till date large scale production of microbial oil has not been reported	[122]
Flash Point	°C	ns	ns	specified for safety during storage, transport and handling Higher value recommended for the safer storage, transport	minimum temperature at which the vapours of a fuel catch fire if in contact with a flame.	[86]
CFPP	°C	ns	+5 to −20	Temperature at which fuel solidifies to plug the filter	-	[34]

(Continued)

Table 14.5 Cont.

Fuel Properties	Unit	Biofuel Standard		Influence in Fuel Properties	Remarks	References
		ASTM D6751	EN 14214			
Cloud point	°C	–3 to 15	ns	First appearance of wax crystal in lower temperature	-	[149]
Pour point	°C	–5 to 10	≤ 0	Minimum temperature at which fuel flows	-	[29]
Allylic position equivalents (APE) and Bis-allylic position equivalents (BAPE)	-	ns	ns	Presence of allylic and bis-allylic carbon in FAME reflects the poly-unsaturation, which makes the fuel prone to auto-oxidation	-	[29]
Oxidative stability	h	3	6	Indicates the quality of gum and macromolecules which are formed when subjecting a sample to certain conditions of pressure in the presence of oxygen. The compounds generated cause an increase in viscosity.	The structure of the FA chains determines this property. While saturated FA chains are problematic for cold flow, unsaturated, especially PUFA chains cause the low oxidative stability of biodiesel fuels. Fatty acid chains with more than three double bonds are even more susceptible to oxidation	[129]
Higher Heating Value	g	ns	~35	the amount of heat energy released during the combustion of one gram of fuel to produce CO_2 and H_2O at its initial temperature.	The calorific value increases with chain length and increase in the ratio of carbon and hydrogen to oxygen and nitrogen while it tends to decrease with increase in level of unsaturation.	[132]

Viscosity	mm^2 / s	1.9 to 6.0	3.5 to 5.0	The fuel viscosity plays a critical role in the fuel spray, mixture formation and combustion process. The injection of fuel into the combustion chamber and fuel atomization is affected by its viscosity.	Reducing viscosity is the major reason for transesterification of oil to biodiesel. The viscosity of biodiesel is many times lower resulting in better fuel atomization in the engine combustion chamber because lower the viscosity of the oil, the easier it is to pump and atomize and achieve finer droplets. The flow properties are also influenced by viscosity making it a critical parameter to consider when using biodiesel at low temperatures as the viscosity decreases exponentially with an increase in temperature	[78]
Density	gm / cm^3	ns	0.86 to 0.9	Influences the performance of the oil in the injectors. It affects the mass of fuel injected at the injection system because a precise fuel quantity is essential for proper combustion.	Along with density, viscosity of the microbial oil also affects the pumping of oil feed into reactor for conversion into biofuel	[132]

References

1. Ageitos, J.M., Vallejo, J.A., Veiga-Crespo, P., Villa, T.G., Oily yeasts as oleaginous cell factories; *Appl. Microbiol. Biotechnol.*, 90, 1219–27, 2011.
2. Aggelis, G., Sourdis, J., Prediction of lipid accumulation-degradation in oleaginous micro-organisms growing on vegetable oils; *Antonie Van Leeuwenhoek* 72, 159–65, 1997.
3. Ahn, Y., Lee, S.H., Kim, H.J., Yang, Y.H., Hong, J.H., Kim, Y.H., Kim, H., Physical state of cellulose in BmimCl: dependence of molar mass on viscoelasticity and sol-gel transition; *Carbohydr. Polym.*, 88, 395, 2012.
4. Aragon, D., Lu, S., Kochergin, V., Conversion of energy cane and sweet sorghum into biofuels and chemicals: a modeling approach; In *New crops: Bioenergy, biomaterials, and sustainability*, (eds.) Janick J, Whipkey A, Von Mark C. Tucson, AZ: The University of Arizona, 2015.
5. Arumugam, N., Anandakumar, S., Mini review on Corncob biomass: A potential resource for value-added metabolites. *Europ. J. Exp. Biology*, 6, 9–13, 2016.
6. Asikainen, A., Availability of wood biomass for biorefining; Cellulose *Chem. Technol.*, 44, 111, 2010.
7. Atik, C., Ates, S., Mass balance of silica in straw from the perspective of silica reduction in straw pulp; *BioResour.*, 7, 3274–82, 2012.
8. Balan, V., Current challenges in commercially producing biofuels from lignocellulosic biomass; *ISRN Biotechnology*; Article ID 463074, 2014.
9. Bals, B., Rogers, C., Jin, M., Balan, V., Dale, B., Evaluation of ammonia fiber expansion (AFEX) pretreatment for enzymatic hydrolysis of switchgrass harvested in different seasons and locations. *Biotechnol. Biofuels*, 3, 1, 2010.
10. Bandhu, S., Dasgupta, D., Akhter, J., Kanaujia, P., Suman, S.K., Agrawal, D., Kaul, S., Adhikari, D.K., Ghosh, D., Statistical design and optimization of single cell oil production from sugarcane bagasse hydrolysate by an oleaginous yeast *Rhodotorula* sp. IIP-33 using response surface methodology. SpringerPlus, 3, 691, 2014.
11. Banerjee, S., Sen, R., Mudliar, S., Pandey, R.A., Chakrabarti, T., Satpute, D., Alkaline peroxide assisted wet air oxidation pretreatment approach to enhance enzymatic convertibility of rice husk; *Biotechnol. Progr.*, 27, 691, 2011.
12. Banerjee, S., Sen, R., Pandey, R.A., Chakrabarti, T., Satpute, D., Giri, B.S., Mudliar, S., Evaluation of wet air oxidation as a pretreatment strategy for bioethanol production from rice husk and process optimization; *Biomas Bioenerg.*, 33, 1680, 2009.
13. Barakat, A., de, Vries, H., Rouau, X., Dry fractionation process as an important step in current and future lignocellulose biorefineries: a review; *Bioresour. Technol.*, 134, 362, 2013.
14. Barros, R.D.R.O., de, Sousa, Paredes, R., Endo, T., da, Silva, Bon, E.P., Lee, S.H., Association of wet disk milling and ozonolysis as pretreatment for enzymatic saccharification of sugarcane bagasse and straw; *Bioresour. Technol.*, 136, 288, 2013.

15. Basu, R., Hazra S., Shanks, M., Paterson, D.I., Oudit, G.Y. Novel mutation in exon 14 of the sarcomere gene MYH7 in familial left ventricular noncompaction with bicuspid aortic valve; *Circulation: Heart Failure* 7, 1059, 2014.
16. Beopoulos, A., Cescut, J., Haddouche, R., Uribelarrea, J.L., Molina-Jouve, C., Nicaud, J.M., *Y. lipolytica* as a model for bio-oil production; *Prog. Lipid. Res.*, 48, 375, 2009.
17. Beopoulos, A., Mrozova, Z., Thevenieau, F., Le, Dall, M.T., Hapala, I., Papanikolaou, S., Chardot, T., Nicaud, J.M., Control of lipid accumulation in the yeast *Y. lipolytica Appl. Environ. Microbiol.*, 74, 7779, 2008.
18. Beopoulos, A., Nicaud, J.M., Gaillardin, C., An overview of lipid metabolism in yeasts and its impact on biotechnological processes. *Appl. Microbiol. Biotechnol.*, 90, 1193, 2011.
19. Binder, J.B., Raines, R.T., Fermentable sugars by chemical hydrolysis of biomass. PNAS 107, 4516, 2016.
20. Bligh, E.G., Dyer, W.J., A rapid method of total lipid extraction and purification. *Can. J. Biochem. Physiol.*, 37, 911, 1959.
21. Bonturi, N., Matsakas, L., Nilsson, R., Christakopoulos, P., Miranda, E.A., Berglund, K.A., Rova, U., Single cell oil producing yeasts *Lipomyces starkeyi* and *Rhodosporidium toruloides*: Selection of extraction strategies and biodiesel property prediction; *Energ., 8,* 5040, 2015.
22. Boulton, C.A., Ratledge, C., Correlation of lipid accumulation in yeasts with possession of ATP:citrate lyase; *Microbiol.*, 127, 169, 1981.
23. Bouvier-Navé, P., Benveniste, P., Oelkers, P., Sturley, S.L., Schaller, H., Expression in yeast and tobacco of plant cDNAs encoding acyl CoA: DAG acyltransferase; *FEBS J.*, 267, 85, 2000.
24. Brodeur, G.,Yau, E., Badal, K., Collier, J., Ramachandran, K.B., Ramakrishnan, S., Chemical and Physicochemical Pretreatment of Lignocellulosic Biomass: A Review; *Enzyme Res.*, 2011, 787532, 2011.
25. Chen, H., Chemical composition and structure of natural lignocellulose. In Biotechnology of lignocellulose; Springer Netherlands p 25, 2014.
26. Chen, X.F., Huang, C., Xiong, L., Ma, L.L., Microbial oil production from corn cob acid hydrolysate by *Trichosporon cutaneum*; *Biotechnol. Letts.*, 34, 1025, 2012.
27. Chouhan, O.P., Bandekar, D., Hazra, M., Baghudana, A., Hazra, S., Biswas, S., Effect of site-directed mutagenesis at the GGEEF domain of the biofilm forming GGEEF protein from *Vibrio cholerae;* AMB Exp 6, 2, 2016.
28. Das, V., Kalyan, G., Pal, M., Hazra, S., Understanding the role of structural integrity and differential expression of integrin profiling to identify potential therapeutic targets in breast cancer; *J. Cell. Physiol.*, https://doi.org/10.1002/jcp.25821.
29. Dasgupta, D., Sharma, T., Bhatt, A., Bandhu, S., Ghosh, D., Cultivation of oleaginous yeast *Rhodotorula mucilaginosa* IIPL32 in split column airlift reactor and its influence on fuel properties; *Biocatalysis. Agri. Biotechnol.*, 10, 308, 2017.

30. Davies, R.J., Yeast oil from cheese whey-process development; In Moreton RS (ed.), *Single cell oil*; Longman Scientific & Technical, New York; USA p. 99, 1988.
31. Davies, R.J., Holdsworth, J.E., Synthesis of lipids in yeasts: Biochemistry, physiology, and production; *Adv. Appl. Lipid Res.*, 1, 119, 1992.
32. Demirbas, A., Products from lignocellulosic materials via degradation processes; *Energ. Sourc.*, Part A 30, 27, 2008.
33. Dujon, B., Sherman, D., Fischer, G., Durrens, P., Genome evolution in yeasts; *Nat.*, 430, 35, 2004.
34. Dunn, R.O., Shockley, M.W., Bagby, M.O., Improving the low-temperature properties of alternative diesel fuels: vegetable oil-derived methyl esters; JAOCS 73, 1719, 1996.
35. Dutt,.D, Tyagi, C.H., Comparison of various eucalyptus species for their morphological, chemical, pulp and paper making characteristics; *Indian J. Chem. Technol.* 18, 145, 2011.
36. El Sabagh, S.M., Keera, S.T., Taman, A.R., The characterization of biodiesel fuel from waste frying oil; Energ Sources, Part A: Recovery Util Environ Effects 33, 401*Energ. Combus. Sci.*, 37, 52, 2011.
37. Folch, J., Lees, M., Sloane-Stanley, GH A simple method for the isolation and purification of total lipids from animal tissues; *J. Biol. Chem.*, 226, 497, 1957.
38. Gabov, K., Fardim, P., da Silva, Júnior, F.G., Hydrotropic fractionation of birch wood into cellulose and lignin: a new step towards green biorefinery. *BioResour.*, 8, 3518, 2013.
39. Galafassi, S., Cucchetti, D., Pizza, F., Franzosi, G., Bianchi, D., Compagno, C., Lipid production for second generation biodiesel by the oleaginous yeast *Rhodotorula graminis*; *Bioresour. Technol.*, 111, 398, 2012.
40. Gao, Q., Cui, Z., Zhang, J., Bao, J., Lipid fermentation of corncob residues hydrolysate by oleaginous yeast *Trichosporon cutaneum*; *Bioresour. Technol.*, 152, 552, 2014.
41. Girard, P., Fallot, A., Review of existing and emerging technologies for the production of biofuels in developing countries; *Energy Sustain. Dev.*, 10, 92, 2006.
42. Gong, Z., Shen, H., Wang, Q., Yang, X., Xie, H., Zhao, Z.K., Efficient conversion of biomass into lipids by using the simultaneous saccharification and enhanced lipid production process. *Biotechnol. Biofuels*, 6, 36, 2013.
43. Gong, Z., Shen, H., Yang, X., Wang, Q., Xie, H., Zhao, Z.K., Lipid production from corn stover by the oleaginous yeast *Cryptococcus curvatus*; *Biotechnol. Biofuels*, 7, 158, 2014.
44. Goodridge, J., Data on California's extreme rainfall from 1862–1995; 1996 *California Weather Symposium*, Sierra College, Rocklin, CA, USA; p 1, 1996.
45. Graham-Rowe, D., Agriculture: Beyond food versus fuel; *Natur* 474, S6, 2011.
46. Groenewald, M., Boekhout, T., Neuvéglise, C., Gaillardin, C., Van, Dijck, P.W., Wyss, M., *Y. lipolytica*: safety assessment of an oleaginous yeast with a great industrial potential. *Crit. Rev. Microbiol.*, 40, 187, 2014.

47. Hazra, S., Hadi, T., Blanchard, J.S., Tanner, M.E., The Structure of MurNAc 6-Phosphate Hydrolase (MurQ) from *Haemophilus influenzae* with Bound Inhibitor; *Biochem.*, 23, 9358, 2013.
48. Hazra, S., Konrad, M., Lavie, A., The sugar ring of the nucleoside is required for productive substrate positioning in the active site of human deoxycytidine kinase (dCK): implications for the development of dCK-activated acyclic guanine analogues; *JMC*, 53, 5792, 2010a.
49. Hazra, S., Kurz, S., Wolff, K., Nguyen, L., Bonomo, R., Blanchard, J.S., Kinetic and Structural Characterization of the Interaction of the 6-methylidene penem 2 with the β-Lactamase from *Mycobacterium tuberculosis; Biochem.*, 54, 5657, 2015.
50. Hazra, S., Ort, S., Konrad, M., Lavie, A., Structural and kinetic characterization of human deoxycytidine kinase variants able to phosphorylate 5-substituted deoxycytidine and thymidine analogues; *Biochem.*, 49, 6784, 2010b.
51. Hazra, S., Sabini, E., Ort, S., Konrad, M., Lavie, A., Extending Thymidine Kinase Activity to the Catalytic Repertoire of Human Deoxycytidine Kinase; *Biochem.*, 48, 1256, 2009.
52. Hazra, S., Szewczak, A., Ort, S., Konrad, M., Lavie, A., Post-translational phosphorylation of serine 74 of human deoxycytidine kinase favors the enzyme adopting the open conformation making it competent for nucleoside binding and release. *Biochem.*, 50, 2870, 2011.
53. Hazra, S., Xu, H., John, S., Blanchard, J.S., Tebipenem, a New Carbapenem Antibiotic is a Slow Substrate that Inhibits the β-Lactamase from *Mycobacterium tuberculosis*; *Biochem.*, 53, 3671, 2014.
54. Heredia, L., Ratledge, C., Simultaneous utilization of glucose and xylose by *Candida curvata* D in continuous culture; *Biotechnol. Letts.*, 10, 25, 1988.
55. Hill, S., Winning, B., Jenner, H., Knorpp, C., Leaver, C., Role of NAD+-dependent 'malic'enzyme and pyruvate dehydrogenase complex in leaf metabolism; *Interaction Photosyn. Respir.*, 24, 743, 1996.
56. Hinkle, J.D., Ciesielski, P.N., Gruchalla, K., Munch, K.R., Donohoe, B.S., Biomass accessibility analysis using electron tomography. *Biotechnol. Biofuels*, 8, 212, 2015.
57. Hoefnagels, R., Smeets, E., Faaij, A., Greenhouse gas footprints of different biofuel production systems; *Renew. Sustain. Energ. Rev.*, 14, 1661, 2010.
58. Hu, Z., Sykes, R., Davis, M.F., Brummer, E.C., Ragauskas, A.J., Chemical profiles of switchgrass; *Bioresour. Technol.*, 101, 3253, 2010.
59. Huang, C., Chen, X.F., Xiong, L., Ma, L.L. Oil production by the yeast *Trichosporon dermatis* cultured in enzymatic hydrolysates of corn cobs; *Bioresour. Technol.*, 110, 711, 2012.
60. Huang, C., Chen, X.F., Yang, X.Y., Xiong, L., Lin, X.Q., Yang, J., Wang, B., Chen, X.D., Bioconversion of corncob acid hydrolysate into microbial oil by the oleaginous yeast *Lipomyces starkeyi*; *Appl. Biochem. Biotechnol.*, 172, 2197, 2014.

61. Huang, C., Zong, M.H., Wu, H., Liu, Q.P., Microbial oil production from rice straw hydrolysate by *Trichosporon fermentans; Bioresour. Technol.*, 100, 4535, 2009.
62. Huang, X., Wang, Y., Liu, W., Bao, J., Biological removal of inhibitors leads to the improved lipid production in the lipid fermentation of corn stover hydrolysate by Trichosporon cutaneum. *Bioresour. Technol.*, 102, 9705, 2011.
63. Huang, Y.F., Huang, G.H., Hu, Z.Y., Maqsood, I., Chakma, A., Development of an expert system for tackling the public's perception to climate-change impacts on petroleum industry. *Expet. Sys. App.*, 29, 817, 2005.
64. Huber, G.W., Iborra, S., Corma, A., Synthesis of Transportation Fuels from Biomass: Chemistry, Catalysts, and Engineering; *Chem. Rev.*, 106, 4044, 2006.
65. Iqbal, H.M.N., Ahmed, I., Zia, M.A., Irfan, M., Purification and characterization of the kinetic parameters of cellulase produced from wheat straw by *Trichoderma viride* under SSF and its detergent compatibility; *Adv. Biol. Biotechnol.*, 2, 149, 2011.
66. Isikgora, F.H., Becer, C.R., Lignocellulosic biomass: a sustainable platform for the production of bio-based chemicals and polymers; *Polym. Chem.*, 6, 4497, 2015.
67. Jacob, Z., Yeast lipids: extraction, quality analysis, and acceptability; *Crit. Rev. Biotechnol.*, 12, 463, 1992.
68. Jain, R.K., Ghosh, D., Agrawal, D., Suman, S.K., Pandey, D., Vadde, V.T., Dixit, A.K., Adhikari, D.K., Dasgupta, D., Ethanol production from rice straw using thermotolerant *Kluyveromyces* sp. IIPE453; *Biomas. Conv. Bioref., 5,* 331, 2015.
69. Kakkad,.H, Khot, M., Zinjarde, S., Ravi, Kumar, A., Biodiesel production by direct *in situ* transesterification of an oleaginous tropical mangrove fungus grown on untreated agro-residues and evaluation of its fuel properties; *BioEnergy, Res.*, 8, 1788, 2015.
70. Kamiryo, T., Nishikawa, Y., Mishina, M., Terao, M., Numa, S., Involvement of long-chain acyl coenzyme A for lipid synthesis in repression of acetyl-coenzyme A carboxylase in *Candida lipolytica*; PNAS 76, 4390, 1979.
71. Katre, G., Joshi, C., Khot, M., Zinjarde, S., RaviKumar, A., Evaluation of single cell oil (SCO) from a tropical marine yeast *Yarrowia lipolytica* NCIM 3589 as a potential feedstock for biodiesel; *AMB Express* 2, 36, 2012.
72. Kautto, J., Realff, M.J., Ragauskas, A.J., Kässi, T., Economic analysis of an organosolv process for bioethanol production; *BioResour.*, 9, 6041, 2014.
73. Khot, M., Ghosh, D., Lipids of *Rhodotorula mucilaginosa* IIPL32 with biodiesel potential: Oil yield, fatty acid profile, fuel properties; *J. Basic Microbiol.*, 57, 345, 2017.
74. Kim, M., Day, D.F., Composition of sugar cane, energy cane, and sweet sorghum suitable for ethanol production at Louisiana sugar mills; *J. Indus Microbiol. Biotechnol.*, 38, 803, 2011.
75. Kim, Y., Mosier, N.S., Ladisch, M.R., Enzymatic digestion of liquid hot water pretreated hybrid poplar; *Biotechnol. Prog.*, 25, 340, 2009.

76. Knittelfelder, O.L., Kohlwein, S.D., Lipid extraction from yeast cells. *Cold Spring Harbor Protocols*; pdb-prot085449, 2017.
77. Knothe, G., Dependence of biodiesel fuel properties on the structure of fatty acid alkyl esters; *Fuel Process Technol.*, 86, 1059, 2005.
78. Knothe, G., Designer Biodiesel: Optimizing Fatty Ester Composition to Improve Fuel Properties; *Energ. Fuels*, 22, 1358, 2008.
79. Knothe, G., Dunn, R.O., A comprehensive evaluation of the melting points of fatty acids and esters determined by differential scanning calorimetry; *JOACS* 86, 843, 2009.
80. Kocaefe, D., Poncsak, S., Boluk, Y., Effect of thermal treatment on the chemical composition and mechanical properties of birch and aspen; *BioResour.*, 3, 517, 2008.
81. Kohlwein, S.D., Analyzing and Understanding Lipids of Yeast: A Challenging Endeavor." *Cold Spr. Har. Prot.*, 5, pdb-top078956, 2017.
82. Kosa, M., Ragauskas, A.J., Lipids from heterotrophic microbes: advances in metabolism research; *Trends Biotechnol.*, 29, 53, 2011.
83. Koutinas, A.A., Chatzifragkou, A., Kopsahelis, N., Papanikolaou, S., Kookos, I.K., Design and techno-economic evaluation of microbial oil production as a renewable resource for biodiesel and oleochemical production; *Fuels* 116, 566, 2014.
84. Koutinas, A.A., Vlysidis, A., Pleissner, D., Kopsahelis, N., Garcia, I.L., Kookos, I.K., Papanikolaou, S., Kwan, T.H., Lin, C.S.K., Valorization of industrial waste and by-product streams via fermentation for the production of chemicals and biopolymers; *Chem. Soc. Rev.*, 43, 2587, 2014.
85. Krishnan, C., Sousa, L.D.C., Jin, M., Chang, L., Dale, B.E., Balan, V., Alkali-based AFEX pretreatment for the conversion of sugarcane bagasse and cane leaf residues to ethanol; *Biotechnol. Bioeng.*, 107, 441, 2010.
86. Krisnangkura, K., Simamaharnnop, R., Continuous transmethylation of palm oil in an organic solvent; *JAOC* 69, 166, 1992.
87. Kumar, P., Barrett, D.M., Delwiche, M.J., Stroeve, P., Methods for pretreatment of lignocellulosic biomass for efficient hydrolysis and biofuel production. *Indus Engg. Chem. Res.*, 48, 3713, 2009.
88. Kurz, S., Hazra, S., Bethel, C., Romagnoli, C., Caselli, E., Prati, F., Blanchard, J.S., Bonomo, R., Inhibiting the β-Lactamase of Mycobacterium tuberculosis (Mtb) with Novel Boronic-Acid-Transition-State-Inhibitors (BATSIs); *ACS Infect. Diseas.*, 1, 234, 2015.
89. Kurz, S., Wolff, K., Hazra, S., Bethel, C., Hujer, A., Smith, K., Xu, Y., Tremblay, L., Blanchard, J.S., Nguyen, L., Bonomo, R., Can inhibitor resistant substitutions in the Mycobacterium tuberculosis β-lactamase BlaC lead to clavulanate resistance? A biochemical rationale for the use of β-lactam β-lactamase inhibitor combinations; *Antimicrob. Agent Chem. Other*, 57, 6085, 2013.
90. Leiva-Candia, D.E., Pinzi, S., Redel-Macías, M.D., Koutinas, A., Webb, C., Dorado, M,P., The potential for agro-industrial waste utilization using oleaginous yeast for the production of biodiesel; *Fuels*, 123, 33, 2014.

91. Li, M., Liu, G.L., Chi, Z., Chi, Z.M., Single cell oil production from hydrolysate of cassava starch by marine-derived yeast *Rhodotorula mucilaginosa* TJY15a; *Biomas. Bioenergy.*, 34, 101, 2010.
92. Li, Q., Du, W., Liu, D., Perspectives of microbial oils for biodiesel production; *Appl. Microbiol. Biotechnol.*, 80, 749, 2008.
93. Li, Z., Sun, H., Mo, X., Li, X., Xu, B., Tian, P., Overexpression of malic enzyme (ME) of *Mucor circinelloides* improved lipid accumulation in engineered *Rhodotorula glutinis*; *Appl. Microbiol. Biotechnol.*, 97, 4927, 2013.
94. Liang, Y., Tang, T., Umagiliyage, A.L., Siddaramu, T., McCarroll, M., Choudhary, R., Utilization of sorghum bagasse hydrolysates for producing microbial lipids; *Appl. Energ.*, 91, 451, 2012.
95. Liu, B., Zhao, Z.K., Biodiesel production by direct methanolysis of oleaginous microbial biomass; *J. Chem. Technol. Biotechnol.*, 82, 775, 2007.
96. Liu, C., Xiao, L., Jiang, J., Wang, W., Gu, F., Song, D., Yi, Z., Jin, Y., Li, L., Biomass properties from different *Miscanthus* species; *Food Energ. Secur.*, 2, 12, 2013.
97. Ma, X.J., Li, H., Wang, D.X., Song, X., Sophorolipid production from delignined corncob residue by *Wickerhamiella domercqiae* var. sophorolipid CGMCC 1576 and *Cryptococcus curvatus* ATCC 96219; *Appl. Microbiol. Biotechnol.*, 98, 475, 2014.
98. Magdouli, S., Yan, S., Tyagi, R.D., Surampalli, R.Y., Heterotrophic microorganisms: a promising source for biodiesel production; *Crit. Rev. Environ. Sci. Technol.*, 44, 416, 2014.
99. Mäki-Arvela, P., Anugwom, I., Virtanen, P., Sjöholm, R., Mikkola, J.P., Dissolution of lignocellulosic materials and its constituents using ionic liquids—A review; *Ind. Crops Prod.*, 32, 175, 2010.
100. Martinez, A., Rodriguez, M.E., York, S.W., Preston, J.F., Ingram, L.O., Effects of $Ca(OH)_2$ Treatments ("Overliming") on the Composition and Toxicity of Bagasse Hemicellulose Hydrolysates; *Biotechnol. Bioengg.*, 69, 526, 2000.
101. Matsakas, L., Sterioti, A.A., Rova, U., Christakopoulos, P., Use of dried sweet sorghum for the efficient production of lipids by the yeast *Lipomyces starkeyi* CBS 1807; *Indus Crops Product*, 62, 367, 2014.
102. McKendry, P., Energy production from biomass (part 1): overview of biomass; *Bioresour. Technol.*, 83, 37, 2002.
103. Meeuwse, P., Sanders, J.P., Tramper, J., Rinzema, A., Lipids from yeasts and fungi: tomorrow's source of biodiesel? *Biofuel Biopro. Bioref.*, 7, 512, 2013.
104. Meng, X., Yang, J., Xu, X., Zhang, L., Nie, Q., Xian, M., Biodiesel production from oleaginous microorganisms; *Renew. Eenerg*, 34, 1, 2009.
105. Mesa, L., González, E., Cara, C., González, M., Castro, E., Mussatto, S.I., The effect of organosolv pretreatment variables on enzymatic hydrolysis of sugarcane bagasse; *Chem. Engg. J.*, 168, 1157, 2011.
106. Michinaka, Y, Shimauchi, T., Aki, T., Nakajima, T., Kawamoto, S., Shigeta, S., Suzuki, O., Ono, K., Extracellular secretion of free fatty acids by disruption

of a fatty acyl-CoA synthetase gene in *Saccharomyces cerevisiae; J. Biosci. Bioengg.*, 95, 435, 2003.

107. Mitchell, R., Vogel, K.P., Uden, D.R., The feasibility of switchgrass for biofuel production; *Biofuels*, 3, 47, 2012.
108. Möller, M., Schröder, U., Hydrothermal production of furfural from xylose and xylan as model compounds for hemicelluloses; *RSC Adv*., 3, 22253, 2013.
109. Montgomery, D.C., Myers, R.H., Response surface methodology; *Design Anal. Exper*., 445, 1995.
110. Moreadith, R.W., Lehninger, A.L., Purification, kinetic behaviour and regulation of NAD(P) + malic enzyme of tumor mitochondria; *J. Biol. Chem*., 259, 6222, 1984.
111. Morin, N., Cescut, J., Beopoulos, A., Lelandais, G., Le, Berre, V., Uribelarrea, J.L., Molina-Jouve, C., Nicaud, J.M., Transcriptomic analyses during the transition from biomass production to lipid accumulation in the oleaginous yeast *Y lipolytica*; PloS one 6, p.e27966, 2011.
112. Moser,B.R., Vaughn, S.F., Efficacy of fatty acid profile as a tool for screening feedstocks for biodiesel production; *Biomas. Bioenerg*., 37, 31, 2012.
113. Naik, S.N., Goud, V.V., Rout, P.K., Dalai, A.K., Production of first and second-generation biofuels: a comprehensive review; *Renew. Sustain. Energy Rev.*, 14, 578, 2010.
114. Nigam, P.S., Singh, A., Production of liquid biofuels from renewable resources; *Prog. Ener. Comb. Sci*., 37, 52, 2011.
115. Noureddini, H., Byun, J., Dilute-acid pretreatment of distillers' grains and corn fiber; *Bioresour. Technol*., 101, 1060, 2010.
116. Ochoa-Estopier, A., Guillouet, S.E., D-stat culture for studying the metabolic shifts from oxidative metabolism to lipid accumulation and citric acid production in *Y. lipolytica; J. Biotechnol*., 170, 35, 2014.
117. Okafor, N., *Modern industrial microbiology and biotechnology*; CRC Press, 2016.
118. Pan, X.J., Sano, Y., Ito, T., Atmospheric acetic acid pulping of rice straw II: behavior of ash and silica in rice straw during atmospheric acetic acid pulping and bleaching; Holzforschung 53, 49, 1999.
119. Papanikolaou, S., Aggelis, G. Modeling lipid accumulation and degradation in *Y. lipolytica* cultivated on industrial fats; *Curr. Microbiol*., 46, 398, 2003a.
120. Papanikolaou, S., Aggelis, G., Selective uptake of fatty acids by the yeast *Yarrowia lipolytica; Eur. J. Lipid Sci. Technol*., 105, 651, 2003b.
121. Papanikolaou, S., Aggelis, G., Lipids of oleaginous yeasts. Part I: Biochemistry of single cell oil production; *Eur. J. Lipid Sci. Technol.*, 113, 1031, 2011.
122. Patel, A., Arora, N., Sartaj, K., Pruthi, V., Pruthi, P.A., Sustainable biodiesel production from oleaginous yeasts utilizing hydrolysates of various non-edible lignocellulosic biomasses; *Renew. Sustain. Energ. Rev*., 62, 836, 2016.
123. Patel, A., Sindhu, D.K., Arora, N., Singh, R.P., Pruthi, V., Pruthi, P.A., Biodiesel production from non-edible lignocellulosic biomass of *Cassia*

fistula L. fruit pulp using oleaginous yeast *Rhodosporidium kratochvilovae* HIMPA1; *Bioresour. Technol.*, 197, 91, 2015.
124. Paul, M., Hazra, M., Barman, A., Hazra, S., Comparative molecular dynamics simulation studies for determining factors contributing to the thermostability of chemotaxis protein CheY; *JBSD* 32, 928, 2013.
125. Pérez, J.A., Ballesteros, I., Ballesteros, M., Sáez, F., Negro, M.J., Manzanares, P., Optimizing liquid hot water pretreatment conditions to enhance sugar recovery from wheat straw for fuel-ethanol production; *Fuel* 87, 3640, 2008.
126. Pointner, M., Kuttner, P., Obrlik, T., Jager, A., Kahr, H., Composition of corncobs as a substrate for fermentation of biofuels; *Agron. Res.*, 12, 391, 2014.
127. Potumarthi, R., Baadhe, R.R., Nayak, P., Jetty, A., Simultaneous pretreatment and saccharification of rice husk by *Phanerochete chrysosporium* for improved production of reducing sugars; *Bioresour. Technol.*, 128, 113, 2013.
128. Probst, K.V., Schulte, L.R., Durrett, T.P., Rezac, M.E., Vadlani, P.V., Oleaginous yeast: a value-added platform for renewable oils; *Crit. Rev. Biotechnol.*, 36, 942, 2016.
129. Pullen, J., Khizer, J., An overview of biodiesel oxidation stability; *Renewab. Sustainab. Energ.Rev.*, 16, 5924, 2012.
130. Quartararo, C.E., Hazra, S., Hadi, T., Blanchard, J.S., Structural, kinetic and chemical mechanism of isocitrate dehydrogenase-1 from *Mycobacterium tuberculosis; Biochem.*, 52, 1765, 2013.
131. Ragauskas, A.J., Williams, C.K., Davison, B.H., Britovsek, Cairney, G.J., Eckert, C.A., Frederick, Jr. W.J., Hallett, J.P., Leak, D.J., Liotta, C.L., Mielenz, J.R., Murphy, R., Templer, R., Tschaplinski, T., The path forward for biofuels and biomaterials; *Sci.*, 311, 484, 2006.
132. Ramirez-Verduzco, L.F., Rodriguez-Rodriguez, J.E., Jaramillo-Jacob, A.R., Predicting cetane number, kinematic viscosity, density and higher heating value of biodiesel from its fatty acid methyl ester composition; *Fuel* 91, 102, 2012.
133. Ramos, M.J., Fernández, C.M., Casas, A., Rodríguez, L., Pérez, Á., Influence of fatty acid composition of raw materials on biodiesel properties; *Bioresour. Technol.*, 100, 261, 2009.
134. Ratledge, C., Yeasts for lipid production; *Biochem. Soc. Trans.* 16, 1088, 1988.
135. Ratledge, C., The role of malic enzyme as the provider of NADPH in oleaginous microorganisms: a reappraisal and unsolved problems; *Biotechnol. letts.*, 36, 1557, 2014.
136. Ratledge, C., Cohen, Z., Microbial and algal oils: do they have a future for biodiesel or as commodity oils?; *Lipid Technol.*, 20, 155, 2008.
137. Ratledge, C., Wynn, J.P., The biochemistry and molecular biology of lipid accumulation in oleaginous microorganisms; *Adv. Appl. Microbiol.*, 51, 1, 2002.
138. Ruenwai, R., Cheevadhanarak, S., Laoteng, K., Overexpression of acetyl-CoA carboxylase gene of *Mucor rouxii* enhanced fatty acid content in *Hansenula polymorpha; Mol. biotechnol.*, 42, 327, 2009.

139. Runguphan, W., Keasling, J.D., Metabolic engineering of *Saccharomyces cerevisiae* for production of fatty acid-derived biofuels and chemicals; *Metabol. Engg.*, 21, 103, 2014.
140. Sabini, E., Hazra, S., Konrad, M., Lavie, A., Structural basis for activation of the therapeutic L-nucleoside analogs 3TC and troxacitabine by human deoxycytidine kinase; *NAR* 35, 186, 2007.
141. Sabini, E., Hazra, S., Ort, S., Konrad, M., Lavie, A., Structural basis for substrate promiscuity of dCK; *JMB* 378, 607, 2008.
142. Sabini, E., Hazra, S., Konrad M., Lavie, A., Elucidation of different binding modes of purine nucleosides to human deoxycytidine kinase; *JMC* 51, 4219, 2008.
143. Saenge, C., Cheirsilp, B., Suksaroge, T.T., Bourtoom, T., Potential use of oleaginous red yeast *Rhodotorula glutinis* for the bioconversion of crude glycerol from biodiesel plant to lipids and carotenoids; *Process Biochem.*, 46, 210, 2011.
144. Salvi, B.L., Panwar, N.L., Biodiesel resources and production technologies–A review. *Renew. Sustainab. Energ. Rev.*, 16, 3680, 2012.
145. Sankh, S., Thiru, M., Saran, S., Rangaswamy, V., Biodiesel production from a newly isolated *Pichia kudriavzevii* strain; *Fuel*, 106, 690, 2013.
146. Sannigrahi, P., Ragauskas, A.J., Tuskan, G.A., Poplar as a feedstock for biofuels: a review of compositional characteristics; *Biofuels Biopro. Bioref.*, 4, 209, 2010.
147. Santos VEN, Ely, R.N., Szklo, A.S., Magrini, A., Chemicals, electricity and fuels from biorefineries processing Brazil's sugarcane bagasse: Production recipes and minimum selling prices; *Renewab Sust Energ. Rev.*, 53, 1443, 2016.
148. Saran, S., Mathur, A., Dalal, J., Saxena, R.K., Process optimization for cultivation and oil accumulation in an oleaginous yeast *Rhodosporidium toruloides* A29; *Fuel* 188, 324–331, 2017.
149. Sarin, A., Arora, R., Singh, N.P., Sarin, R., Malhotra, R.K., Kundu, K., Effect of blends of *Palm-Jatropha-Pongamia* biodiesels on cloud point and pour point; *Energ.*, 34, 2016, 2009.
150. Scharnewski, M., Pongdontri, P., Mora, G., Hoppert, M., Fulda, M., Mutants of *Saccharomyces cerevisiae* deficient in acyl-CoA synthetases secrete fatty acids due to interrupted fatty acid recycling; *FEBS J.*, 275, 2765, 2008.
151. Schneiter,R., Daum, G., Extraction of yeast lipids; *Yeast Protoc* 41, 2006.
152. Schnitzer, M., Monreal, C.M., Powell, E.E., Wheat straw biomass: A resource for high-value chemicals; *J. Environ. Sci. Health*, Part B 49, 51, 2014.
153. Shang, C.H., Zhu, S.N., Yuan, Z.H., Wang, Z.M. Molecular Cloning and Characterization Analysis of Malic Enzyme Gene from *Dunaliella parva; Adv. Mat. Res.*, 347, 2536, 2012.
154. Shankarappa, T.H., Geeta, G.S., Alkali and autohydrolysis pretreatments for effective delignification and recovery of cellulose and hemicellulose in selected agro residues; Karnataka *J. Agri. Sci.*, 26, 67, 2013.

155. Sheng, J., Feng, X., Metabolic engineering of yeast to produce fatty acid-derived biofuels: bottlenecks and solutions; *Front Microbiol.*, 6, 554, 2015.
156. Sindhu, R., Kuttiraja, M., Binod, P., Janu, K.U., Sukumaran, R.K., Pandey, A., Dilute acid pretreatment and enzymatic saccharification of sugarcane tops for bioethanol production; *Bioresour. Technol.*, 102, 10915, 2011.
157. Slininger, P.J., Dien, B.S., Kurtzman, C.P., Moser, B.R., Bakota, E.L., Thompson, S.R., O'Bryan, P.J., Cotta, M.A., Balan, V.J.M., Sousa, L.D.C., Comparative lipid production by oleaginous yeasts in hydrolyzates of lignocellulosic biomass and process strategy for high titers; *Biotechnol. Bioengg.*, 113, 1676, 2016.
158. Soccol, C.R., Neto, C.J.D., Soccol, V.T., Sydney, E.B., Ferreira, da Costa, E.S., Medeiros, A.B.P., Vandenberghe, L.Pd.S., Pilot scale biodiesel production from Rhodosporidium toruloides DEBB 5533 microbial oil using sugarcane juice: Performance in diesel engine and preliminary economic study; *Bioresour. technol.*, 223, 259, 2017.
159. Somerville, C., Youngs, H., Taylor, C., Davis, S.C., Long, S.P., Feedstocks for lignocellulosic biofuels; *Sci.*, 329, 790, 2010.
160. Sun, N., Rodriguez, H., Rahman, M., Rogers, R.D., Where are ionic liquid strategies most suited in the pursuit of chemicals and energy from lignocellulosic biomass? *Chem. Commun.*, 47, 1405, 2011.
161. Sun, Y., Cheng, J., Hydrolysis of lignocellulosic materials for ethanol production: a review; *Bioresour. Technol.*, 83, 1, 2002.
162. Taarning, E., Osmundsen, C.M., Yang, X., Voss, B., Andersen, S.I., Christensen, C.H., Zeolite-catalyzed biomass conversion to fuels and chemicals; *Energ. Environ. Sci.*, 4, 793, 2011.
163. Tai, M., Stephanopoulos, G., Engineering the push and pull of lipid biosynthesis in oleaginous yeast *Y. lipolytica* for biofuel production; *Metabol. Engg.*, 15, 1, 2013.
164. Tamano, K., Bruno, K.S., Karagiosis, S.A., Culley, D.E., Deng, S., Collett, J.R., Umemura, M., Koike, H., Baker, S.E., Machida, M., Increased production of fatty acids and triglycerides in *Aspergillus oryzae* by enhancing expressions of fatty acid synthesis-related genes; *Appl. Microbiol. Biotechnol.*, 97, 269, 2013.
165. Tan, Z., Liu, S., Bliss, N., Tieszen, L.L., Current and potential sustainable corn stover feedstock for biofuel production in the United States; *Biomass Bioenerg.*, 47, 372, 2012.
166. Tang, W., Zhang, S., Tan, H., Zhao, Z.K., Molecular cloning and characterization of a malic enzyme gene from the oleaginous yeast *Lipomyces starkeyi*; *Mol. Biotechnol.*, 45, 121, 2010.
167. Tanimura, A, Takashima, M., Sugita, T., Endoh, R., Ohkuma, M., Kishino, S., Ogawa, J., Shima, J., Lipid production through simultaneous utilization of glucose, xylose, and L-arabinose by *Pseudozyma hubeiensis*: a comparative screening study; AMB Express 6, 58, 2016.
168. Theresa, M.S., Ort, S., Hazra, S., Lavie, A., Konrad, M., Mimicking phosphorylation of Ser-74 on human deoxycytidine kinase selectively increases catalytic activity for dC and dC analogues; *FEBS Letter,* 582, 720–4, 2008.

169. Thiru, M., Sankh, S., Rangaswamy, V., Process for biodiesel production from *Cryptococcus curvatus; Bioresour. Technol.*, 102, 10436, 2011.
170. Tsigie, Y.A., Wang, C.Y., Kasim, N.S., Diem, Q.D., Huynh, L.H., Ho, Q.P., Truong, C.T., Ju, Y.H., Oil production from *Y. lipolytica* Po1g using rice bran hydrolysate; *J. Biomed. Biotechnol.*, 2012.
171. Tsigie, Y.A., Wang, C.Y., Truong, C.T., Ju, Y.H., Lipid production from *Y. lipolytica* Po1g grown in sugarcane bagasse hydrolysate; *Bioresour. Technol.*, 102, 9216, 2011.
172. U.S. Department of Energy Office of Science. *Breaking the Biological Barriers to Cellulosic Ethanol: A Joint Research Agenda.* Report from the December 2005 Workshop, June 2006.
173. Van, Soest, P.J., Rice straw, the role of silica and treatments to improve quality; *Animal Feed Sci. Technol.*, 130, 137, 2006.
174. Viljoen, M., Subden, R.E., Krizus, A., van, Vuuren, H.J., Molecular analysis of the malic enzyme gene (mae2) of *Schizosaccharomyces pombe*; Yeast 10, 613, 1994.
175. Wan, C., Li, Y., Fungal pretreatment of lignocellulosic biomass; *Biotechnol. Adv.*, 30, 1447, 2012.
176. Wang, M.D., Wu, H., Fu, G.B., Zhang, H.L., Zhou, X., Tang, L., Dong, L.W., Qin, C.J., Huang, S., Zhao, L.H., Zeng, M., Wu, M.C., Yan, H.X., Wang, H.Y., Acetyl-coenzyme A carboxylase alpha promotion of glucose-mediated fatty acid synthesis enhances survival of hepatocellular carcinoma in mice and patients; *Hepatology* 63, 1272–86, 2016.
177. Ways, P., Hanahan, DJ., Characterization and quantification of red cell lipids in normal man; *J. Lipid Res.*, 5, 318, 1964.
178. Wenzl, H., *The chemical technology of wood*; Academic Press; New York, 2012.
179. Wijaya, Y.P., Putra, R.D.D., Widyaya, V.T., Ha, J.M., Suh, D.J., Kim, C.S., Comparative study on two-step concentrated acid hydrolysis for the extraction of sugars from lignocellulosic biomass; *Bioresour. Technol.* 164, 221, 2014.
180. Woli, K.P., David, M.B., Tsai, J., Voigt, T.B., Darmody, R.G., Mitchell, C.A., Evaluating silicon concentrations in biofuel feedstock crops *Miscanthus* and *Switchgrass*; *Biomass Bioenerg.*, 35, 2807, 2011.
181. Xu, H., Hazra, S., Blanchard, J.S., NXL104 Irreversibly inhibits the β-Lactamase from *Mycobacterium tuberculosis*; *Biochem.*, 51, 4551, 2012.
182. Ykema, A., Verbree, E.C., Kater, M.M., Smit, H., Optimization of lipid production in the oleaginous yeast *Apiotrichum curvatum* in whey permeate; *Appl. Microbiol. Biotechnol.*, 29, 211, 1988.
183. Yu, X., Zheng, Y., Dorgan, K.M., Chen, S., Oil production by oleaginous yeasts using the hydrolysate from pretreatment of wheat straw with dilute sulfuric acid; *Bioresour. Technol.*, 102, 6134, 2011.
184. Yu, X., Zheng, Y., Xiong, X., Chen, S., Co-utilization of glucose, xylose and cellobiose by the oleaginous yeast *Cryptococcus curvatus*; *Biomass Bioenerg.*, 71, 340, 2014.

185. Yüksel, F., Yüksel, B., The use of ethanol–gasoline blend as a fuel in an SI engine. *Renewab. energy.*, 29, 1181, 2004.
186. Zhang, B., A possible connection between fast radio bursts and g-ray bursts; *Astrophys J. Letts.*, 780, L21, 2013.
187. Zhang, H., Zhang, L., Chen, H., Chen, Y.Q., Ratledge, C., Song, Y., Chen, W., Regulatory properties of malic enzyme in the oleaginous yeast, *Yarrowia lipolytica*, and its non-involvement in lipid accumulation; *Biotechnol. Letts.*, 35, 2091, 2013.
188. Zhang, Y., Adams, I.P., Ratledge, C., Malic enzyme: the controlling activity for lipid production? Overexpression of malic enzyme in *Mucor circinelloides* leads to a 2.5-fold increase in lipid accumulation; *Microbiol.*, 153, 2013, 2007.
189. Zhao, X., Kong, X., Hua, Y., Feng, B., Zhao, Z.K., Medium optimization for lipid production through co-fermentation of glucose and xylose by the oleaginous yeast *Lipomyces starkeyi*; *Eur. J. Lip. Sci. Technol.,* 110, 405, 2008.
190. Zhou, C.H., Beltramini, J.N., Fan, Y.X., Lu, G.Q., Chemoselective catalytic conversion of glycerol as a biorenewable source to valuable commodity chemicals; *Chem. Soc. Rev.*, 37, 527, 2008.
191. Zhou, C.H., Xia, X., Lin, C.X., Tong, D.S., Beltramini, J., Catalytic conversion of lignocellulosic biomass to fine chemicals and fuels; *Chem. Soc. Rev.*, 40, 5588, 2011.
192. Zhu, J., Wan, C., Li, Y., Enhanced solid-state anaerobic digestion of corn stover by alkaline pretreatment. *Bioresour. Technol.*, 101, 7523, 2010.
193. Zhu, L.Y., Zong, M.H., Wu, H., Efficient lipid production with *Trichosporon fermentans* and its use for biodiesel preparation; *Bioresour. Technol.*, 99, 7881, 2008.
194. Zhu, Y., Lee, Y.Y., Elander, R.T., Optimization of dilute-acid pretreatment of corn stover using a high-solids percolation reactor; *Appl. Biochem. Biotechnol.*, 124, 1045, 2005.

15

Pre-Treatment of Lignocellulose for the Production of Biofuels

Biva Ghosh[1], Debalina Bhattacharya[2] and Mainak Mukhopadhyay[1]*

[1]*Department of Biotechnology, JIS University, Kolkata, West Bengal, India*
[2]*Department of Biochemistry, University of Calcutta, Kolkata, West Bengal, India*

Abstract

Lignocellulose, which is produced from plant resources are the most abundant and cheap resources available in our earth that can be used for biofuel production. It can be produced from all plant related wastes such as agricultural wastes, vegetables wastes, terrestrial plants etc. Lignocellulosic biofuels are one of the best supplements to fossil fuels for the production of energy. It is eco-friendly, renewable and acts as sustainable source of energy. The pre-treatment of lignocellulose includes many different processes such as enzymatic and chemical treatment for the production of different biofuels such as bioethanol, butanol etc. As lignocellulosic biomass has the capacity to produce different biofuels thus it can be studied for the production of more different biofuels which is easy to transport, handle and feasible to be used in place of fossil fuels such as coal, petroleum etc. Many more studies are also done to enhance the production of biofuels in efficient way for more yield and consistence quality. Thus, the chapter of this book deals with the different pre-treatment processes of lignocellulose for high yield as well as consistence quality of different biofuels. It also puts light on different possible biofuels that can be produced from the lignocellulose biomass.

Keywords: Lignocellulose; biofuels; pre-treatment

15.1 Introduction

As population is increasing, the demand of energy is also increasing. From the ancient times, fossil fuel is being used as the energy resource. But combustion

**Corresponding author*: m.mukhopadhyay85@gmail.com

Arindam Kuila and Vinay Sharma (eds.) Principles and Applications of Fermentation Technology, (307–350) © 2018 Scrivener Publishing LLC

of fossil fuel leads to pollution which is a major environmental problem [1]. As well as fossil fuel is also not sustainable resource. Thus to fulfil the increasing demand of energy, biofuel is the best supplement of fossil fuels [2].

Biofuels are the biochemical resources which can be used as fuels but generated from photosynthetic living biomass such as plant resources, or microbial resources such as vegetable wastes, food industry wastes, algae, bacterial etc. [2]. Due to the local availability of photosynthetic biomass and its renewable nature sustainable development of industry depending on renewable resource occurs, which leads to economic development in developing countries [3–5]. Biofuels can be categorized into primary and secondary biofuels. Primary biofuels are produced directly from unprocessed plant and animal waste whereas secondary biofuels are produced from processed biomass. Biofuels are further divided into first generation, second generation, and third generation biofuels. First generation biofuels are produced from sugar and oil rich crops that are generally edible crops whereas second generation biofuels are produced from waste biomass and third generation biofuel such as biodiesel is produced from microorganism such as algae [6–8]. Photosynthetic plants not only help to produce biofuel but also help in fixing carbon dioxide in the nature. Photosynthetic process occurs in two stages i.e., light dependent and light independent [4]. In light dependent pathways, the light energy is directly absorbed by the chlorophyll and converted to Adenosine triphosphate (ATP) and Nicotinamide adenine dinucleotide phosphate with hydrogen (NADPH). During this process electron transport chain is activated by the electron carriers present in thylakoids such as Ferredoxin, Plastoquinone, Plastocyanin, and Cytochrome C. Ferredoxin Nicotinamide adenine dinucleotide phosphate (NADP) oxidoreductase reduces $NADP^+$ to NADPH in presence of water. In oxygenic condition of photosynthesis, water molecules are broken into electrons and protons. These protons are sometimes converted to molecular hydrogen by hydrogenase rather than accepted by Ferredoxin. Thus, oxygenic condition of photosynthesis produces bio-hydrogen biofuel. On the other hand, light independent photosynthesis produces ATP and NADPH which further converts to sugar molecules which help to synthesize bio-alcohol, biodiesel and on further fermentation produce bio-hydrogen as biofuels [2, 4].

Thus, photosynthetic biomass is storage of natural energy, which can be used to produce sustainable biofuels. As there are lot of agricultural waste, nonedible parts of plant, waste biomass in different industries such as food industries, textile industries etc. which could be efficiently used to produce different biofuels. This process will not only solve the air pollution problem but also decrease the waste load on the environment [1]. These wastes thus can increase the country's economy without much investment, which is very helpful for developing countries. These plant or photosynthetic biomass mainly consists of huge amount of lignocellulose. Lignocellulose is very robust in nature, consisting of microcrystalline cellulose covered with hemicellulose and lignin. Thus, degradation of lignocellulose to simpler molecules are very tough [9, 10]. Lignin is the most resisting molecule toward converting to simpler molecules. Thus, pre-treatment of lignocellulose is needed, depending on the degree of lignification. This pre-treatment process includes chemical, mechanical, or biological treatment [11]. As the process of converting lignocellulose to simpler molecules is costly, thus its use in replace of fossil fuel is still limited. Many modifications and researches are done to improve the pre-treatment process, which are discussed in this chapter. This chapter also puts light on other biofuels rather than focussing only on the traditional one as well as new upcoming sources of biofuels.

15.2 Lignocellulose

Lignocellulose is very robust in nature, consisting of microcrystalline cellulose covered with hemicellulose and lignin and is found in plants and photosynthetic organisms [9, 10]. The main sources of lignocelluloses, which can be used for biofuel production, are forestland residue, agricultural waste, municipal and industrial waste, marine algae [2, 12–14].

The main composition of lignocellulose consists of cellulose (30–50%), hemicellulose (15–35%) and lignin (10–20%), which are linked by covalent and hydrogen bonding that makes them rigid and strong [1, 15]. Cellulose and hemicellulose present in lignocellulose are fermentable, thus it is used to produce biofuels. But their chemical characteristics and interaction between the molecules makes them robust and firm rejecting biodegradation into small molecules by enzymes and chemical alone [13]. Cellulose is one of the most abundant components of cell wall. It is composed of linear chain of glucose molecules clustery known as cellobiose. These cellobioses are linked together by strong β-1–4 glycosidic bond [9, 16]. As cellulose consist of hydroxyl groups, which provides hydrogen bonding and Van der Waals forces between inter and intra molecules, providing it a microfibrillar structure [17]. The different orientation of cellulose gives them different crystallinity. Cellulose microfibrils further can be divided into amorphous (low crystallinity) and crystalline (high crystallinity) regions. The higher the crystallinity tougher to degrade in simpler molecules [14, 18]. The crystallinity of cellulose can be measured by crystallinity index. These cellulose microfibrils are further wrapped with hemicellulose and then lignin [15]. In contrast with cellulose, hemicellulose consists of amorphous regions and heterogeneous polysaccharides such as pentose, hexose, and acids. This nature of hemicellulose helps to form network like structure with cellulose forming cellulose- hemicellulose- lignin rigid structure [19, 20]. The branched nature of hemicellulose leads to more susceptibility to biological, chemical, thermal hydrolysis under certain physical conditions [15, 21].

The second most abundant natural polymer present in lignocellulose is lignin. Lignin is a large, complex aromatic, hydrophobic heteropolymer. The heteropolymer consist of phenylpropane units such as coniferyl and sinapyl alcohol with hydroxyl, methoxyl, carbonyl as functional group [22]. These structural properties of lignin make it as a perfect polymer to cross-link cellulose and hemicellulose providing it rigidity [23, 24]. Though lignin is not soluble in water, but it dissolves in water at high temperature at neutral pH [24]. Other than this, lignin may dissolve in acidic or alkaline nature depending on precursors of the lignin. These properties of lignin make the cellulose -hemicellulose - lignin more robust and hard in nature. As the lignin quantities increase in the biomass, recalcitrant nature increases and resistance toward biodegradation increases [20, 21]. It is known that soft woods has more lignin than hard woods which comprises mostly feedstocks, agricultural waste etc. [15]. Thus, as hydrolysis of lignocellulose is tough and need pre-treatment to improve and easy the digestibility of lignocellulose and increase the efficiency of bioconversion. Improvement in pre-treatment will also lower the cost of biofuel generation making it more affordable.

15.3 Parameters Effecting the Hydrolysis of Lignocellulose

Lignocelluloses are very robust and recalcitrant in nature. Thus, pre-treatment is done to degrade lignocellulose into simple reducing sugar which can be easily hydrolysed by enzymes or can be easily fermented [9]. There are many types of pre-treatment such as physical, chemical, physio-chemical, biological pre-treatments [13]. The pre-treatment of the lignocellulosic biomass depends on the nature of biomass. The nature of the biomass depends on crystallinity of cellulose, degree of polymerization (DP), moisture, and lignin content and available surface area.

15.3.1 Crystallinity of Cellulose

Cellulose microfibrils consist of two regions crystalline component and amorphous component. The crystallinity depends on the relative amount of these components [9]. As amorphous region of cellulose microfibrils are more accessible, thus enzymes could easily hydrolyse amorphous region of cellulose microfibrils whereas crystalline region is not easy accessible. The more the crystalline regions in the lignocellulosic biomass the more recalcitrant they are [25].

15.3.2 Cellulose Degree of Polymerization

Degree of polymerization (DP) of cellulose depends on the strength of cellulose fibre, accessible surface area, pore size, and encapsulation of cellulose by hemicelluloses–lignin matrix but mainly depends on the crystallinity of cellulose [26]. The enzymes such as Endoglucanases and cellobiases acts upon β,1–4 glycosidic bond at internal site of cellulose chain connecting two sugar molecules. Degradation of these β,1–4 glycosidic bond decreases the DP cellulose. This also releases the recalcitrant nature of cellulose crystals. Many chemical, physical or biological pre-treatment can be done to decrease DP of cellulose [27].

15.3.3 Effect of Accessible Surface Area

Accessible surface area of the biomass decides the yield after enzymatic hydrolysis or fermentation. This accessible surface area depends on two factors (i) external and (ii) internal [28]. External factors are size and shape of the biomass fibers, which can be managed by physical pre-treatment such as size reduction by chipping, shredding, grinding, and milling. However, internal factors depend upon capillary structure of cellulosic fibers. Depending on these factors, pre-treatment must be deigned to get better yield [29].

15.3.4 Encapsulation by Lignin

Lignin is the most robust and recalcitrant part of lignocellulosic biomass. Lignocellulosic biomass consists of cellulose fibrils covered with hemicellulose and further protected with lignin. Thus, without degradation of lignin digestible part of biomass is not exposed. So to expose the digestive part which consist of cellulose and hemicellulose which can be converted to simple reducing sugar molecules lignin has to be degraded [30]. It is known that enzymes such as cellulases are inhibited by lignin whereas xylanases and glucosidase are less affected. Softwoods are more robust than hardwoods because of the fact that soft wood consist of more amount of guaiacyl lignin whereas hardwood consist of mix of guaiacyl and syringyl lignin. Guaiacyl lignin reduces biomass swelling than syringyl lignin resulting in more robust nature of soft wood [31]. Delignification causes increase in internal surface area, biomass swelling, disruption of lignin structure, and increased accessibility of cellulolytic enzymes, which results in better enzymatic saccarification and increased sugar yield [9, 30].

15.3.5 Hemicellulose Content

Hemicellulose when degraded by the pre-treatment produces simple sugar molecules such as pentose and hexose and some value added products such as (5-hydroxymethylfurfural, xylitol, ethanol, butanediol, butanol) and polymers (polyhydroxyalkanoates, polylactates). Some of these such as 5-hydroxymethylfurfural are inhibitors for enzymatic hydrolysis [30]. Hexose and pentose on further treatment such as fermentation produces bioethanol. As hemicellulose degrade porosity of the biomass increases and cellulose become more accessible for conversion to reducing sugars [32]. Degree of acetylation is another factor that effect porosity. Acetyl group of hemicellulose is bonded with lignin, which are very tough to break. Thus, the content of acetylated hemicellulose determines the porosity of the biomass[33].

15.3.6 Porosity

Porosity of the biomass mainly effects the enzymatic saccharification. Size of the enzyme to the size of the pore effects the enzyme activity. Cellulase gets trapped in a pore where the internal pore size is bigger than external pore size seen in many lignocellulosic biomasses. Other factors such as drying of biomass decreased hydrolysability. All these factors affect the enzyme activity over the biomass. Pre-treatment causes increase in pore size which increases the efficiency of enzymatic hydrolysis [30, 33].

15.4 Pre-Treatment of Lignocellulose

As discussed above lignocellulose degradation to simpler molecules is very tough due to their rigid nature. Pre-treatment is needed to make them more feasible for production of different biofuels. Many bio-refinery systems have developed for the production of different biofuels and chemicals from lignocellulosic biomass. There are mainly two primary type of bio-refinery route. They are biological conversion route and thermochemical route. In case of thermochemical bio-refinery, through gasification biomass is converted to biogas or through pyrolysis and catalytic hydrothermal treatment leads to bio-oil formation or other liquid fuels [34]. Whereas biological conversion includes production of sugar molecules by biodegradation of biomass using biological or chemical means. This sugar is further fermented to form biofuels such as bioethanol (Figure 15.1). Pre-treatment is one of the most important steps to increase the accessibility of carbohydrate polymer to hydrolytic enzymes which is one of the most important steps toward biofuel production [35]. Hydrolysis conversion of sugar makes them accessible for fermentation by extracting hemicellulose, disrupting lignin and finally releasing cellulose for fermentation. Depending on the type of biofuel to be formed, pre-treatment is divided into biological, physical, physiochemical, and chemical treatment (Figure 15.2) [11] .

15.4.1 Physical Pre-Treatment

15.4.1.1 Milling

Mechanical grinding or milling includes chipping, grinding, and/or milling techniques for the reduction the crystallinity of cellulose. Chipping reduces the size of biomass to 10–30 mm whereas milling or grinding reduces the size to 0.2 mm [36]. Due to the shearing force involved in milling and

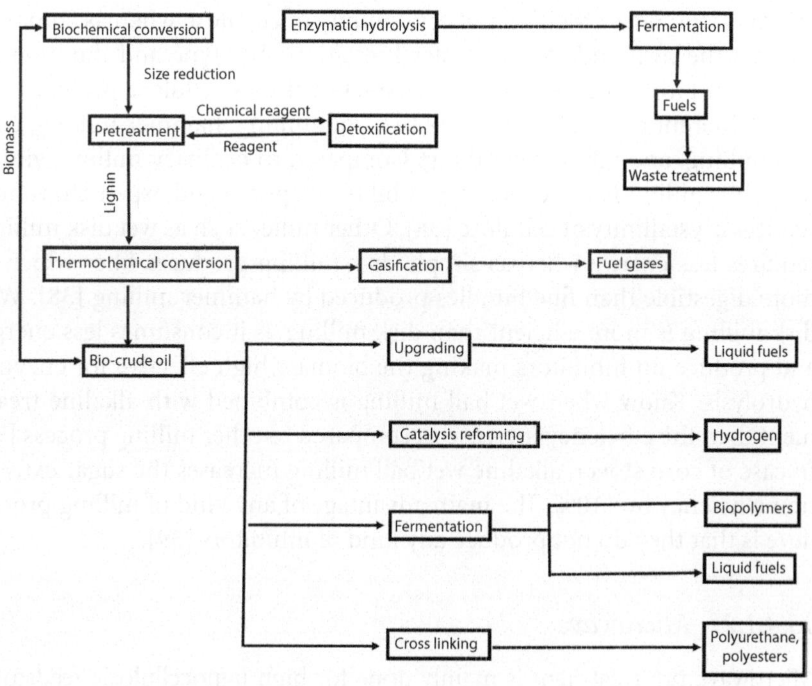

Figure 15.1 Schemetic diagram of pretreatment process leading to different biofuel formation.

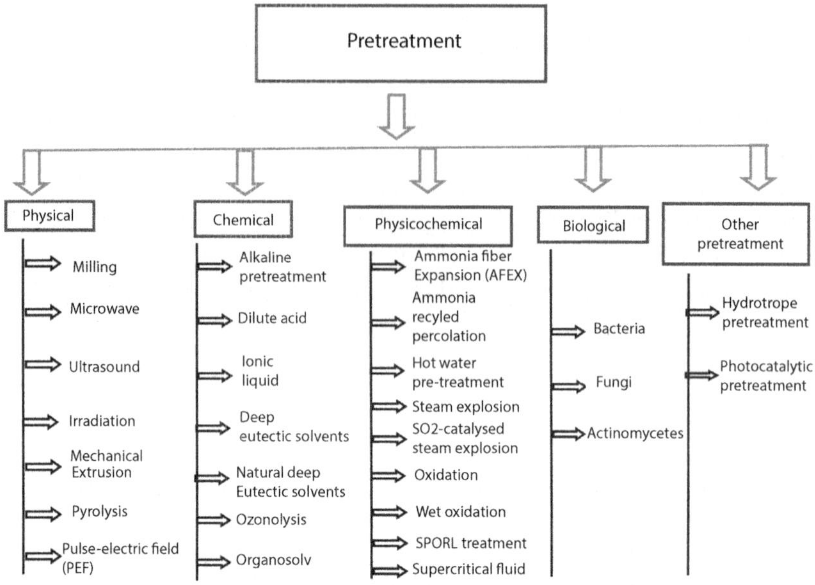

Figure 15.2 Types of pretreatment.

grinding, it reduces the size and crystallinity of cellulose whereas chipping reduces the heat and mass transfer limitation. The type and duration of milling of biomass determine the characteristics of cellulose present in it [37]. Different types of milling are two-roll milling, hammer milling, colloid milling, and vibratory milling. Compared to ordinary milling, vibratory ball milling is better for digestibility of spruce and aspen chips and reduces crystallinity of cellulose [38]. Other miller such as wet disk milling requires less energy whereas simple disk milling produces fiber which is more digestible than fine bundles produced by hammer milling [38]. Wet disk milling is more efficient than disk milling as it consumes less energy and produce no inhibitors making the biomass high effective for enzyme hydrolysis. Know when wet ball milling is combined with alkaline treatment then the efficiency increase as compared to other milling process [9]. In case of corn stover, alkaline wet ball milling increases the sugar extraction efficiency by 110%. The main advantage of any kind of milling procedure is that they do not produce any kind of inhibitors [39].

15.4.1.2 Microwave

Microwave pre-treatment is mainly done for high lignocellulosic feedstock because of main advantages such as (i) it is easy to operate as well as consumes low energy (ii) produces high heat at short duration of time (iii) high heat degrades crystalline structure of cellulose into small molecules (iv) produces negligible inhibitors [40]. Microwave technique is made more efficient by combining with mild alkaline treatment which increases its efficiency [41]. In recent studies, it is mentioned that mild alkaline treatment combined with microwave resulted in 70–90% cellulose extraction from switch grass biomass. In case of microwave pre-treatment combined with alkaline addition, the biomass is heated at 150 °C for 5 mins where as in case of acid treatment combination such as H_2SO_4 the biomass is heated at 180 °C for 20 min [41].

15.4.1.3 Ultrasound

Ultrasound or sonication is comparatively new but effective pre-treatment for lignocellulosic biomass as it has both physical and chemical effect on the biomass. Ultrasound changes the morphology of cellulose and hemicellulose by formation of small cavitation bubbles. This allows enzymes to easily access on the biomass surface and penetrate to convert cellulose and hemicellulose into simple reducing sugars [42]. The major factors affecting this pre-treatment are ultrasonic frequency, duration, reactor geometry and its type and solvent. Other factors are biomass type and reactor configuration and kinetics. All these factors play vital role for production of maximum yield. Alkali use as a solvent increases the delignification but the main factor that effects the delignification is power and duration of sonicate [43]. Maximum researchers sonicate at 10–100 kHz for better results. Higher sonication than this cause formation of bubbles in front of transducer and the power of sonication cannot reach the liquid medium. Other than this, if the power is increased to 400 W, oxidation of cellulose is seen. Thus, optimization of power and duration is needed for maximum yield of reducing sugar [44].

15.4.1.4 Irradiation

Irradiation includes delignification of biomass using gamma rays and electron beams to improve enzymatic hydrolysis. However, irradiation with high power that is 100 MR may lead to decomposition of oligosaccharides and glucose ring. In case of rice straw, it yields 52% reducing sugar and in case of wheat straw the yield was 13% after enzyme hydrolysis [45].

15.4.1.5 Mechanical Extrusion

In mechanical extrusion, pre-treatment process both heat (> 300 °C) and shearing process is used for delignification. This pre-treatment process mainly produces gaseous and char as product [46]. Due to the application of high temperature in the reaction barrel and shearing force by the rotating screw blades, the amorphous and crystalline structures of cellulose are disrupted. Mechanical extrusion defibrillate the lignocellulosic biomass as well as shorten the fibril size. It also increases the carbohydrate content in the mixture, which results in easy accessibility of enzymes for hydrolysis. The factors affecting biomass delignification are screw design, compression ratio, screw speed, and barrel temperature [47]. The maximum yield of sugar can be extracted from the lignocellulosic biomass by optimizing the system to 75 rpm and 125 °C and with the combination of enzymes cellulase and β-glucosidase in the ratio of 1:4. In this process, production of inhibitor are very lower than other pre-treatment process [46]. The biomass containing high lignin such as soybean hulls does not produce better yield of sugar whereas less lignin content biomass such as wheat bran produces high reducing sugar. This may be because cellulose degrading enzymes avidly and irreversibly bind to lignin, which make them unavailable for cellulose degradation. Though this pre-treatment has high advantage and high yield with low inhibitor production, but is not cost efficient as it requires significant amount of high energy making it tough to scale up for industries [48].

15.4.1.6 Pyrolysis

Pyrolysis pre-treatment is an endothermic process where lignocellulosic biomass is decomposed into simpler molecules by thermochemical process. In this process, the temperature is generally raised greater than 300 °C in which the fibrous structure of biomass is disrupted, and H_2 and CO_2 are released and some residual char are left. After this pre-treatment, the caloric value of the biomass is increased as well as it gains hydrophobic nature which improves stability of the biomass while storage. In this pre-treatment at low temperature low volatile products are produced [49]. This residual char contains enough carbon source for microbial growth for fermentation [50]. Treatment of residual char with water and mild acid converts 80–85% of cellulose into reducing sugar consisting of more than 50% glucose [51]. Pyrolysis is of two types, slow and fast depending on the heating rate. Depending on the type of pyrolysis, biomass characteristic and reaction parameters different products are produced. As it produces high caloric products and it is adapted by thermal industries. Pyrolysis in presence of oxygen is more efficient than in presence of nitrogen. When pyrolysis of biomass is done in presence of oxygen at 25 °C, 7.8 × 109 bonds/min/g cellulose is cleaved where as in presence of nitrogen, 1.7 × 108 bonds are cleaved [52].

15.4.1.7 Pulse Electric Field (PEF)

In pulse-electric field, pre-treatment of biomass is done by subjecting the biomass to a sudden change of high voltage between 5.0–20.0 kV/cm for short durations (nano to milliseconds). In this process due to sudden electric pulse the pores of the cell member increases and allows the agents to enter and convert cellulose into reducing sugar. PEF is a simple instrument that needs very low energy [53]. It is very efficient for the production of biogas such as methane from sludge and manure. About two fold increase in methane production from sludge and 80% increase of methane from manure was reported [13]. This pre-treatment process is also useful for softwood and hardwood biomass.

15.4.2 Chemical Pre-Treatment

15.4.2.1 Alkaline Pre-Treatment

It includes chemical treatment such as sodium hydroxide, lime, ammonium fibre expansion, ammonium recycle percolation. It is known that alkaline solution improves digestibility of cellulose and also degrades lignin [11].

15.4.2.1.1 Sodium Hydroxide

Among alkaline solution, sodium hydroxide is best for pre-treatment to digest lignocellulose because of its high alkalinity than others. Depending on the concentration of NaOH cellulose is extracted [54]. It is mentioned by Zhang and Shahbazi (2011) that 4% of NaOH can yield 78% of cellulose from raw cattails and increases its fermentation rate. NaOH with other chemicals such as H_2O_2 and urea also increases lignin degradability. 5% NaOH and 5% H_2O_2 yield 80% sugar at 80 °C and 7 % NaOH and 12% urea solution yield 70% sugar from wood chips at -15 °C [11]. This treatment also increases the enzymatic hydrolysis efficiency of the lignocellulose biomass. NaOH treatment doesnot require high temperature and pressure leading to cost effectiveness. NaOH is one of the best pre-treatment which increases biomass porosity, reduces cellulose crystallinity, increases DP and removes lignin and hemicellulose [55]. One example of NaOH pre-treatment was done on *Ipomoea carnea* by Sharma *et al.* (2015) where 1%, 3% and 5% NaOH was used as pre-treatment. Maximum yield of fermentable sugar (202.38 μg/μl) was obtained by 3% NaOH treatment at 50 °C for 5 days incubation [55]. According to Sambusiti *et al.* (2012) 10% NaOH pre-treatment on sorghum forage and wheat straw at 40 °C for 24 hrs increased the digestibility of lignocellulose and increased the fermentable sugar to 5 times higher [56]. Another example of genetically modified Alamo switchgrass *(Panicum virgatum L.)* where expression of 4-coumarate-CoA ligase (4CL) was reduced leading to 5.8% reduction in lignin content. When this grass was pre-treated with 0.5%, 1%, and 2% (w/v) NaOH for 15, 30, and 60 min at 121 °C glycan and xylan conversion was higher as 16% and 18% than other conventional method. Thus leading to higher sugar yield for fermentation with low alkali treatment[57]. Though pre-treatment with NaOH does not require high temperature and pressure but it requires long time duration for pre-treatment and moreover recovery of NaOH is also very complex [58]. NaOH being costly it a huge disadvantage to overcome with different combination of other pre-treatments to reduce the cost effect [11].

15.4.2.1.2 Lime

Lime pre-treatment is similar to NaOH pre-treatment. But it is less expensive and easy to recover rather than NaOH. It is reported in Zhang *et al.* (2011) that when 0.5g/ g biomass $Ca(OH)_2$ was used to pre-treat corn stover at 25- 55 °C lignin and hemicellulose was removed but cellulose remain unaffected [11]. Though the degree of crystallinity increases to 43% to 60% after delignification. Sierra *et al.* (2009, 2010, 2011) when combined lime pre-treatment with oxygen pressure the delignification and digestibility of cellulose to fermentable sugar increase [59–61]. When lime pre-treatment was continued with fermentation by microorganism then carboxylic acid like acetic, propionic or butyric acids are produced. These were neutralized with calcium carbonate, which resulted in carboxylate salts such as calcium acetate, propionate and butyrate. These salts were further dried and thermal treatment done to convert into ketone such as acetone. Further hydrogenation leads to different secondary alcohols [11]. The carboxylic acid used is further recovered from fermentation solution by reacting with tertiary amines to form tertiary amine carboxylates and calcium carbonate. These salts were further regenerated by cracking into carboxylic acid and tertiary amines. The secondary alcohols produced are further oligomerized to form biodiesel, gasoline, jet fuel etc. In this way, reuse of lime is done. Though it has some demerits such as neutralization need lot of wash water and time as lots of salts are produced and lime recovery process is also very costly. Thus if a better lime recovery process can be found then it can be a cheapest and effective pre-treatment process [11].

15.4.2.2 Dilute-Acid Pre-Treatment

In this process, 0.5–1% sulphuric acid is used at temperature of 140–190 °C at 3–15 atm pressure to remove and recover dissolve hemicellulose. Along with this lignin is also disrupted and removed from the biomass thus increasing the enzyme susceptibility to pre-treated biomass [62]. Dilute acid hydrolysis leads to two types of chemical reaction (i) conversion of cellulose extraction to sugar (ii) to convert sugar into other chemicals. The main disadvantage of this method is the production of inhibitory compounds such as organic acids, furans, and phenols, which affects the downstream fermentation by microbes. The other disadvantages are the due to high corrosive costly construction materials are required and requirement for acid neutralizer [11]. In this treatment process the biomass is first preheated then followed by heating to final temperature and finally cooling down of the reaction system is done. Here the reactors used are batch, continuous, percolation reactors [13]. In conventional batch reactor, the biomass is indirectly heated by heaters or directly heated by stem injection. In continuous pre-treatment, the reactor used is plug-flow reactors. Here, first, the biomass is steamed in a separate reactor and the non-condensable gases are separated from the biomass before sending it to the main reactor. The reactor consists of a moving paddle on central shaft and continuous steam pressure on which solid biomass is passed though. In this system time consumed for reaction is short and high solid biomass can be loaded [63]. In percolation reactor, dilute acid is passed through the bed of biomass, which solubilizes the monomers from the media and removes it thus minimises the decomposition of sugars. As the acid used is low in concentration thus to increase the activity time of acid on the biomass can be increase by increasing the retention time or reducing solid loading [64]. Generally, the solid loading done is 2–4 % for 12–24 mins at 190–200 °C temperature and 2–24 atm pressure. The major factors affecting the yield of sugar are retention time, neutralization and decomposition of sugar produced [64]. The main drawback of this pre-treatment is low solid to liquid ratios and dissolved sugar in the liquid that requires costly methods for purification. In this acid catalysed reaction two types of product are form (i) solid and (ii) liquid. The solid part includes cellulose and lignin and is called hydrocellulose. On the other hand, the liquid part consists of solubilized hemicellulose such as xylooligomers and xyloses and lignin. The solid part is neutralized by washing but in case of liquid part bases such as NaOH or $Ca(OH)_2$ are used to neutralize. This causes this system to be costly [34] . The reaction mechanism involved in system are breaking of hydrogen bonding to decrystallize cellulose and beak the glycosidic bonds of hemicellulose thus the molecules are loosening and dissolves in the solution. In severe condition, degradation of cellulose and lignin occurs producing inhibitory products such as furfural, 5-hydroxymethylfurfural (HMF), and organic acids such as uronic, formic, levulinic, and acetic acid.

These inhibitory products hamper, the fermentation and enzymatic hydrolysis. The other reasons for the loss of glucose molecules are (i) formation of glucose-lignin complex (ii) reverse reaction causes glucose to form oligosaccharides. Thus, the reaction mixture is taken care of by adjusting different parameters to reduce the production of inhibitory products [64, 65].

15.4.2.3 Ionic Liquids

Ionic liquid are salts in liquid form with low boiling point and consist of short-lived ion pairs. These liquids are environment friendly and can replace other organic solvents. As their volatility is low thus it has less effect on environment and the reagents used in pre-treatment process [66]. Some of the well-known ionic liquids are 1-n-butyl-3-methylimidazolium chloride ([Bmim]Cl), 1-allyl-3 methylimidazoliumchloride ([Amim]Cl), N-methyl morpholine N-oxide (NMMO), 1-buthyl-3-methylimidazolium acetate [BMIM][OAc], 1-ethyl-3-methylimidazolium acetate ([Emim]Ac) and 1-ethyl-3- methylimidazolium diethyl phosphate ([Emim]Dep) which low volatility and low boiling point [67–70]. These ionic liquids consist of small anions and large cations and are liquid at room temperature with very low vapour pressure. This property of these liquids helps to play with the anions and cations to dissolve different lignocellulosic biomass such as switchgrass, cotton, bagasse, wheat straw, corn stover and pine, poplar, and oak of different hardness. NMMO is known to dissolve lignocellulosic biomass with greater than 99% solvent recovery rate and with no chemical modification in it. It is known as Lyocell solvent. It increases rate of cellulose hydrolysis by disrupting cellulose crystalline structure and thus improves enzymatic hydrolysis over the biomass [70]. Though ionic liquids are the most studied ones in this sector but it has few disadvantage which needs to be overcome such as high cost of ionic liquid, inhibitors generations which needs to be taken care of [13].

15.4.2.4 Deep Eutectic Solvents

Deep eutectic solvents (DES) are kind of ionic liquids. These solvents consist of two or three cheap and safe components that self-interact through hydrogen bonding to form eutectic mixture. DES has melting point lower than the melting point of each component present in it [71]. DES can be generally described by the formula $Cat^+ X^- zY$ where Cat^+ symbolizes ammonium, phosphonium, or sulfonium cation, X symbolizes Lewis base mainly halide anions, Y symbolizes Lewis or Brønsted acid where z symbolizes number of Y molecules that interacts with the halide anions.

Choline chloride (ChCl), one of the most widely used DES solvent because of its low cost, biodegradable and non-toxic ammonium salt, which can be easily extracted from the biomass. ChCl is a hydrogen donor like urea, carboxylic acids, and polyols [72]. Though DES acts as ionic liquids, but it is not always composed of entirely ionic compounds, it can be formed of non-ionic compounds. Rather DES mainly functions on hydrogen bond interactions [13].

15.4.2.5 Natural Deep Eutectic Solvents

In the recent years, many natural products are included in ionic liquid and DES categories. Compounds such as choline, urea, sugars, amino acids, and several other organic acids are considered as natural deep eutectic solvents (NDES). These solvents are cost effective, can easily be synthesize, non-toxic, biodegradable, eco-friendly, easy to recover for reuse. The principle behind NDES is the complex formation between hydrogen acceptor and hydrogen bond donor [73, 74]. These solvents have low melting point then each component present in it during their initial state. This phenomenon occurs due to charge delocalization of each component. NDES has high affinity toward lignin solubilisation and has multiple diverse functions. Though it has multiple advantage and is a very good solvent to dissolve lignin and hemicellulose during pre-treatment, its viscosity is the biggest disadvantage. The viscosity of the solvent is due to intense hydrogen bonding between the components of solvent. To reduce the viscosity, the solvent is diluted with 50% (v/v) water and the intense bonding almost disappears. It also has many pharmaceutical, food processing and enzyme industries applications [74, 75].

15.4.2.6 Ozonolysis

Ozonolysis pre-treatment is mainly done for delignification and it negligible affect hemicellulose and cellulose. It is done in ambient temperature and pressure and does not produce any toxic inhibitors. Thus, this pre-treatment is eco-friendly and it also does not affect the post pre-treatment treatments such as enzyme hydrolysis and fermentation by microbes. Various lignocellulosic biomasses such as bagasse, green hay, peanut, pine, wheat stalk is treated by ozonolysis and it increases the post pre-treatment treatments efficiency. The condition need for ozonolysis are moisture content of the biomass. If the moisture of the biomass are more then, less lignin are oxidised [76]. Thus, more amount of ozone is needed for pre-treatment, which makes this process costlier for use in industrial sector.

To overcome this problem research has been done to produce industrial feasible zone concentration and generation of different reactors such as packed bed, fixed-bed, and stirred tank semi-batch reactors that are capable of loading low moisture biomass nearly less than 30 % with particle size 1- 200 nm [13].

15.4.2.7 Organosolv

In this process, different aqueous organic solvents such as ethanol, methanol, ethylene glycol, acetone etc. are used for delignification under specific temperature, pressure and catalysts. Depending on the type of lignocellulosic biomass the temperature, pressure changes and catalysts used are acid, base or salt. This process mainly used to extract lignin, but cellulose and hemicellulose of C5 and C6 carbons are also extracted. Due to effective delignification cellulose are exposed for better enzymatic hydrolysis thus improving its efficiency. Depending on the temperature, pressure and catalyst and retention time of pre-treatment the nature of pre-treated biomass is different [77]. High temperature and acid concentration leads to formation of inhibitory products, which hampers the fermentation process. In a study of pine biomass delignification H_2SO_4, NaOH, and $MgSO_4$ was used as catalyst where H_2SO_4 has shown to be better catalyst for ethanol production but with respect to digestibility of biomass NaOH is the best one [78]. Though these solvents are costly and this is the biggest disadvantage of organosolv (OV) to be used as industrial scale purification, but it can be solved by recycling and reuse of the solvents used. Recovery of these solvents is very important because they will otherwise effect the fermentation and hydrolysis process. OV are rarely used for delignification because of the high-risk factors involved in the handling of harsh organic solvents. Other than this these solvents are highly inflammable and corrosive in nature. Handling these solvent at high temperature and high pressure is a high-risk job. Some of the most commonly used oranosolvs are Battelle, formasolv, ethanosolv. Battelle is a combination of phenol, HCl, and water where acid moiety that is HCL depolymerizes lignin as well as hydrolyses hemicellulose. Lignin is further dissolves in phenol and while the sugar can be obtained from the aqueous phase after cooling. This whole process is conducted under 100 °C temperature at 1 atm pressure. Another solvent formasolv is a combination of formic acid, water, and HCl where lignin is dissolved in formic acid and sugars are extracted from aqueous solvent under low temperature and pressure. Ethanosolv are another solvent composed of ethanol, water and HCL under high temperature and pressure [79]. This solvent also functions in the same way as other mentioned OVs.

Ethanol is less toxic than other solvents but it hinders the hydrolysis process. Thus, complete extraction of ethanol is needed. This problem can be overcome by diluting ethanol with water where majority is water and less ethanol and this diluted solvent is easy to extract for reuse making it industrial feasible and cost effective [13].

15.4.3 Physicochemical Pre-Treatment

15.4.3.1 Ammonia Fiber Expansion (AFEX)

Ammonia fiber expansion is a process in which the lignocellulosic biomass is pre-treated with high-pressure ammonium gas and heat for a period of time and then the pressure is released rapidly to explosive decompression which leads to decomposition of the lignocellulosic biomass into simpler molecules. Most of the ammonia used are recovered and the rest is used in the downstream by microorganism during fermentation. As ammonia is highly and ammonia decomposes into NH_4^+ and OH^- in water thus leading to easy recovery of ammonia [80]. This process decrystallizes cellulose and depolarised hemicellulose and lignin and separates cellulose from lignin and hemicellulose. It is one of the best methods of pre-treatment before enzymatic hydrolysis of biomass. The parameters of AFEX, which affects the yield of reducing sugar produced, are temperature, pressure, ammonia content, time for pre-treatment. With proper condition nearly 80 % of reducing sugar can be recovered [58]. The different reactors systems used for AFEX are Conventional batch reactors, Plug-flow reactor (PF-AFEX), Packed-bed reactor (PB-AFEX), Fluidized gaseous reactors (FG-AFEX) and extractive reactors (E-AFEX). It is most effective for agricultural waste but is not good for soft and hard wood. Enzymatic treatment such as xylase, hemicellulose are recommended to get reducing sugar [81].

15.4.3.2 Ammonia Recycled Percolation (ARP) and Soaking in Aqueous Ammonia

In ARP 5–15 wt % of aqueous ammonia is used to pre-treat lignocellulosic biomass through flow-through column reactor [11]. The aqueous ammonia follows through the column which is prepacked with lignocellulosic biomass under heat (150–180 °C) and pressure (9–17 atm) to avoid evaporation of ammonia as it is volatile in nature [16, 80]. The reaction continues for 10–90 mins with a flow rate of aqueous ammonia of 1–5ml/min [80]. After the reaction is complete, the solid biomass is separated from the liquid. This liquid is further sent to steam heated evaporator to ammonia recovery and the residual sugar is also extracted. The solid biomass rich in cellulose and hemicellulose is further sent for crystallization. After crystallization, it is washed to separate reducing sugar molecules and further send for hydrolysis and fermentation for producing biofuels. The ammonia is resend to the reactor chamber for next batch of lignocellulosic biomass [80].

Another way of lignocellulosic treatment for lignin removal is soaking biomass into ammonia solution at comparatively low temperature. In this way the interaction between lignin and hemicellulose is release increasing the surface area and pore size for the interaction of enzymes to hydrolyse the biomass [16]. The temperature preferred for this pre-treatment is 24–60 °C for 12 hrs for several days or 60–120 °C for 1–24 hrs. Addition of H_2O_2 increases the delignification but hydrogen peroxide is expensive as well as unstable [16, 80].

15.4.3.3 Hot Water Pre-Treatment

Hot water pre-treatment is also called autohydrolysis, hydrothermolysis, hydrothermal pre-treatment, aqueous fractionation, solvolysis or aquasolv [16, 58, 82, 83]. It has many advantages such as less expensive, low by-products including inhibitory products, low corrosion to the equipment and less xylose degradation [84]. Hot water has property to penetrate inside the cell wall. Thus, hot water hydrates the cellulose and loosens the hemicellulose and lignin from the system. Generally depending on the type of lignocellulosic biomass temperature varies [34]. In case of aquatic plants such as cattails, 190 °C for 10 mins is used to dissolve the lignocellulosic biomass. Solvolysis when done with a temperature of 200–230 °C for 15 mins with hot pressurized water, 90 % of the hemicellulose is recovered as monomeric sugar molecules. Even 35 – to 60% lignin is also dissolve in the solution. Few by-products are produced which includes acetic acid formation which further catalysis the polysaccharide fermentation [79]. This results in the formation of monomeric sugar, which on further decomposition results into furfural, which is a fermentation inhibitor. This pre-treatment mostly involves batch mode of operation or percolation reactors. Other types of reactors such as co-current and counter current reactors are also used in some cases. In cases of percolation reactor low solid form of biomass (2–4 %) is used and physical conditions includes high temperature of 190–230 °C and high pressure of 20–24 atm for 12–24 mins. In this case the product of the reaction is also in liquid or semi solid form where all the oligomers, hemicellulose lignin, cellulose are dissolved. Further, the solid mass and liquid is separated and treated to collect maximum reducing sugar. Whereas in case of batch mode of operation, higher solid loading (5–30%) is possible and the reaction conditions are comparatively low temperature of 160–190 °C and pressure of 6–14 atm for 10- 30 mins. Here pH is controlled between 4–7 by the addition of KOH or NaOH in the reaction mixture [85]. In autohydrolysis, mainly hydrolysis of hemicellulose occurs whereas moderate change occurs in lignin structure and glucan are not affected. In hemicelluloses, hemicaetal linkages are broken and acetyl group is released in the solution in the form of acetic acid. As water acts, as acidic in nature in high temperature as well as acetic acid is present in the solution is the driving force to the formation of monomers and oligomers from the hemicellulose. Hot water pre-treatment increases the accessible surface area of the biomass by removing lignin and hemicellulose for enzyme digestibility of lignocellulose [58].

15.4.3.4 Steam Explosion

Steam explosion can be done with or without chemicals and is one of the most investigated one. The process can be demonstrated in two ways (i) batch and (ii) continuous mode of operation [86]. This process is in use from early days and the basic operation principle involved here are lignocellulosic biomass is pre-treated with elevated temperature of 160–260 °C at pressure of 0.7–4.8 MPa for 30 s to 20 mins [79]. After this step, the biomass is discharge explosively from the system in vessel of low pressure. Then sudden change in pressure and autohydrolysis by water caused the biomass to disintegrate and solubilise in the medium. The liquid media consist of separated and disintegrated lignin, hemicellulose, cellulose and inhibitors [15]. Further different process is followed to extract and purify the reducing sugar for further fermentation. On the completion of the reaction, three components are generated from the system those are (i) a solid biomass containing less recalcitrant cellulose and lignin (ii) a liquid component containing solubilize form of hemicellulose, some degrading lignin, pentose and phenolic compounds (iii) vapour component contain all the volatile compounds mainly furfural (60–70%). The drawback of the steam explosion is low saccharification of the yield along with loss of carbohydrates due solubilisation and formation of inhibitory products. To overcome this problem, two step steam explosion is done where firstly the lignocellulosic biomass is treated at low temperature as first steam explosion and then again, the biomass is treated at high temperature (greater than 210 °C). In this case, comparatively yield is high. Other than this, methods include first pre-treatment of lignocellulosic biomass with alkali or SO_2, ionic liquid, OV and pre-treatment of the same biomass with steam explosion. This increase the yield of glucose to 90% approximately [58].

15.4.3.5 SO_2-Catalyzed Steam Explosion

This pre-treatment process is one of the most cost-effective processes studied generally. In this case, SO_2 catalyst is used to pre-soak the lignocellulosic biomass similarly as done in AFEX. This pre-treatment process increases the hemicellulose hydrolysis and increases the efficiency of enzymatic activity on it [11]. In general, 0–5% of SO_2 is used to treat the biomass at the high temperature of 190–210 °C for hard woods and 200–220 °C for softwoods. Here the biomass is not diluted using excess water rather the SO_2 steam penetrates the biomass and due to explosive pressure release the biomass degrades into simple molecules. It solubilizes the acid soluble lignin and break the glycosidic bonds of hemicellulose [87]. Thus, separating cellulose from lignin and hemicellulose. The explosive discharge of the pre-treated biomass at the ends results in disruption of biomass fibres into small and simple molecules thus increasing the specific surface area for enzymes to act on it. The main advantage of this pre-treatment is that due to explosive discharge most of the inhibitory products formed are removed from the system through the steam. Though the blowdown vessel is expensive and the vapour produced is highly corrosive and has to be treated in waste water management system but the cellulose product yield is high [88]. Similarly, as single steam explosion system, single acid catalysed steam explosion system efficiency can be increased by pre-treating the lignocellulosic biomass with acid or alkali and then treat with SO_2. While treating with acid first the liquid media generated after pre-treatment has to be removed to improve the saccharification [89].

15.4.3.6 Oxidation

Oxidative pre-treatment involves pre-treatment of lignocellulosic biomass by oxidative agents such as hydrogen peroxide, oxygen or air, ozone. In this pre-treatment system, oxidative agents causes many chemical reactions such as electrophilic substitution, side chain displacements, and oxidative cleavage of aromatic ring ether linkages [9]. Delignification is done by converting lignin to acids, which acts as an inhibitor for hydrolysis and fermentation. Thus, these acids need to be removed. This oxidative pre-treatment also damages hemicellulose leading to loss of significant amount of hemicellulose for fermentation [36]. Hydrogen peroxide is commonly used as oxidative agent in maximum cases. Hydrogen peroxide on hydrolysis produces hydroxyl radicals which causes degradation of lignin and other low molecular weight products. As the lignin is removed, cellulose and hemicellulose is exposed for enzymatic hydrolysis and fermentation [53].

15.4.3.7 Wet Oxidation

Wet oxidation pre-treatment is a process in which lignocellulosic biomass is treated in presence of gaseous and liquid oxidative agents such as air or oxygen with combination of water or hydrogen peroxide at high temperature such as 120 °C for 30 min [90]. This method is also used for waste water treatment and soil remediation [91]. High lignin content biomass is best suited for this type of pre-treatment. The factors affecting this type of reaction are oxygen content, temperature, which may rise to 170 °C, retention time. At high temperature, water acts as acid and leads to hydrolytic reaction, which converts hemicellulose into smaller pentose monomers and lignin is degraded by oxidation whereas cellulose remains unaffected. Addition of some other chemicals such as sodium carbonate and alkaline peroxide reduces the temperature and retention time of the reaction. It also reduced the production inhibitory products such as furfurals and furfuraldehydes and improves hemicellulose hydrolysis [92]. The major drawback that does not make this pre-treatment feasible for industrial use are high cost of hydrogen peroxide and high combustibility of pure oxygen which causes high risk [93].

15.4.3.8 SPORL Treatment

Sulfite pre-treatment to overcome recalcitrance of lignocellulose (SPORL) is a popular and efficient pre-treatment of lignocellulose by using sulphite salts with combination of mechanical milling process [94]. In this process, two steps are involved (i) biomass are treated with calcium or magnesium sulphite to remove lignin and hemicellulose (ii) than using mechanical disk miller to reduce the size of biomass [95]. Formation of inhibitors such as hydroxymethyl furfural (HMF) and furfural were reduced by the increase of bisulfite use. This pre-treatment process is one of the most popular treatment because of its versatility, efficiency, and simplicity [13]. The energy consumed for the size reduction of biomass is 1/10 as compare to other pre-treatment process. Other advantages are high cellulose conversion to sugar molecules and degradation of hemicellulose and lignin. It has excellent scalability and retrofitting for processing different biomass for commercial purpose of biofuel production. Though it has some disadvantage which limits its large scale use in industries such as high cost for recovery of chemical used, degradation of useful sugar and needs lots of water for washing post pre-treatment [93].

15.4.3.9 Supercritical Fluid

In this process, supercritical CO_2 is used to treat the lignocellulosic biomass. The supercritical fluid is passed the chamber with high pressure where biomass is present. The biomass is treated under high pressure and high constant temperature for several minutes in that chamber [96]. The basic reaction mechanism involved in it is CO_2 under high pressure and temperature form carbonic acid, which hydrolyses hemicellulose. When the pressure is released due to sudden change in pressure, it disrupts the biomass into small size and also increase the accessible surface area for hydrolysis. In this pre-treatment higher the moisture content of the biomass higher the hydrolysis of biomass [96]. This pre-treatment has some major advantages which includes low cost of CO_2, low temperature requirement compared to other pre-treatments, high solid loading capacity, low toxin formation making it an excellent pre-treatment process. But the reactor used to hold the high pressure is costlier which limits its application [79].

15.4.4 Biological Pre-Treatment

Biological pre-treatment involves digestion of lignocellulosic biomass by microbes such as white, brown and soft-rot fungi, bacteria and actinomycetes. During other pre-treatment, most of the cellulose remains unaffected by the treatment, which is converted to simpler molecules by biological pre-treatment. It requires less energy and simple maintenance. It is cost efficient and does not produce any toxic compounds. But in this pre-treatment the rate of hydrolysis is low and need long duration which is its demerit [9].

15.4.4.1 White-Rot Fungi

Lignolytic basidiomycetes is a saprophytic fungus which is commonly known as white-rot fungi. From the past decades, it is known to mineralization of lignin. Some of the species such as *Phanerochaete chrysosporium* [97], *Pycnoporus cinnabarinus* [98], *Phlebia spp.* [99], *Echinodontium taxodii* [100], *Irpex lacteus* [101] and *Pycnoporus sanguineus* [102] are used for the degradation of lignin. Other species like *Ceriporiopsis subvermispora* [103], *Phlebia brevispora, P. Floridensis, P. radiate* [104], *Echinodontium taxodii* [100], *Euc-1* [105], *Gonoderma sp.* [106], *Oxysporus sp.* [107], *Trametes versicolor* [100], *Pleurotus sajor-caju* [108], and *Trichoderma reesei* [109], are known to degrade carbohydrates. Biodelignification produces reducing sugar, which can be directly used for fermentation and produces high yield in case of ethanol production. When biodelignification is done prior to chemical pre-treatment reducing sugar yield is increased to 80% [100, 107].

15.4.4.2 *Brown-Rot Fungi*

Brown-Rot Fungi are generally known to degrade cellulose and hemicellulose as compared to lignin. Thus causes brown rotted wood due to incomplete degradation of lignin and known as brown-rot fungi [9]. Some of the species used for biodegradation of cellulose and hemicellulose are *Serpula lacrymans, Coniophora puteana, Meruliporia incrassata, Laetoporeus sulphureus* and *Gleophyllum trabeum* [110].

15.4.4.3 *Soft-Rot Fungi*

There are two types of soft-rot fungi, type I consist of biconical or cylindrical cavities formed within secondary walls whereas type II is known to be erosion form of degrading fungi [111]. *Daldinia concentric*, the most effective type II fungus known to degrade 53% weight loss in 2 months [112]. Other fungus such as *Paecilomyces sp.* [113] and *Cadophora sp.* [114] are also known to rapidly biodelignification of biomass. Though in early classification wood-rotting fungi, *Xylariaceous ascomycetes* from genera *Daldinia, Daldinia* and Xylaria are classified under white-rot fungi but now it is classified under soft-rot fungi as it causes type II soft rot on woods [9].

15.4.4.4 *Bacteria and Actinomycetes*

Though fungi are best delignification microorganism as bacteria are poor producer of lignolytic enzyme but in some case some bacteria also cause delignification. *Bacillus sp. AS3, Bacillus circulans* and *Sphingomonas paucimobilis* [115], *Cellulomonas* and *Zymomonas sp.* [109] are some known bacteria used for biodegradation. Bacteria causes 50% delignification which is similar to fungal treatment. Bacterial delignification is known to be done by xylanases which can be enhanced by the addition of MnP, pectinase or a-L-arabinofuranosidase [116, 117]. High lignin degradation can cause unnneccessary cellulose degradation as seen in *Bacillus macerans, Cellulomonas cartae, C. uda* degrades cellulose to 31–51%. To overcome this issue purified enzyme treatment or cellulase free extracts can be used on the biomass which not only shorten incubation duration but also increases delignification to 20% as compared to whole cell [9].

15.4.5 Other Pre-Treatment Process

15.4.5.1 *Hydrotrope Pre-Treatment*

The hydrotrope pre-treatment was patented by McKee (1946) [118]. In this pre-treatment, lignocellulosic biomass is treated with concentrated solution of benzene derived hydrotrope salts. Other salts used in this process are amphiphilic molecules such as sodium and potassium salts of an alkyl group substituted benzoic and aryl sulfonic acids are known as hydrotropes [119]. These hydrotropes easily dissolves sparingly soluble organic compounds into solution. Hydrotropes consist of amphiphilic compounds, which exhibits both hydrophilic and hydrophobic functional group. The main advantage of hydrotrope is easy recovery of solute from the solution by dilution with water [120]. Lignin is hydrophobic in nature thus insoluble in water, thus diluted hydrotrope pretreatment is a green process, as it is nonhazardous and ecofriendly. This can be alternative method for harsh chemical pre-treatment such as acid treatment, alkaline treatment etc. which produces harmful effluents [121]. Lignin extracted after hydrotrope pre-treatment can be used for different phenolic compound production. Hydrotrope solution can be reused until it gets saturated with lignin. Sodium xylene sulfonate (SXS) is most commonly used hydrotrope solution for lignin extraction. Except lignin other barrier to reach cellulose is hemicellulose which cover cellulose inside it. Hydrotrope solution such as SXS solubilizes lignin as well as degrades hemicellulose and expose cellulose for enzymatic hydrolysis [122]. It degrades hemicellulose into organic compounds such as acetic acid and formic acid. Moreover, presence of formic acid enhances degradation of hemicellulose. Another factor effecting hydrolysis of cellulose is crystallization of cellulose which is also significantly decrease [121].

15.4.5.2 Photocatalytic Pre-Treatment

Corro *et al.* (2014) describe the use of photocatalytic pretreatment for delignification and hydrolysis of coffee pulp for biogas production [123]. From the past studies, it is known to be used for degrading organic pollutants. Photocatalytic pretreatment involves photocatalytic oxidation also known as advanced oxidation processes (AOP). It generates highly reactive oxidizing species such as surface-bound hydroxyl radical (·OH), superoxide ($\cdot O_2^-$), hydroperoxy radical ($HO_2\cdot$), and free holes which degrades absorbed molecules [124–126]. Among photocatalysts, TiO_2 in anatase phase are the most efficient one because of its high stability, high photocatalytic efficiency, low toxicity, and low cost and high optical absorbance in the near-UV region. Corro *et al.* (2014) used photocatalysts as Cu/TiO_2, which absorbs strong absorbance in between 235 and 400 nm, and an intense absorption band in the visible region (400–800 nm). A quartz reactor of inner diameter 10cm was used with 10% Cu/TiO_2 as catalyst, 100g of coffee pulp, water 500ml as reaction mixture. The reactor was exposed to sunlight for UV irradiation and ambient air at the rate of 0.5 L/min was supplied by compressor. Small transparent sachet was used for catalyst contained to separate catalyst from solution and prevent leakage [123]. The whole system was set to run for 9 to 16 hours for 30 days where sun intensity was nearly 1000 W m^{-2}. After the reaction is completed, it is transferred for fermentation and production of biogas. The superoxides produced during the reaction causes hydroxylation of lignin and degradation. Delignification is most tough factor for lignocellulosic biomass hydrolysis, which could be easily obtained by this process. The major disadvantage known in this process TiO_2 if soluble in the reaction mixture cannot be separated but that also was solved by using sachet as catalyst container [123].

15.5 Case Studies of Biofuels

The global demand for energy consumption has increase with the increase in technology. The main sources of energy include fossil fuels such as coal, oil, natural gas etc., which are nonrenewable resources of energy as well as combustion causes pollution to the environment. Thus, to combat with problem green energy sources can be used which are collectively called as biofuels. But production and processing of biofuels are costlier. Thus, different pre-treatment process is invented by many researchers to reduce cost of biofuel production. As lignocellulose biomass consist of lots of sugar molecules, which can be converted to different biofuels thus lignocellulosic is of high demand. This all helps to reuse the waste lignocellulosic biomass and increase the economic value. Different biofuels consist of ethanol, butanol, bio-hydrogen, biogas etc. Some case studies of pretreatment process used to produce high yield biofuels are discussed here.

15.5.1 Ethanol Production

With the depletion of fossil fuel, need for alternative fuel has arisen. As lignocellulosic biomass is abundant in nature thus, biofuels such bioethanol can be produced by fermentation of biomass. For enhanced production of bioethanol, pre-treatment of biomass is needed. Depending on the nature of biomass, different pre-treatment process can be applied for high yield of bioethanol. Table 15.1 list the comparisons of bioethanol production from various lignocellulosic biomass at different pretreatment procedures. Thakur *et al.* (2012) pre-treated wheat straw (WS) and banana stem (BS) with biologically (fungus - *Pleurotus ostreatus* HP-1) and chemically (mild acid or dilute alkali) pre-treatment and further subjected to enzymatic saccharification. Pre-treatment resulted into removal of 4.0–49.2 % lignin, cellulose (0.3–12.4 %) and for hemicellulose (0.7–21.8 %) from both WS and BS. Enzymatic hydrolysis resulted into the production of 64–306.6 mg/g (1.5–15 g/L) of reducing sugar which further produced 0.15–0.54 g/g ethanol on fermentation by *Saccharomyces cerevisiae* NCIM 3570. Wet oxidation such as with acetone/water oxidation treatment is an attractive method for delignification with less degradation products [127]. Beech wood residual biomass pre-treatment using acetone/water oxidation was studied by Katsimpouras *et al.* (2017) [128] . In this study, the optimum reaction mixture conditions were acetone/water mixture in 1:1 ratio with 40 atm initial pressure of 40 volume % oxygen gas and final reaction pressure of 64 atm at 175 °C for 2 hrs incubation. Further enzymatic hydrolysis of biomass was done using custom designed free-fall mixer at 50 °C for 6–12 hrs with enzyme loading of 9 mg/g dry matter at 20 wt% initial solids content. After fermentation of this biomass high ethanol concentration of 75.9 g/L was obtained. In this method lignin isolated can be used for conversion of different value added phenolic compounds [128]. *Prosopis juliflora* (a leguminous plant) is wildly found in barren lands as well as it can grow

Table 15.1 Comparisons of bioethanol production from various lignocellulosic biomass at different pretreatment procedures.

Sl. No.	Lignocellulose biomass	Solid content %	Ethanol production %	Pretreatment methods	Ref.
1	Wheat straw	25	56.9	Steam explosion	[148]
2	Rice straw	13.8	73.4	Dilute acid-dilute alkali	[149]
3	Sweet sorghum bagasse	18	70.4	Hydrothermal	[150]
4	Corn stover	24	51.3	Acetic acid-catalysed hydrothermal	[151]
5	Sugarcane bagasse	20	82.7	Formiline	[152]
6	Corncob residue	20	89.4	Dilute acid hydrolysis-alkaline extraction	[153]
7	Empty palm fruit bunch	30	70.6	Alkali	[154]

in variety of soil and in harsh condition is a source of lignocellulose. As feedstocks and agricultural waste are also used for feeding animal in developing country so only depending on the feedstock and agricultural waste for lignocellulose source is not enough. *Prosopis juliflora* is widely available in nature and can be easily grown at any condition can be good alternative. Pre-treatment of *Prosopis juliflora* for delignification and ethanol production is studied by Naseeruddin *et al.* (2016) [129]. Delignification of the biomass was studied in three different quantities of $Na_2S_2O_4$ are 10g, 100g, 1kg level with 1:10 ratio using 2% (w/v) $Na_2S_2O_4$ at 30 ± 2 °C for 18 hrs which resulted in 82.16 ± 0.34%, 81.82 ± 0.36% and 79.23 ± 0.25% lignin removal respectively. Further the biomass when hydrolysed with biphasic dilute acid, 51.4 ± 0.47%, 51.2 ± 0.52% and 48.1 ± 0.18% of holocellulose was hydrolysed and released 1.94 ± 0.03 g/L, 2.16 ± 0.10 g/L and 1.68 ± 0.05 g/L phenolics and 1.17 ± 0.02 g/L, 1.10 ± 0.03 g/L and 1.07 ± 0.04 g/L of furans, respectively. On detoxification of the pre-treated biomass 85.83 ± 2.8% of phenolics, 87.85 ± 2.4% of furans was removed with loss of 4.74 ± 0.12% sugar. Enzymatic hydrolysis produced 39.37 ± 0.92 g/L, 37.37 ± 0.8 g /L and 30.07 ± 0.48 g/L of sugars, respectively. Further fermentation of the hydrolysate by *Saccharomyces cerevisiae* VS3 and *Pichia stipitis* NCIM 3498 produced 87.34 ± 0.28%, 0.445 ± 1.32 g/g and 0.301 ± 0.011 g/L/h, respectively with total yield of 10.85 g/L of ethanol [129]. Agave bagasse (AGB) is a drought tolerant feedstock used for biofuel production. A comparative analysis of ionic liquid (IL) and OV on AGB was done by Pérez-Pimienta *et al.* (2016) [130]. Pre-treatment of AGB with OV removed 86% of xylan and 45% of lignin whereas IL removed 28% lignin and 50% xylan. On sequential enzymatic saccharification and fermentation (SESF) with cellulolytic enzymes and the ethanologenic *Escherichia coli* strain MS04 improved ethanol production by IL (82%) and OV (85%) [130].

15.5.2 Butanol

Fermentation of lignocellulose produces butanol with other alcohols such as ethanol, acetone etc. Butanol has high demand in present scenario due to its properties. It is less volatile than ethanol and methanol. It contains 25 % more energy than low carbon methanol and ethanol. Butanol has low auto-ignition temperature than methanol and ethanol. It has better inter-solubility quality than low carbon alcohols. Thus, it can easily blend with gasoline and diesel or any co-solvent. As it has high kinematic viscosity, it makes butanol suitable fuel for fuel pump. With low vapour pressure, butanol is suitable for high temperature processes. Butanol is less corrosive and can be transported through pipelines where as low carbon methanol and ethanol needs vehicles for transportation [131]. Butanol is also used as raw material for hydrogen generation or other low carbon fuel generation. Production of butanol is done by fermentation. For better yield of butanol pre-treatment of lignocellulosic biomass must be done. However, inhibitors produced during pre-treatment hamper fermentation process and it need to be de toxified. Table 15.2 list the Comparisons of acetone butanol ethanol (ABE) production from various lignocellulosic biomass at different pretreatment procedures. Qureshi *et al.* (2010) produced butanol from agricultural waste by pre-treating biomass using $Ca(OH)_2$ [132]. Clostridium beijerinckii P260 was used for fermentation process. It produced 26.27 g /L ABE after removal of inhibitors. Further distillation process is used for purification [132]. Another example of inhibitors produced by acid treatment cause inhibition of cell growth and butanol production is explained by Qureshi *et al.* (2008) [133]. Fermentation of sulfuric acid treated corn fiber by Clostridium beijerinckii BA101 produced 1.7 ± 0.2 g/L ABE and also inhibition of cell growth occurred. When inhibitors were removed by XAD-4 resin, ABE yield on fermentation increased to 9.3 ± 0.5 g/L. fermentation of corn fiber with enzymatic hydrolysis lead to 8.6 ± 1.0 g/L ABE production on fermentation [133].

15.5.3 Biohydrogen

Biohydrogen production from lignocellulosic biomass has gained a lot of attraction due to eco-friendly nature and it is a renewable resource. The

Table 15.2 Comparisons of acetone butanol ethanol (ABE) production from various lignocellulosic biomass at different pretreatment procedures.

Sl no.	Lignocellulose biomass	ABE yield g/L	Pretreatment method	Ref.
1	Bagasse	18.1	Alkali pretreated + enzyme hydrolysed	[155]
2	Rice straw	13.0	Alkali pretreated + enzyme hydrolysed	[155]
3	Wheat straw	17.7	Alkali pretreated + enzyme hydrolysed	[156]
4	Corn fiber	9.3	Dilute sulfuric acid	[157]
5	sugarcane molasses	18.90	gas stripping system with nitrogen supplement	[158]
6	spent mushroom substrate	30.21	organosolv pretreatment	[159]

different technology involves for the production of biohydrogen from lignocellulosic biomass are direct biophotolysis, indirect biophotolysis, photo, and dark fermentations [134]. Hydrogen has high energy yield of 122 kJ/g which is 2.75 times greater than hydrocarbon fuels [134]. It can be used in machines, vehicles, aeroplanes etc. as clean energy as its combustion produce only water as product. Hydrogen are generally obtained from fossil fuels, but it can be produced from nuclear power or solar energy which can be stored and transported. Hydrogen produces by conventional methods such as steam reforming of natural gas, thermal cracking or coal gasification are not eco-friendly (Figure 15.3). Whereas biological process of hydrogen production is cost effective and eco-friendly. There are two biological methods of biohydrogen production (i) photofermentative processes by photosynthetic microorganisms such as bacteria, algae (ii) dark fermentation processes by fermenting microorganism [135]. Production of hydrogen from biomass by fermentation has advantage of valorization of residual biomass. Dark fermentation process has potential to use waste biomass as well as wastewater for hydrogen production and is more popular than photofermentative process. Lignocellulosic biomass is most abundant resource in nature. Lignocellulosic biomass consists of glucose and xylose, which can be converted to biohydrogen by microorganisms. Direct conversion of hydrogen from lignocellulosic biomass needs pre-treatment of biomass for delignification and hydrolysis of heterogeneous and crystalline cellulose for fermentation. Pre-treatment process has to be optimised to reduce inhibitory products formation as inhibitory products will hamper

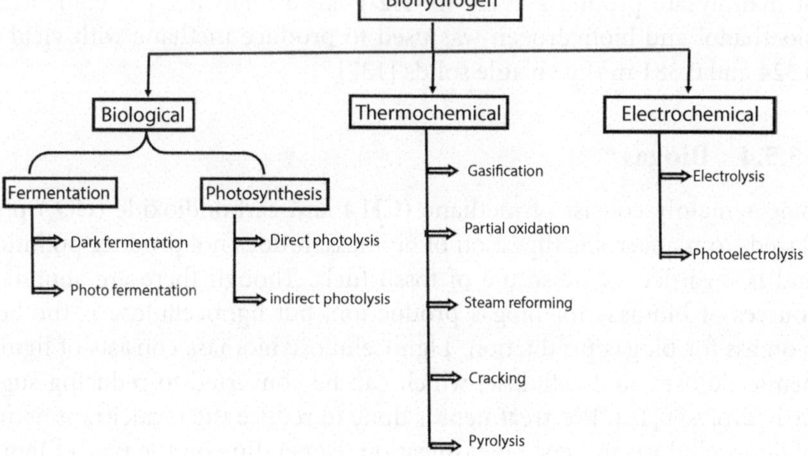

Figure 15.3 Types of process leading to biohydrogen formation.

Table 15.3 Comparisons of biohydrogen production from various lignocellulosic biomass at different pretreatment procedures.

Sl no.	Lignocellulosic biomass	Biohydrogen yield (mmol/g sugar)	Pretreatment methods	Ref.
1	Corn stover	4.17	Acid	[160, 161]
2	Cornstalk	12	Wet stream explosion	[162, 163]
3	Corncob	31	High-pressure autohydrolysis	[164]
4	Wheat straw	2.62	Acid, stream explosion, and enzyme	[165]
5	Wood chopsticks	195	Alkaline and enzyme	[166]
6	Rice straw	0.44	Acid	[167]
7	Soybean straw	47.65	Acid	[169]

fermentation process and yield [136]. Table 15.3 list the Comparisons of biohydrogen production from various lignocellulosic biomasses at different pre-treatment procedures. According to Kaparaju *et al.* (2009), wheat straw was pre-treated by hydrothermally into cellulosic rich fiber fraction and hemicellulose rich liquid fraction (hydrolysate) [137]. Cellulosic fraction was further treated with enzymatic hydrolysis and subsequent fermentation produce 0.41 g-ethanol/g-glucose. Whereas, dark fermentation of hydrolysate produces 178.0 ml-H2/g-sugars. Further the effluents of bioethanol and biohydrogen was used to produce methane with yield of 0.324 and 0.381 m^3/kg volatile solids [137].

15.5.4 Biogas

Biogas mainly consist of methane (CH_4) and carbondioxide (CO_2) produced from anaerobic digestion of biomass. It does not produce pollution and is regarded as substitute of fossil fuels. Though there are abundant sources of biomass for biogas production, but lignocellulose is the best biomass for biogas production. Lignocellulose biomass consists of lignin, hemicelluloses, and cellulose, which can be converted to reducing sugar by hydrolysis [16]. Pre-treatment is done to reduce the recalcitrant nature of lignocellulose and fast biogasification. Depending on the type of lignocellulosic biomass, different pre-treatment is decided for delignification.

The main aim of pre-treatment is to convert lignocellulose into reducing sugar that can be easily hydrolysed. Physical pre-treatment such mechanical extrusion, pyrolysis, milling reduces the size of biomass and combined with other pre-treatment such as alkaline treatment, steam explosion, acid treatment delignify biomass by breaking the bonds between them leading to easy hydrolysis of biomass. According to Sambusiti *et al.* (2012) sodium hydroxide pre-treatment of ensiled sorghum forage and wheat straw by 1–10 % NaOH per gram total solid (TS) of biomass with TS of biomass is 160 gTS/L at 40 °C for 24 hrs in closed improved methane production [138]. Cellulose, hemicellulose, lignin extracted from biomass at 1–10% NaOH pre-treatment are 31, 66 and 44%, and 13, 45 and 3% for sorghum and wheat straw respectively. NaOH pre-treatment favoured methane generation and increased methane production by 14 to 31% for ensiled sorghum forage and 17 to 47% for wheat straw. NaOH pre-treatment process produces inhibitory product which need to be optimized for high yield [137]. Many pre-treatments were studied from the past decades to produces biogas but hydrothermal pre-treatment is one of the most cost effective and high yield pre-treatment process [16]. Hydrothermal pre-treatment is also known as liquid hot water pre-treatment, hot compressed water, autohydrolysis, hydrothermolysis process, thermal hydrolysis, pressure- cooking in water, or aqueous pre-treatments [139–144]. During this process of pre-treatment, biomass is exposed to high temperature and pressure for delignification. Cellulose degrades at greater than 200 °C and forms liquids, gases or char. As some inhibitory products can be formed, thus the reaction system needs to be controlled. The factors affecting biogasification for methane production are temperature, retention time, pressure, solid content, particle size and pH. Mainly temperature and retention time are the most important parameter for enhanced methanogenesis [145, 146]. If temperature increase above 200 °C, phenolic compounds are produced as inhibitors which inhibits growth of anaerobic microorganisms [146]. Thus temperature is better to be controlled between 100–230 °C [16]. Temperature and retention time are related to each other by the relation $R_0 = t \cdot \exp[(T\text{-}100)/14.75]$. Where t is the pre-treatment time (min), T is the temperature (°C), 100 is the base temperature (°C), 14.75 is the conventional energy of activation assuming the overall reaction is hydrolytic and the overall conversion is first order. The severity factor is defined by logarithm of R_0 (log R_0) which depends on time and temperature. Thus, controlling the severity factors controls the production of inhibitory products [144]. Table 15.4 list some of the comparisons of biogas production from various lignocellulosic biomass at different pre-treatment procedures. Another example of pre-treatment of lignocellulosic biomass is use of OV.

Table 15.4 Comparisons of biogas production from various lignocellulosic biomass at different pretreatment procedures.

Sl no.	Lignocellulosic biomass	Biogas yield	Pretreatment methods	Ref.
1	Winter rye	96 %	Wet oxidation	[1]
2	Oilseed rape	85%	Wet oxidation	[1]
3	Faba bean	75%	Wet oxidation	[1]
4	Bamboo	215 ml / g bamboo	Steam explosion	[2]
5	Biofibers	66% increase + 34% increase	(physical +chemical + biological) + steam explosion pretreatment	[3]
6	Wheat straws	87.5% increase	NaOH pretreated	[4]
7	Rice straw	225.6%	Hydrothermal +5% NaOH	[5]

According to Mirmohamad Sadeghi *et al.*, (2014), lignocellulosic biomass when treated with ethanol and sulfuric acid as catalyst at 150–180 °C for 30–60 mins, 152.7, 93.7, and 71.4 liter per kg carbohydrates, methane is produced which significant enough to be used for methane production in industrial scale [147].

15.6 Conclusion

With depletion of fossil fuels and increasing demand of energy production, biofuels are the best alternative. Biofuels are eco-friendly and renewable makes them prime solution to pollution caused due to use of fossil fuels. Lignocellulose biomass is the most abundant resource available in the nature. It also includes all the plant based waste material. Production biofuels from lignocellulosic biomass not only reduces use of fossil fuels but also helps in waste reuse or management. Lignocellulosic biomass consists of cellulose covered with hemicellulose and lignin. Degradation of lignocellulose into smaller molecules is a robust mechanism. The recalcitrant nature of lignocellulose, mainly lignin makes them hard to digest. Pre-treatment are done to digest lignocellulosic biomass into reducing sugars. Due to pre-treatment delignification and conversion of hemicellulose and cellulose to reducing sugar by breaking of inter molecular hydrogen bonds. Pre-treatment increases the efficiency of enzymatic hydrolysis and fermentation. Pre-treatment includes chemical, physical, physicochemical, biological pre-treatment. During pre-treatment, inhibitory products mainly furans are produced which are inhibitory product for enzymatic hydrolysis and fermentation. Detoxification and optimization of reaction system for less inhibitory product formation is necessary. Physical pre-treatment process produces less inhibitory products than chemical pre-treatment but is costly. However, biological pre-treatment processes produce no inhibitory products but are slow in process. Thus, mix match of physical, chemical or biological process is done to enhance the yield of reducing sugar keeping in check the cost, time and inhibitory products. Biofuels such as bioethanol, butanol, biohydrogen and biogas are green energy and are primary biofuels produced from plant sources by fermentation process. Alcohols such as ethanol, methanol, and butanol are used during the production of biodiesel from waste or nonedible vegetable oils by transesterification. Further studies are needed for better pre-treatment of lignocellulosic biomass for more cost effective and increase yield and efficiency of biofuel production, as they are the future fuels of the world.

Reference

1. Rodionova, M.V., *et al.*, Biofuel production: Challenges and opportunities. *Int. J. Hydrog. Energ.*, 42(12), 8450–8461, 2017.
2. Roy, L., Garlapati, V., Banerjee, R., Challenges in Harnessing the Potential of Lignocellulosic Biofuels and the Probably Combating Strategies, in Bioenergy. *Apple Academic Press.*, 171–203, 2015.
3. Allakhverdiev, S.I., *et al.*, Hydrogen photoproduction by use of photosynthetic organisms and biomimetic systems. *Photochem. Photobiol. Sci.*, 8(2), 148–56, 2009.
4. Voloshin, R.A., *et al.*, Photoelectrochemical cells based on photosynthetic systems: a review. *Biofuel Res. J.*, 2(2), 227–235, 2015.
5. Razzak, S.A., *et al.*, Integrated CO2 capture, wastewater treatment and biofuel production by microalgae culturing—A review. *Renew. Sust. Energ. Rev.*, 27, 622–653, 2013.
6. Alam, F., Mobin, S., Chowdhury, H., Third Generation Biofuel from Algae. *Procedia Eng.*, 105, 763–768, 2015.
7. Dragone, G., *et al.*, Current research, technology and education topics in applied microbiology and microbial biotechnology. 2010.
8. Abdelaziz, A.E., Leite, G.B., Hallenbeck, P.C., Addressing the challenges for sustainable production of algal biofuels: II. Harvesting and conversion to biofuels. *Environ. Technol.*, 34(13–16), 1807–36, 2013.

9. Akhtar, N., *et al.*, Recent advances in pretreatment technologies for efficient hydrolysis of lignocellulosic biomass. *Environ. Prog. Sustain. Energy.*, 35(2), 489–511, 2016.
10. Zhang, M., *et al.*, Metabolic Engineering of a Pentose Metabolism Pathway in Ethanologenic Zymomonas mobilis. *Science.*, 267(5195), 240–3, 1995.
11. Zhang, B., Shahbazi, A., Recent Developments in Pretreatment Technologies for Production of Lignocellulosic Biofuels. *J. Pet. Environ. Biotechnol.*, 02(02), 108, 2011.
12. Zhu, J.Y., Pan, X., Zalesny, R.S., Jr., Pretreatment of woody biomass for biofuel production: energy efficiency, technologies, and recalcitrance. *Appl. Microbiol. Biotechnol.*, 87(3), 847–57, 2010.
13. Kumar, A.K., Sharma, S., Recent updates on different methods of pretreatment of lignocellulosic feedstocks: a review. *Bioresour. Bioprocess.*, 4(1) 7, 2017.
14. Ju, X., Engelhard, M., Zhang, X., An advanced understanding of the specific effects of xylan and surface lignin contents on enzymatic hydrolysis of lignocellulosic biomass. *Bioresour. Technol.*, 132, 137–145, 2013.
15. Zheng, Y., *et al.*, Pretreatment of lignocellulosic biomass for enhanced biogas production. *Prog. Energy Combust. Sci.*, 42, 35–53, 2014.
16. Leilei, H., *et al.*, A Review of Hydrothermal Pretreatment of Lignocellulosic Biomass for Enhanced Biogas Production. *Curr. Org. Chem.*, 19(5), 437–446, 2015.
17. Frey-Wyssling, A., The Fine Structure of Cellulose Microfibrils. *Science*, 119(3081), 80–82, 1954.
18. Atalla, R.H., Vanderhart, D.L., Native cellulose: a composite of two distinct crystalline forms. Science, 223(4633), 283–285, 1984.
19. Ademark, P., *et al.*, Softwood hemicellulose-degrading enzymes from Aspergillus niger: purification and properties of a beta-mannanase. *J. Biotechnol.*, 63(3), 199–210, 1998.
20. Bobleter, O., Hydrothermal degradation of polymers derived from plants. *Prog. Polym. Sci.*, 19(5), 797–841, 1994.
21. Garrote, G., Domínguez, H., Parajó, J.C., Hydrothermal processing of lignocellulosic materials. Holz als Roh- und Werkstoff, 57(3), 191–202, 1999.
22. Stamatelatou, K., *et al.*, The Effect of Physical, Chemical, and Biological Pretreatments of Biomass on its Anaerobic Digestibility and Biogas Production, in Biogas Production. John Wiley & Sons, Inc. p. 55–90, 2012.
23. Palmqvist, E., Hahn-Hägerdal, B., Fermentation of lignocellulosic hydrolysates. II: inhibitors and mechanisms of inhibition. *Bioresour. Technol.*, 74(1), 25–33, 2000.
24. Grabber, J.H., How Do Lignin Composition, Structure, and Cross-Linking Affect Degradability? A Review of Cell Wall Model Studies This paper was originally presented at the Lignin and Forage Digestibility Symposium, 2003 CSSA Annual Meeting, Denver, CO. *Crop. Science.*, 45(3): 820–831, 2005.

25. Taherzadeh, M.J., Karimi, K., Pretreatment of Lignocellulosic Wastes to Improve Ethanol and Biogas Production: A Review. *Int. J. Mol. Sci.*, 9(9), 1621–1651, 2009.
26. Alvira, P., *et al.*, Pretreatment technologies for an efficient bioethanol production process based on enzymatic hydrolysis: A review. *Bioresour. Technol.*, 101(13), 4851–4861, 2010.
27. Mansfield, S.D., Mooney, C., Saddler, J.N., Substrate and Enzyme Characteristics that Limit Cellulose Hydrolysis. *Biotechnol. Prog.*, 15(5), 804–816, 1999.
28. Palmowski, L.M., Muller, J.A., Influence of the size reduction of organic waste on their anaerobic digestion. *Water Sci. Technol.*, 41(3), 155–62, 2000.
29. Sun, Y., Cheng, J., Hydrolysis of lignocellulosic materials for ethanol production: a review. *Bioresour. Technol.*, 83(1), 1–11, 2002.
30. Chang, V.S., Holtzapple, M.T., Fundamental Factors Affecting Biomass Enzymatic Reactivity, in Twenty-First Symposium on Biotechnology for Fuels and Chemicals: Proceedings of the Twenty-First Symposium on Biotechnology for Fuels and Chemicals Held May 2–6, 1999, in Fort Collins, Colorado, M. Finkelstein and B.H. Davison, (Eds.), 2000, Humana Press: Totowa, NJ. p. 5–37.
31. Ramos, L.P., Breuil, C., Saddler, J.N., Comparison of steam pretreatment of eucalyptus, aspen, and spruce wood chips and their enzymatic hydrolysis. *Appl. Biochem. Biotechnol.*, 34(1), 37, 1992.
32. Grous, W.R., Converse, A.O., Grethlein, H.E., Effect of steam explosion pretreatment on pore size and enzymatic hydrolysis of poplar. *Enzyme Microb. Technol.*, 8(5), 274–280, 1986.
33. Grous, W.R., Converse, A.O., Grethlein, H.E., Effect of steam explosion pretreatment on pore size and enzymatic hydrolysis of poplar. *Enzyme Microb. Technol.*, 8, 1986.
34. Mosier, N., *et al.*, Features of promising technologies for pretreatment of lignocellulosic biomass. *Bioresour. Technol.*, 96(6), 673–686, 2005.
35. Yang, B., Wyman, C.E., Pretreatment: the key to unlocking low-cost cellulosic ethanol. *Biofuels, Bioprod. Biorefin.*, 2(1), 26–40, 2008.
36. Sun, Y., Cheng, J., Hydrolysis of lignocellulosic materials for ethanol production: a review. Bioresour Technol., 83, 2002.
37. Millett, M.A., Baker, A.J., Satter, L.D., Physical and chemical pretreatments for enhancing cellulose saccharification. *Biotechnol. Bioeng. Symp.*, (6): p. 125–53, 1976.
38. Zhu, J.Y., *et al.*, Specific surface to evaluate the efficiencies of milling and pretreatment of wood for enzymatic saccharification. *Chem. Eng. Sci.*, 64(3), 474–485, 2009.
39. Taherzadeh, M.J., Karimi, K.m Pretreatment of lignocellulosic wastes to improve ethanol and biogas production: a review. *Int. J. Mol. Sci.*, 9, 2008.
40. Hu, Z., Wen, Z., Enhancing enzymatic digestibility of switchgrass by microwave-assisted alkali pretreatment. *Biochem. Eng. J.*, 38(3), 369–378, 2008.

41. Xu, J., *et al.*, Optimization of microwave pretreatment on wheat straw for ethanol production. *Biomass Bioenergy.*, 35(9). 3859–3864, 2011.
42. Yachmenev, V., *et al.*, Acceleration of the Enzymatic Hydrolysis of Corn Stover and Sugar Cane Bagasse Celluloses by Low Intensity Uniform Ultrasound. *J. Biobased Mater. Bioenergy.*, 3(1), 25–31, 2009.
43. Bussemaker, M.J., Zhang, D., Effect of Ultrasound on Lignocellulosic Biomass as a Pretreatment for Biorefinery and Biofuel Applications. *Ind. Eng. Chem. Res.*, 52(10), 3563–3580, 2013.
44. Gogate, P.R., Sutkar, V.S., Pandit, A.B., Sonochemical reactors: Important design and scale up considerations with a special emphasis on heterogeneous systems. *Chem. Eng. J.*, 166(3), 1066–1082, 2011.
45. Yang, C., *et al.*, Effect and aftereffect of gamma radiation pretreatment on enzymatic hydrolysis of wheat straw. *Bioresour. Technol.*, 99(14), 6240–5, 2008.
46. Chinnadurai, K., Muthukumarappan, K., Julson, J.L., Influence of High Shear Bioreactor Parameters on Carbohydrate Release from Different Biomasses, in 2008 Providence, Rhode Island, June 29 – July 2, 2008, ASABE: St. Joseph, MI, 2008.
47. Zheng, J., *et al.*, Enzymatic hydrolysis of steam exploded corncob residues after pretreatment in a twin-screw extruder. *Biotechnol. Rep.*, 3, 99–107, 2014.
48. Zheng, J., Rehmann, L., Extrusion pretreatment of lignocellulosic biomass: a review. *Int. J. Mol. Sci*, 15, 2014.
49. Sánchez, M.E., *et al.*, Effect of pyrolysis temperature on the composition of the oils obtained from sewage sludge. *Biomass Bioenergy.*, 33(6), 933–940, 2009.
50. Sarkar, N., *et al.*, Bioethanol production from agricultural wastes: An overview. *Renew. Energy*, 37(1), 19–27, 2012.
51. Melo, E., Kennedy, J.F., Cellulose hydrolysis (biotechnology monographs, Vol. 3) edited by L.-T. Fan, M. M. Gharpuray and Y.-H. Lee, Springer-Verlag, Berlin, Heidelberg, New York, London, Paris and Tokyo, 1987. 198, *British Polymer J.*, 20(6), 532–532, 1988.
52. Shafizadeh, F., Bradbury, A.G.W., Thermal degradation of cellulose in air and nitrogen at low temperatures. *J. Appl. Polym. Sci.*, 23(5), 1431–1442, 1979.
53. Kumar, P., *et al.*, Methods for Pretreatment of Lignocellulosic Biomass for Efficient Hydrolysis and Biofuel Production. *Ind. Eng. Chem. Res.*, 48(8), 3713–3729, 2009.
54. Zhang B., Wang L.S.A., Diallo O., Whitmore A., Alkali Pretreatment and Enzymatic Hydrolysis of Cattails from Constructed Wetlands. Am *J. Eng. Appl. Sci.*, 3, 328–332, 2010.
55. Sharma, A.R.K.B.A.R.G.S., Proximate chemical analysis of ipomoea carnea and its sodium hydroxide pretreatment to obtain fermentable sugars. Int. *J. Pharm. Bio. Sci.*, 6(3B), 1247- 1255, 2015.

56. Sambusiti, C., *et al.*, Sodium hydroxide pretreatment of ensiled sorghum forage and wheat straw to increase methane production. *Water Sci. Technol.*, 66(11), 2447–52, 2012.
57. Wang, Z., *et al.*, Sodium hydroxide pretreatment of genetically modified switchgrass for improved enzymatic release of sugars. *Bioresour. Technol.*, 110, 364–370, 2012.
58. Shafiei, M., Kumar, R., Karimi, K., *Pretreatment of Lignocellulosic Biomass, in Lignocellulose-Based Bioproducts*, K. Karimi, (Ed.) Springer International Publishing: Cham. p. 85–154, 2015.
59. Sierra, R., Garcia, L.A., Holtzapple, M.T., Selectivity and delignification kinetics for oxidative and nonoxidative lime pretreatment of poplar wood, part III: long-term. *Biotechnol. Prog.*, 26(6), 1685–94, 2010.
60. Sierra, R., Granda, C., Holtzapple, M.T., Short-term lime pretreatment of poplar wood. *Biotechnol. Prog.*, 25(2), 323–32, 2009.
61. Sierra, R., Holtzapple, M.T., Granda, C.B., Long-term lime pretreatment of poplar wood. *AIChE J.*, 57(5), 1320–1328, 2011.
62. Yang, B., Wyman, C.E., Effect of xylan and lignin removal by batch and flowthrough pretreatment on the enzymatic digestibility of corn stover cellulose. *Biotechnol. Bioeng.*, 86(1), 88–98, 2004.
63. Diedericks, D., van Rensburg, E., Görgens, J.F., Enhancing sugar recovery from sugarcane bagasse by kinetic analysis of a two-step dilute acid pretreatment process. *Biomass. Bioenergy.*, 57, 149–160, 2013.
64. Wyman, C.E., *et al.*, Coordinated development of leading biomass pretreatment technologies. *Bioresour. Technol.*, 96(18), 1959–1966, 2005.
65. Palmqvist, E., Hahn-Hägerdal, B., Fermentation of lignocellulosic hydrolysates. I: inhibition and detoxification. *Bioresour. Technol.*, 74(1), 17–24, 2000.
66. Behera, S., *et al.*, Importance of chemical pretreatment for bioconversion of lignocellulosic biomass. *Renew. Sustain. Energy Rev.*, 2014. 36, 91–106, 2014.
67. Moulthrop, J.S., *et al.*, High-resolution 13C NMR studies of cellulose and cellulose oligomers in ionic liquid solutions. *Chem. Commun. (Camb)*, (12), 1557–9, 2005.
68. Dadi, A.P., Varanasi, S., Schall, C.A., Enhancement of cellulose saccharification kinetics using an ionic liquid pretreatment step. *Biotechnol. Bioeng.*, 95(5), 904–10, 2006.
69. Liu, L., Chen, H., Enzymatic hydrolysis of cellulose materials treated with ionic liquid [BMIM] Cl. *Chin. Sci. Bull.*, 51(20), 2432–2436, 2006.
70. Kuo, C.-H., Lee, C.-K., Enhanced enzymatic hydrolysis of sugarcane bagasse by N-methylmorpholine-N-oxide pretreatment. *Bioresour. Technol.*, 100(2), 866–871, 2009.
71. Zhang, S., Xu, Y., Hanna, M.A., Pretreatment of Corn Stover with Twin-Screw Extrusion Followed by Enzymatic Saccharification. *Appl. Biochem. Biotechnol.*, 166(2), 458–469, 2012.
72. Smith, E.L., Abbott, A.P., Ryder, K.S. Deep Eutectic Solvents (DESs) and Their Applications. *Chem. Rev.*, 114(21), 11060–11082, 2014.

73. Paiva, A., *et al.*, Natural Deep Eutectic Solvents – Solvents for the 21st Century. *ACS Sustain. Chem. Eng.*, 2(5), 1063–1071, 2014.
74. Kumar, A.K., Parikh, B.S., Pravakar, M., Natural deep eutectic solvent mediated pretreatment of rice straw: bioanalytical characterization of lignin extract and enzymatic hydrolysis of pretreated biomass residue. *Environ. Sci. Pollut. Res. Int.*, 23(10), 9265–75, 2016.
75. Dai, Y., *et al.*, Tailoring properties of natural deep eutectic solvents with water to facilitate their applications. *Food Chem.*, 187, 14–19, 2015.
76. Quesada, J., Rubio, M., Gomez, D., Ozonation of lignin rich solid fractions from corn stalks. *J. Wood Chem. Technol.*, 19, 1999.
77. Agbor, V.B., *et al.*, Biomass pretreatment: Fundamentals toward application. *Biotechnology Adv.*, 29(6), 675–685, 2011.
78. Park, J.-y., *et al.*, A novel lime pretreatment for subsequent bioethanol production from rice straw – Calcium capturing by carbonation (CaCCO) process. *Bioresour. Technol.*, 101(17), 6805–6811, 2010.
79. Agbor, V.B., *et al.*, Biomass pretreatment: fundamentals toward application. *Biotechnol. Adv.*, 2011. 29.
80. Harmsen, P., *et al.*, *Literature review of physical and chemical pretreatment processes for lignocellulosic biomass*, 2010.
81. Chundawat, S.P.S., *et al.*, Primer on Ammonia Fiber Expansion Pretreatment, in Aqueous Pretreatment of Plant Biomass for Biological and Chemical Conversion to Fuels and Chemicals. John Wiley & Sons, Ltd. p. 169–200, 2013.
82. Maurya, D.P., Singla, A., Negi, S., An overview of key pretreatment processes for biological conversion of lignocellulosic biomass to bioethanol. *3 Biotech*, 5(5), 597–609, 2015.
83. Limayem, A., Ricke, S.C., Lignocellulosic biomass for bioethanol production: Current perspectives, potential issues and future prospects. *Prog. Energy Combust. Sci.*, 38(4), 449–467, 2012.
84. Yang, B., Wyman, C.E., Effect of xylan and lignin removal by batch and flowthrough pretreatment on the enzymatic digestibility of corn stover cellulose. *Biotechnol. Bioeng.*, 86(1), 88–95, 2004.
85. Kohlmann, K.L., *et al.*, Enhanced Enzyme Activities on Hydrated Lignocellulosic Substrates, in Enzymatic Degradation of Insoluble Carbohydrates. *American Chemical Society*. p. 237–255, 1996.
86. Li, B.Z., *et al.*, Process optimization to convert forage and sweet sorghum bagasse to ethanol based on ammonia fiber expansion (AFEX) pretreatment. *Bioresour. Technol.*, 101(4), 1285–92, 2010.
87. Bura, R., *et al.*, SO2-catalyzed steam explosion of corn fiber for ethanol production. *Appl. Biochem. Biotechnol.*, 98–100, 59–72, 2002.
88. Bura, R., *et al.*, Optimization of SO2-catalyzed steam pretreatment of corn fiber for ethanol production. *Appl. Biochem. Biotechnol.*, 106(1), 319–335, 2003.
89. Kumar, L., *et al.*, Can the same steam pretreatment conditions be used for most softwoods to achieve good, enzymatic hydrolysis and sugar yields? *Bioresour. Technol.*, 101(20), 7827–7833, 2010.

90. Varga, E., *et al.*, Pretreatment of corn stover using wet oxidation to enhance enzymatic digestibility. *Appl. Biochem. Biotechnol.*, 104(1), 37–50, 2003.
91. Chaturvedi, V., Verma, P., An overview of key pretreatment processes employed for bioconversion of lignocellulosic biomass into biofuels and value added products. *3 Biotech.*, 3(5), 415–431, 2013.
92. Banerjee, S., *et al.*, Alkaline peroxide assisted wet air oxidation pretreatment approach to enhance enzymatic convertibility of rice husk. *Biotechnol. Prog.*, 27(3), 691–7, 2011.
93. P, B., In Pretreatment of lignocellulosic biomass for biofuel production. Springer Briefs in Molecular Science. p. 17–70, 2016.
94. Xu, H., Li, B., Mu, X., Review of Alkali-Based Pretreatment To Enhance Enzymatic Saccharification for Lignocellulosic Biomass Conversion. *Ind. Eng. Chem. Res.*, 55(32), 8691–8705., 2016
95. Zhu, J.Y., *et al.*, Sulfite pretreatment (SPORL) for robust enzymatic saccharification of spruce and red pine. *Bioresour. Technol.*, 100(8), 2411–2418, 2009.
96. Kim, K.H., Hong, J., Supercritical CO2 pretreatment of lignocellulose enhances enzymatic cellulose hydrolysis. *Bioresour. Technol.*, 77(2), 139–144, 2001.
97. Shi, J., Chinn, M.S., Sharma-Shivappa, R.R., Microbial pretreatment of cotton stalks by solid state cultivation of Phanerochaete chrysosporium. *Bioresour. Technol.*, 99(14), 6556–6564, 2008.
98. Meza, J.C., *et al.*, New process for fungal delignification of sugar-cane bagasse and simultaneous production of laccase in a vapor phase bioreactor. *J. Agric. Food Chem.*, 54(11), 3852–8, 2006.
99. Arora, D.S., Sharma, R.K., Comparative ligninolytic potential of *Phlebia* species and their role in improvement of *in vitro* digestibility of wheat straw. *J. Ani. Feed Sci.*, 18(1): p. 151–161, 2009.
100. Yu, H., *et al.*, The effect of biological pretreatment with the selective white-rot fungus Echinodontium taxodii on enzymatic hydrolysis of softwoods and hardwoods. *Bioresour. Technol.*, 100(21), 5170–5175, 2009.
101. Xu, C., Ma, F., Zhang, X., Lignocellulose degradation and enzyme production by Irpex lacteus CD2 during solid-state fermentation of corn stover. *J. Biosci. Bioeng.*, 108(5), 372–375, 2009.
102. Lu, C., *et al.*, An efficient system for pre-delignification of gramineous biofuel feedstock *in vitro:* Application of a laccase from Pycnoporus sanguineus H275. *Process Biochem.*, 45(7), 1141–1147, 2010.
103. Wan, C., Li, Y., Microbial delignification of corn stover by Ceriporiopsis subvermispora for improving cellulose digestibility. *Enzyme Microb. Technol.*, 47(1), 31–36, 2010.
104. Sharma, R.K., Arora, D.S., Changes in biochemical constituents of paddy straw during degradation by white rot fungi and its impact on *in vitro* digestibility. *J. Appl. Microbiol.*, 109(2), 679–86, 2010.
105. Tripathi, M.K., *et al.*, Selection of white-rot basidiomycetes for bioconversion of mustard (Brassica compestris) straw under solid-state fermentation into

energy substrate for rumen micro-organism. *Lett. Appl. Microbiol.*, 46(3), 364–370, 2008.
106. Dias, A.A., *et al.*, Enzymatic saccharification of biologically pre-treated wheat straw with white-rot fungi. *Bioresour. Technol.*, 101(15), 6045–6050, 2010.
107. Haddadin, M.S., *et al.*, Bio-degradation of lignin in olive pomace by freshly-isolated species of Basidiomycete. *Bioresour. Technol.*, 82(2), 131–137, 2002.
108. Kannan, K., Oblisami, G., Loganathan, B.G., Enzymology of ligno-cellulose degradation by Pleurotus sajor-caju during growth on paper-mill sludge. *Biol. Wastes*, 33(1), 1–8, 1990.
109. Singh, P., *et al.*, Biological pretreatment of sugarcane trash for its conversion to fermentable sugars. *World J. Microbiol. Biotechnol.*, 24(5), 667–673, 2008.
110. Monrroy, M., *et al.*, Structural change in wood by brown rot fungi and effect on enzymatic hydrolysis. *Enzyme Microb. Technol.*, 49(5), 472–477, 2011.
111. Blanchette, R.A., A review of microbial deterioration found in archaeological wood from different environments. *Int. Biodeterior. Biodegradation.*, 46(3), 189–204, 2000.
112. Narayanaswamy, N., *et al.*, Biological Pretreatment of Lignocellulosic Biomass for Enzymatic Saccharification, in *Pretreatment Techniques for Biofuels and Biorefineries*, Z. Fang, (ed.) Springer Berlin Heidelberg: Berlin, Heidelberg. p. 3–34, 2013.
113. Kluczek-Turpeinen, B., *et al.*, Lignin degradation in a compost environment by the deuteromycete Paecilomyces inflatus. *Appl. Microbiol. Biotechnol.*, 61(4), 374–379, 2003.
114. Chandel, A.K., *et al.*, Biodelignification of lignocellulose substrates: An intrinsic and sustainable pretreatment strategy for clean energy production. *Crit. Rev. Biotechnol.*, 35(3), 281–293, 2015.
115. Kurakake, M., Ide, N., Komaki, T., Biological pretreatment with two bacterial strains for enzymatic hydrolysis of office paper. *Curr. Microbiol.*, 54(6), 424–428, 2007.
116. Bezalel, L., Shoham, Y., Rosenberg, E. Characterization and delignification activity of a thermostable α-l-arabinofuranosidase from Bacillus stearothermophilus. *Appl. Microbiol. Biotechnol.*, 40(1), 57–62, 1993.
117. Kaur, A., *et al.*, Application of cellulase-free xylano-pectinolytic enzymes from the same bacterial isolate in biobleaching of kraft pulp. *Bioresour. Technol.*, 101(23), 9150–9155, 2010.
118. McKee, R.H., Use of Hydrotropic Solutions in Industry. *Ind. Eng. Chem.*, 38(4), 382–384, 1946.
119. Devendra, L.P., Gaikar, V.G. Is sodium cinnamate a photoswitchable hydrotrope? *J. Mol. Liq.*, 165, 71–77, 2012.
120. Mata, J., *et al.*, Aggregation behavior of a PEO–PPO–PEO block copolymer + ionic surfactants mixed systems in water and aqueous salt solutions. Colloids and Surfaces A: *Physicochem. Eng. Asp.*, 247(1), 1–7, 2004.

121. Devendra, L.P., Kiran Kumar, M., Pandey, A., Evaluation of hydrotropic pretreatment on lignocellulosic biomass. *Bioresour. Technol.*, 213, 350–358, 2016.
122. Mou, H.-Y., *et al.*, Topochemical pretreatment of wood biomass to enhance enzymatic hydrolysis of polysaccharides to sugars. *Bioresour. Technol.*, 142, 540–545, 2013.
123. Corro, G., Pal, U., Cebada, S., Enhanced biogas production from coffee pulp through deligninocellulosic photocatalytic pretreatment. *Energy Sci. Eng.*, 2(4), 177–187, 2014.
124. Ghanbari, F., *et al.*, Comparison of electron beam and gamma ray irradiations effects on ruminal crude protein and amino acid degradation kinetics, and *in vitro* digestibility of cottonseed meal. *Radiat. Phys. Chem.*, 81(6), 672–678, 2012.
125. Fox, M.A., Dulay, M.T., Heterogeneous photocatalysis. *Chem. Rev.*, 93(1), 341–357, 1993.
126. Huang, G., *et al.*, Fluorination of ZnWO4 Photocatalyst and Influence on the Degradation Mechanism for 4-Chlorophenol. *Environ. Sci. Technol.*, 42(22), 8516–8521, 2008.
127. Thakur, S., *et al.*, Degradation and selective ligninolysis of wheat straw and banana stem for an efficient bioethanol production using fungal and chemical pretreatment. *3 Biotech.*, 3(5), 365–372, 2013.
128. Katsimpouras, C., *et al.*, Production of high concentrated cellulosic ethanol by acetone/water oxidized pretreated beech wood. *Biotechnol. Biofuels*, 10(1), 54, 2017.
129. Naseeruddin, S., Desai, S., Venkateswar Rao, L., Ethanol production from lignocellulosic substrate Prosopis juliflora. *Renew. Energy*, 103, 701–707, 2017.
130. Pérez-Pimienta, J.A., *et al.*, Sequential enzymatic saccharification and fermentation of ionic liquid and organosolv pretreated agave bagasse for ethanol production. *Bioresour. Technol.*, 225, 191–198, 2017.
131. Kumar, B., Kumar, S., Kumar, S., Butanol reforming: an overview on recent developments and future aspects, in *Rev. Chem. Eng.*, 2017.
132. Qureshi, N., *et al.*, Production of butanol (a biofuel) from agricultural residues: Part II – Use of corn stover and switchgrass hydrolysates. *Biomass Bioenergy*, 34(4), 566–571, 2010.
133. Qureshi, N., *et al.*, Butanol production by Clostridium beijerinckii. Part I: use of acid and enzyme hydrolyzed corn fiber. *Bioresour. Technol.*, 99(13), 5915–22, 2008.
134. Kapdan, I.K., Kargi, F., Bio-hydrogen production from waste materials. *Enzyme Microbial. Technol.*, 38(5), 569–582, 2006.
135. Ghimire, A., *et al.*, A review on dark fermentative biohydrogen production from organic biomass: Process parameters and use of by-products. *Appl. Energy*, 144, 73–95, 2015.
136. Singh, A., *et al.*, Biohydrogen Production from Lignocellulosic Biomass: Technology and Sustainability. *Energies*, 8(11), 12357, 2015.

137. Kaparaju, P., *et al.*, Bioethanol, biohydrogen and biogas production from wheat straw in a biorefinery concept. *Bioresour. Technol.*, 100(9), 2562–2568, 2009.
138. Sambusiti, C., *et al.*, Sodium hydroxide pretreatment of ensiled sorghum forage and wheat straw to increase methane production. *Water Sci. Technol.*, 66(11), 2447–2452, 2012.
139. López González, L.M., *et al.*, Effect of liquid hot water pre-treatment on sugarcane press mud methane yield. *Bioresour. Technol.*, 169, 284–290, 2014.
140. Sun, P., *et al.*, Direct liquefaction of paulownia in hot compressed water: Influence of catalysts. *Energy*, 35(12), 5421–5429, 2010.
141. Suryawati, L., *et al.*, Effect of hydrothermolysis process conditions on pretreated switchgrass composition and ethanol yield by SSF with Kluyveromyces marxianus IMB4. *Process Biochem.*, 44(5), 540–545, 2009.
142. Ferreira, L.C., *et al.*, Biomethane potential of wheat straw: Influence of particle size, water impregnation and thermal hydrolysis. *Chem. Eng. J.*, 242, 254–259, 2014.
143. Weil, J.R., *et al.*, Pretreatment of corn fiber by pressure cooking in water. *Appl. Biochem. Biotechnol.*, 73(1), 1–17, 1998.
144. Fractionation of lignocellulosics by steam-aqueous pretreatments. Philosophical Transactions of the Royal Society of London. Series A, *Math. Phys. Sci.*, 321(1561), 523–536, 1987.
145. Minowa, T., *et al.*, Liquefaction of Cellulose in Hot Compressed Water Using Sodium Carbonate: Products Distribution at Different Reaction Temperatures. *J. Chem. Eng. Japan*, 30(1), 186–190, 1997.
146. Pérez, J.A., *et al.*, Effect of process variables on liquid hot water pretreatment of wheat straw for bioconversion to fuel-ethanol in a batch reactor. *J. Chem. Technol. Biotechnol.*, 82(10), 929–938, 2007.
147. Mirmohamadsadeghi, S., *et al.*, Enhanced Solid-State Biogas Production from Lignocellulosic Biomass by Organosolv Pretreatment. *BioMed. Res.Int.*, p. 6, 2014.
148. Alvira, P., *et al.*, Improving the fermentation performance of Saccharomyces cerevisiae by laccase during ethanol production from steam-exploded wheat straw at high-substrate loadings. *Biotechnol Prog*, 29(1): p. 74–82, 2013.
149. Sun, W.L. and W.Y. Tao, Simultaneous Saccharification and Fermentation of Rice Straw Pretreated by a Sequence of Dilute Acid and Dilute Alkali at High Dry Matter Content. Energy Sources, Part A: Recovery, Utilization, and Environmental Effects, 35(8): p. 741–752, 2013.
150. Katsimpouras, C., P. Christakopoulos, and E. Topakas, Acetic acid-catalyzed hydrothermal pretreatment of corn stover for the production of bioethanol at high-solids content. *Bioprocess and Biosystems Engineering*, 39(9): p. 1415–1423, 2016.
151. Matsakas, L. and P. Christakopoulos, Fermentation of liquefacted hydrothermally pretreated sweet sorghum bagasse to ethanol at high-solids content. *Bioresource Technology*, 127: p. 202–208, 2013.

152. Zhao, X., *et al.*, Batch and multi-step fed-batch enzymatic saccharification of Formiline-pretreated sugarcane bagasse at high solid loadings for high sugar and ethanol titers. *Bioresource Technology*, 135: p. 350–356, 2013.
153. Lei, C., *et al.*, An alternative feedstock of corn meal for industrial fuel ethanol production: delignified corncob residue. *Bioresour Technol*, 167: p. 555–9, 2014.
154. Park, J.M., *et al.*, Efficient production of ethanol from empty palm fruit bunch fibers by fed-batch simultaneous saccharification and fermentation using Saccharomyces cerevisiae. *Appl Biochem Biotechnol*, 170(8): p. 1807–14, 2013.
155. Soni, B.K., K. Das, and T.K. Ghose, Bioconversion of agro-wastes into acetone butanol. *Biotechnology Letters*, 4(1): p. 19–22, 1982.
156. Marchal, R., M. Rebeller, and J.P. Vandecasteele, Direct bioconversion of alkali-pretreated straw using simultanesous enzymatic hydrolysis and acetone-butanol fermentation. *Biotechnology Letters*, 6(8): p. 523–528, 1984.
157. Qureshi, N., *et al.*, Butanol production by Clostridium beijerinckii. Part I: Use of acid and enzyme hydrolyzed corn fiber. *Bioresource Technology*, 99(13): p. 5915–5922, 2008.
158. Wechgama, K., L. Laopaiboon, and P. Laopaiboon, Enhancement of batch butanol production from sugarcane molasses using nitrogen supplementation integrated with gas stripping for product recovery. *Industrial Crops and Products*, 95: p. 216–226, 2017.
159. Zhu, H.-J., *et al.*, Conversion of spent mushroom substrate to biofertilizer using a stress-tolerant phosphate-solubilizing Pichia farinose FL7. *Bioresource Technology*, 111: p. 410–416, 2012.
160. Zhang, K., N.-Q. Ren, and A.-J. Wang, Enhanced biohydrogen production from corn stover hydrolyzate by pretreatment of two typical seed sludges. *International Journal of Hydrogen Energy*, 39(27): p. 14653–14662, 2014.
161. Zhang, Z., *et al.*, *Photo-fermentative Bio-hydrogen Production from Agricultural Residue Enzymatic Hydrolyzate and the Enzyme Reuse*. Vol. 9. 2014.
162. Liu, C.-M., *et al.*, Biohydrogen production from rice straw hydrolyzate in a continuously external circulating bioreactor. *International Journal of Hydrogen Energy*, 39(33): p. 19317–19322, 2014.
163. Liu, Z., *et al.*, Effects of operating parameters on hydrogen production from raw wet steam-exploded cornstalk and two-stage fermentation potential for biohythane production. *Biochemical Engineering Journal*, 90: p. 234–238, 2014.
164. Nasr, N., *et al.*, Biohydrogen production from pretreated corn cobs. *International Journal of Hydrogen Energy*, 39(35): p. 19921–19927, 2014.
165. Patel, A.K., *et al.*, Biohydrogen production from a novel alkalophilic isolate Clostridium sp. IODB-O3. *Bioresource Technology*, 175: p. 291–297, 2015.
166. Phummala, K., *et al.*, Delignification of disposable wooden chopsticks waste for fermentative hydrogen production by an enriched culture from a hot spring. *Journal of Environmental Sciences*, 26(6): p. 1361–1368, 2014.

167. Liu, C.-M., *et al.*, Biohydrogen production evaluation from rice straw hydrolysate by concentrated acid pre-treatment in both batch and continuous systems. *International Journal of Hydrogen Energy*, 38(35): p. 15823–15829, 2013.
168. Han, H., *et al.*, Optimization of biohydrogen production from soybean straw using anaerobic mixed bacteria. *International Journal of Hydrogen Energy*, 37(17): p. 13200–13208, 2012.
169. Petersson, A., *et al.*, Potential bioethanol and biogas production using lignocellulosic biomass from winter rye, oilseed rape and faba bean. *Biomass and Bioenergy*, 31(11): p. 812–819, 2007.
170. Kobayashi, F., *et al.*, Methane production from steam-exploded bamboo. *Journal of Bioscience and Bioengineering*, 97(6): p. 426–428, 2004.
171. Bruni, E., A.P. Jensen, and I. Angelidaki, Comparative study of mechanical, hydrothermal, chemical and enzymatic treatments of digested biofibers to improve biogas production. *Bioresource Technology*, 101(22): p. 8713–8717, 2010.
172. Chandra, R., *et al.*, Improving biodegradability and biogas production of wheat straw substrates using sodium hydroxide and hydrothermal pretreatments. *Energy*, 43(1): p. 273–282, 2012.
173. Chandra, R., H. Takeuchi, and T. Hasegawa, Hydrothermal pretreatment of rice straw biomass: A potential and promising method for enhanced methane production. *Applied Energy*, 94: p. 129–140, 2012.

16

Microalgal Biomass as an Alternative Source of Sugars for the Production of Bioethanol

Maria Eugenia Sanz Smachetti, Lara Sanchez Rizza, Camila Denise Coronel, Mauro Do Nascimento and Leonardo Curatti*

Instituto de Investigaciones en Biodiversidad y Biotecnología (INBIOTEC-CONICET), Mar del Plata, Argentina and Fundación para Investigaciones Biológicas Aplicadas, Mar del Plata and Argentina

Abstract

Ethanol is mostly produced by fermentation. The demand of ethanol as a renewable transportation fuel has increased dramatically during the last decades. The current feedstocks for first generation bioethanol are sugar cane and corn kernels, in Brazil and USA, respectively. More recently, concerns regarding food security and environment conservation promoted R+D+i of second generation (2G) bioethanol from lignocellulosic plant feedstocks. However, the complex structure of these materials poses a difficult-to-overcome barrier to a wider production of 2G bioethanol. Thus, production of third generation (3G) bioethanol from photosynthetic microorganisms such as microalgae and cyanobacteria cultivated in aquatic farms is increasingly considered a viable alternative according to higher productivities and simpler biochemical composition and structure in comparison with terrestrial crops. Strategies for cultivation, downstream processing for conversion into biofuels, with emphasis in 3G bioethanol, and main drawbacks for technology developments are discussed in this chapter.

Keywords: Renewable energy, biofuels, bioethanol, microalgae, cyanobacteria, sugars

**Corresponding author*: lcuratti@inbiotec.conicet.gov.ar

Arindam Kuila and Vinay Sharma (eds.) Principles and Applications of Fermentation Technology, (351–386) © 2018 Scrivener Publishing LLC

16.1 Overview

Second-generation biofuels present clear advantages over first-generation ones, mostly related to the availability, low cost and non-competition with food production of lignocellulose as a feedstock and its reduced environmental impact [1]. However, they face hard-to-overcome disadvantages due to the composition and structure of the biomass, requiring quite intensive mechanical and physicochemical pre-treatments, and expensive saccharification enzymes for its conversion into the desired biofuel [2].

Lignocellulose pre-treatments frequently result in the generation of fermentation inhibitors such as weak acids, furans and phenolic compounds formed or released during hydrolysis. Although some alternatives for detoxification have been shown (such as additional treatments with alkali, sulfite or enzymes, pre-fermentation by a fungus, removal of non-volatile compounds, extraction with ether or ethyl acetate, and improved fermentation technology), implementing them increases production costs [3]. Furthermore, it has been reported that cellulases impact the most in the total cost of production of second-generation biofuels. They also represent one of the most uncertain parameters in techno-economic analyses mostly due to assumptions on future prices and the heterogeneity in the way results are presented in the literature, making it difficult to cross-compare studies [4].

In view of these difficulties, researchers have envisioned changing both its composition and/or structure by genetically modifying the lignocellulose synthetic pathway in plants. The modification of lignin content in plants has been attempted before for other reasons such as to increase digestibility of feed production and to decrease the need of bleaching in the paper industry. Thus, although some details of the metabolic pathway for lignin biosynthesis are still not completely understood [5], some very promising genetic modifications have already been demonstrated. For example, down regulation of cinnamyl alcohol dehydrogenase (ADH) in poplar resulted in improved lignin solubility in an alkaline medium, decreasing the need for pre-treatment before saccharification [6]. Also, downregulation of 4-coumarate CoA ligase in the lignin biosynthesis pathway in aspen resulted in a 45% decrease in lignin content and a concomitant 15% increase in cellulose content [7] and, when coniferaldehyde 5-hydroxylase was additionally down regulated, the lignin content was further reduced to 52% and cellulose increased by 30% [8]. More recently, a proof-of-principle study conducted in alfalfa, in which six different genes for the lignin biosynthetic pathway were down regulated, showed reduction or elimination of need for chemical pre-treatment in the production of fermentable sugars [9]. Although modification of lignocellulosic biomass composition and/or structure by genetic engineering is very promising, further research toward plant structural integrity and defense against pathogens and insects should also be addressed to continue improving lignocellulosic biomass from genetically modified crop plants [5].

Other genetic modification approaches had also been pursued to improve yield, pre-treatment's efficiency and saccharification: *i*) increasing the overall biomass productivity by modifying plant growth regulators and other factors such as carbon allocation; efficiency of uptake and use of nutrients, among others; *ii*) increasing cell-wall polysaccharide content by modifying the expression of genes that are involved in both cellulose and hemicellulose biosynthesis; *iii*) expressing microbial hydrolases in specific cellular compartments of the plants. The latter is a very appealing approach considering the possibility that hydrolytic enzymes could be produce on-site and at a very low cost by the same crop plant. The apoplast accumulation of heterologous hydrolases is often a selected target. However, the expression of thermophilic enzymes would be preferred to avoid premature degradation of the plant cell walls before lignification at cultivation temperatures [5].

Thus, despite the improvements in the technology for converting lignocellulosic biomass into biofuels achieved over the last years, the structural nature of this feedstock still represents a remarkable challenge. It is presumed that expanding the search for alternative feedstock for biofuels by looking into natural biodiversity would be a reasonable approach. Among non-conventional crops, aquatic photosynthetic species such as macro and microalgae and cyanobacteria, arise as an alternative source of low-cost sugars for biofuels and other applications.

16.2 Aquatic Species as Alternative Feedstocks for Low-Cost-Sugars

16.2.1 Seaweed

Macroalgae can be classified into three major classes according to the presence or absence of pigments other than chlorophyll: brown algae (Phaeophyceae), red algae (Rhodophyceae), and green algae (Chlorophyceae) [10].

16.2.1.1 Seaweed Biomass

Brown macroalgae display their characteristic olive-green to dark-brown color due to the abundance of the yellow-brown pigment fucoxanthin that masks the green color of chlorophyll. Kelp (Laminaria) is a well-known marine macroalga from this group that is found at depths below the low-tide level in temperate and polar regions and frequently reaches lengths of 10–50 m and grows as much as 50 cm/day. Laminaria biomass contains up to 55% (w/w) of carbohydrates mostly as laminarin and mannitol and can be extracted from milled biomass at low pH and high temperature [11]. Both laminarin [11] and mannitol [12] can be quite easily hydrolyzed into hexoses by laminarase (endo-1,3[4]-b-glucanase) or mannitol dehydrogenase, respectively [10].

Red macroalgae owe their color to the pigments phycocyanin and, especially, phycoerythrin. They are mostly found from the intertidal zones to depths down to 250 m, at which their light-harvesting pigments are crucial. Red algae biomass contains 30– 60% (w/w) of carbohydrates [13], frequently in the form of cellulose, glucan, and galactan. In addition to cellulose, the cell wall also contains the long-chain polysaccharides agar and carrageenan that are valued for their gel-forming properties in the food industry [11].

Green macroalgae comprise mostly freshwater species that thrive in shallow waters. They have evolutionary and biochemical similarity with higher plants [14] accumulating 25–50% (w/w) of carbohydrates [13], preferentially as starch (a carbon reserve) while their cell wall contains cellulose and pectin as the main structural polysaccharide [15].

16.2.1.2 Seaweed Cultivation

Approximately 90% of the global market of seaweed is currently focused on food products for human consumption, while the rest corresponds to other products extracted from macroalgal biomass, such as alginate, agar, and carrageenan [14, 16].

The bulk production of macroalgal biomass is obtained by aquaculture, which reached a global productivity of 3.1 million dry metric tons in 2006, whereas harvesting of wild seaweed was only 22,000 dry metric tons for the same period. Seaweed aquaculture is more developed in Asia, especially in China, which accounts for about 72% of global annual production, mostly of *Laminaria japonica*, *Plantae aquaticae*, *Undaria pinnatifida*, *Porphyra tenera*, and *Gracilaria verrucosa*. Instead, harvesting of wild seaweed is more geographically distributed throughout the world [14].

Seaweed can be cultivated either vegetatively or by a separate reproductive cycle. For vegetative propagation, small algal pieces are grown in a suitable aquatic environment until they reach a harvestable size. Harvesting can be complete or partial, leaving smaller pieces for the next growing season. Cultivation by a separate reproductive cycle is much more complex and expensive but is mandatory for some commercial species, such as *Laminaria*, which cannot be cultivated by the vegetative method. Thus, seed production and raising of young seedlings is to be conducted in land-based facilities under controlled conditions [17].

Seaweed farming can take place offshore, in coastal regions and in land-based settings [14, 17]. Nearshore farms are currently in use for culturing macroalgae in China and Japan. However, concerns toward the environmental impact of this option have prevented its implementation in the United States and Europe. Offshore farming has been tested in the North Sea with success, but current capital and operating cost might still be prohibitive for the production of low-cost feedstocks for biofuels. Lastly, land-based pond strategies might improve nutrients management, oversight of the meteorological conditions, disease and predation and can be coupled with fish farming and recycling of waste/nutrients. Nevertheless, pond construction and operation costs need to be improved in order to achieve inexpensive production of biofuels feedstocks [14].

16.2.1.3 Seaweed as a Biofuels Feedstock

In addition to an increased production potential, seaweed biomass is very attractive as a feedstock for biofuels due to the absence of lignin and a low content of cellulose, making the biomass easier to convert in comparison to land plants [14, 18]. Currently, seaweed biomass can be used as a source of bioenergy through thermochemical alternatives that include direct combustion, gasification, pyrolysis and liquefaction, and fermentation technology, mostly to produce biogas or bioethanol. Although direct combustion and pyrolysis would be the simplest options, the mineral content of seaweed makes these alternatives less attractive. Furthermore, the water content in macroalgae is higher than in terrestrial biomass (80–85%), making seaweed biomass more suitable for microbial conversion than for direct combustion or thermo-chemical conversion processes [18].

Almost four decades ago, the US Marine Biomass Program stated the feasibility of cultivating and converting kelp biomass into methane by anaerobic digestion [19, 20]. This process is carried out by a complex community of microorganisms displaying specific metabolic activities for the deconstruction of the seaweed (or other) biomass by hydrolysis, intermediate fermentations and final production of methane [18]. At the time, it was concluded that, although methane yield from kelp exceeded that produced from terrestrial biomass or waste feedstocks by over three-fold, the reliability of using large, open-ocean growth structures to provide the feedstock supply was inconclusive and mainly demonstrated the inadequacies of the available technology for techno-economic feasibility [20]. More recently, the Tokyo Gas Company and the New Energy and Industrial Technology Development Organization of Japan demonstrated the production of electricity and heat from biogas derived from anaerobic digestion of macroalgae collected from shorelines and processed at the considerably large scale of one-metric ton macroalgae per day [14].

Additionally, seaweed carbohydrates may be converted into a wide range of fuels and chemicals by microbial fermentation. However, unlike anaerobic digestion, the biomass is to be pretreated and saccharified before fermentation with specific microorganisms. Since seaweed lacks lignin and normally contains low levels of cellulose, these processes are simpler and less energy-intensive than current pre-treatments for saccharification of lignocellulosic biomass. Fresh harvested brown seaweed contains about 15–20% carbohydrates of the total wet weight, which is an appropriate substrate concentration for microbial fermentation. Thus, although drying improves further downstream processing of the biomass, it is not strictly necessary. Alginates in seaweed biomass are normally slowly released, even during hydrolysis, which may constitute a rate-limiting step for saccharification of macroalgal biomass. After a first step of mechanical disruption through grinding, chemically and/or enzymatic processes are normally carried out by modification and adaptation of existing technologies, such as those developed for pre-treatment of wood biomass, and by combination of acid or alkali with steam treatment [18].

The potential ethanol production from saccharified seaweeds biomass can be calculated assuming a carbohydrate content of 60% of dry weight and a 90% conversion rate to ethanol. Thus, an optimal bioethanol productivity of 19,000 liters/ha/year has been estimated, which is approximately two times higher than the ethanol productivity from sugarcane and 5 times higher than the ethanol productivity from corn [21]. However, in brown macroalgae, hexose-based polysaccharides constitute only about 30–40% of the total carbohydrates. The remaining fraction is composed of C-5 sugars released from alginates that are very poor substrates for yeast fermentation and other most naturally occurring microorganisms. In recent years, some breakthrough demonstrations have paved the way to ethanol production from brown macroalgae using genetically engineered *Escherichia coli*. For example, a recombinant strain was developed by integrating *Zymomonas mobilis* ethanol production genes into the *pflB* gene, which was able to ferment a mixed sugar solution containing glucose, galactose, xylose, L-arabinose, and mannitol with an ethanol yield of about 0.4 g/g total sugar [22]. Another strain was obtained by genomic integration of a DNA fragment from *Vibrio splendidus* encoding enzymes for alginate transport and metabolism, together with engineered systems for extracellular alginate depolymerization and for ethanol synthesis. A bioethanol production of 0.281 g ethanol/g dry weight of kelp was achieved in this consolidated process [21]. Other successful examples have been reviewed for brown or other seaweed biomass conversion into ethanol and butanol [10].

Seaweed biomass can also be converted into liquid biofuels such as bio-oil, and chemicals, through thermo-chemical conversion processes (e.g. pyrolysis and hydrothermal liquefaction). Bio-oil production aims at breaking the biomass macromolecules (carbohydrate, protein, and lipid) bonds to form a more homogeneous organic liquid phase. Although biomass from most origins can be converted into bio-oil [23], seaweed biomass differs from lignocellulosic biomass in their constitutional compounds and higher nitrogen and sulfur content and in its considerably higher contents of ash- up to 40%. Hence, the main compounds in bio-oils not only vary between macroalgae species but are also significantly different from those of land biomass. Furthermore, a high ash content is not a desirable aspect for bio-oil production although acid-washing pre-treatment has allowed some improvements in bio-oil recovery [24]. For example, demineralization of *Sargassum* spp. biomass with citric acid (among other acids) greatly reduced the ash content from 27.46 to 7% and allowed bio-oil yields of 22.2% (ash-free dry basis) by hydrothermal liquefaction [25].

Since some of the other products that can be extracted from seaweed, especially those that are not intended for bioenergy purposes (adipinic or citric acid, among others), display a more cost-efficient production than liquid biofuels, it is anticipated that biofuel production in the frame of a biomass biorefinery would largely improve the economic aspects of seaweed-to-biofuel production and commercialization [18].

16.2.2 Microalgae

Microalgae are photosynthetic microorganisms that exist as individual cells or chains of cells. Along with other microorganisms that accumulate copious amounts of carbon reserves (especially lipids, but also carbohydrates), they are called feedstocks for third generation biofuels [26]. Even though microalgae and cyanobacteria proliferate considerable slower and often under more complex cultivation systems than other microorganisms (see below), they allow a direct production of biofuels feedstocks from CO_2 due to their photosynthetic life-style [27]. While there are probably more than 100,000 species of microalgae, the most important classes in terms of abundance and biotechnological significance are diatoms (Bacillariophyceae), green algae (Chlorophyceae), and golden algae (Chrysophyceae). On the other hand, cyanobacteria (blue-green algae) (Cyanophyceae) are a taxonomic and phylogenetic divergent group from those previously mentioned but biotechnologically speaking are commonly also referred to as microalgae. Diatoms are the dominant life form in phytoplankton and probably represent the largest group of biomass producers [28].

Although the O_2-evolving and photosystem II-dependent photosynthetic mechanism and the carbohydrates metabolism of microalgae are similar to those in land-based plants [29, 30], the microalgae photosynthetic efficiency generally spans from 6% to 10%, surpassing the 1–2% of land plants [31]. Additionally, their larger surface-to-volume ratio allows the uptake of larger amounts of nutrients, especially when submerged in an aqueous environment. This, coupled with their simpler cellular structure, makes microalgae more efficient in converting solar energy into biomass [32].

Microalgae display some additional features that make them very attractive as feedstocks for biofuels in comparison to land plants and even macroalgae: *i*) higher productivity of biomass per unit of surface and time (for example a 4- to 5- fold higher oil productivity than the most productive crop plants currently used as biodiesel feedstock has been demonstrated [33]); *ii*) non competitiveness for land or food market with crops, since it can be produced even on non-arable land; *iii*) better economy of water and nutrients through effective recycling; *iv*) possibility of using industrial residues as a source of inexpensive nutrients, especially CO_2, N and P, and/or thriving in saline or freshwater environments and converting sunlight, CO_2 and water to algal biomass [34].

16.2.2.1 Microalgae Biomass as a Biofuel Feedstock

Since many microalgal strains, especially diatoms and green algae, accumulate around 30% of their dry biomass as lipids (exceptionally up to 90%) [27], most research and development regarding microalgal biomass for biofuels has been directed toward assessment of possibilities for biodiesel production, which is extensively covered in the literature [32, 35]. However, there are microalgal strains that accumulate up to 70% of their dry biomass as carbohydrates, mostly water-insoluble structural and energy-reserve polysaccharides that make them also attractive as a feedstock for a variety of fermentation processes toward biofuels (Table 16.1) [34, 36–41].

While the main reserve polysaccharide of cyanobacteria is glycogen (α-1,4 linked glucan), which accumulates in the cytosol, green algae

Table 16.1 Biomass composition of some representative microalgal strains.

Microalgae	**Carbohydrates (dwt %)**	**Proteins (dwt %)**	**Lipids (dwt %)**	**Induction method**	**Reference**
Chlorophyta strain SP_2–3	70.35 ± 8.17	17.58 ± 8.60	10.30 ± 2.60	Nutrient starvation	[34]
***Spirogyra* sp.**	33–64	6–20	11–21	NA	[48]
***Chlorella* sp. strain MI**	57.84 ± 16.62	17.45 ± 8.20	11.04 ± 2.19	Nutrient starvation	[34]
Porphyridium cruentum	40–57	28–39	9–14	NA	[48]
Porphyridium cruentum	40–57	8–39	9–14	NA	[49]
Dunaliella salina	55.50	12.50	9.20	Deficient nutrients, salt stress	[50]
***Desmodesmus* sp. strain FG**	53.47 ± 14.18	16.25 ± 6.93	18.76 ± 2.40	Nutrient starvation	[34]
***Scenedesmus* sp. strain SP_2–9**	52.93 ± 4.14	13.85 ± 1.90	14.26 ± 0.51	Nutrient starvation	[34]
Chlorophyta strain C_1C	52.85 ± 6.04	19.40 ± 3.35	nd	Nutrient starvation	[34]
Chlorophyta strain C_1	51.45 ± 4.45	17.50 ± 2.12	nd	Nutrient starvation	[34]
***Ankistrodesmus* sp. strain LP_1**	51.25 ± 8.99	28.87 ± 3.63	nd	Nutrient starvation	[34]
***Chlorella vulgaris* *(dry-ash-free)**	51.00	6.00	43.00	Nutrient starvation	[51]
***Chlorella sorokiniana* strain RP**	49.8 ± 2.40	14.75 ± 1.76	nd	Nutrient starvation	[34]

(*Continued*)

Table 16.1 Cont.

Microalgae	Carbohydrates (dwt %)	Proteins (dwt %)	Lipids (dwt %)	Induction method	Reference
***Ankistrodesmus* sp. strain SP$_2$–15**	47.25 ± 9.54	31.61 ± 2.28	nd	Nutrient starvation	[34]
***Scenedesmus* sp. strain SP$_1$–20**	46.11 ± 6.40	24.95 ± 4.80	nd	Nutrient starvation	[34]
***Scenedesmus* sp. strain PL**	45.66 ± 10.08	17.97 ± 7.49	nd	Nutrient starvation	[34]
***Chlorella* sp. strain SP$_2$–1**	44.37 ± 2.23	30.80 ± 5.37	nd	Nutrient starvation	[34]
***Scenedesmus* sp. strain RD**	41.87 ± 13.15	11.94 ± 4.26	nd	Nutrient starvation	[34]
Dunaliella bardawil	40.40	9.70	10.40	Nutrient starvation, salt stress	[50]

NA: not available. nd: not determined

synthesize and accumulate amylopectin-like polysaccharides (starch) in their chloroplasts [42]. As in plants and macroalgae, cellulose is an integral constituent of the cell wall of eukaryotic microalgae and its levels remain mostly constant for a given strain [41]. More interestingly from a biotechnological point of view, microalgal cell wall not only lacks lignin but also hemicellulose. Furthermore, some eukaryotic microalgae lack a cell wall altogether, which largely facilitates access to the storage polysaccharides that are normally preferred for downstream conversion technologies. In contrast to structural polysaccharides, starch and glycogen accumulation levels largely fluctuate due to the environmental and overall growth conditions such as nutrient availability, salt stress, light intensity, and temperature [42]. For example, starch content in *Chlorella vulgaris* increased about 8-fold, to 41% of the algal dry weight, after N starvation [40]. In addition to N starvation, deficiency of other nutrients such as P and S also triggered carbohydrates accumulation in a variety of microalgal strains [34]. However, most times the final carbohydrates (or lipids) productivity is offset by the lower yields in biomass resulting from the imposed nutrients limitations [34, 40, 42]. Therefore, optimization of nutrient concentration is very important to balance biomass production and desired biomass composition for overall carbohydrates (or lipids) yield. It has been shown that optimized limited media are superior to the absolute deprivation of nutrients for increased yield of target compounds, likely because they may favor some of the conversion processes [43].

Alternative to the use of synthetic fertilizers for an optimized nutrients composition, microalgae can be cultivated at the expense of wastewaters, especially those from the agro-industrial sector which are rich in inorganic pollutants such as nitrogen and phosphorus. This dual-purpose strategy is interesting for the simultaneous management of environmental pollutants and the production of microalgae biomass with a composition that can be tuned according to some operational variables. However, the specific composition of wastewater and its variations make it difficult to take advantage of the optimized limited medium strategy [42, 44].

Glucose is the main sugar in microalgal biomass and, in some cases, can represent up to 90% of the total sugars, when cells are appropriately induced [42, 45]. However, rhamnose, fucose, ribose, arabinose, xylose, mannose, and galactose are frequently detected in varying proportions [46].

As it happens with most organisms, microalgae produce high yields of biomass of a desired composition in laboratory shaker-flask cultures under conditions of constant temperature and light intensity and do not necessarily perform as satisfactorily under environmental conditions and/or in large-scale cultivation systems [47].

16.2.2.2 Microalgal Biomass Production Technology

Different cultivation systems have been tested for microalgal biomass production at large scale (Figure 16.1) [52]. Currently, most commercial-scale algae cultivation systems are open ponds due to their relative low-cost and simple scaling-up [53, 54, 55]. Additionally, open ponds can be installed in areas with marginal crop production potential, avoiding competition for land with existing agricultural crops [56], have lower energy input requirements during operation [56, 57], and easier regular maintenance and cleaning than alternative culture systems [58]. There are four major types of open-culture systems currently in use: shallow big ponds, tanks, circular ponds and raceway ponds [59].

Raceway ponds consist of a closed-loop recirculation channel that is typically between 0.2 and 0.5 m deep, built in concrete, or compacted soil, and may be lined with plastic materials. Mixing and circulation is done by a paddlewheel and baffles that guide the flow around bends [32, 55]. CO_2 requirements are usually supplied from the surface air, but supplementation from CO_2 addition-sumps close to the paddlewheels are often installed to enhance CO_2 absorption, resulting in increased cells proliferation by promoting C-assimilation and by aiding in the pH control of the culture [60]. However, open ponds also have many disadvantages since being exposed to the air makes extremely difficult to control the cells environment. A common problematic issue is water evaporation at temperatures that favor cell proliferation. Ponds depth should be balance between the need to maintain an adequate water depth for mixing and to avoid large changes in ionic composition due to evaporation with the need to provide adequate light to the algal cells. Contamination and/or predation are also critical aspects to sustain productivity [53, 59].

Typical closed reactors include bag systems, vertical bubbled columns, other tubular photobioreactors (PBRs) and flat plate reactors (Figure 16.1) [59, 61]. In contrast to open pond systems, PBRs allow a better control over water evaporation, contamination and predation, and higher productivities [32].

Bag systems are perhaps the simplest setting and consist of large plastic bags of approximately 0.5 m in diameter fitted with aeration systems. This system is being extensively used for aquaculture feed production [59]. Maintenance of this culture system is labor intensive and algal cultures usually crash due to inadequate mixing [53].

Vertical-column PBRs are characterized by their high volumetric gas transfer coefficients. The bubbling of gas from the bottom of the column enables efficient CO_2 utilization and optimal O_2 removal that ameliorates

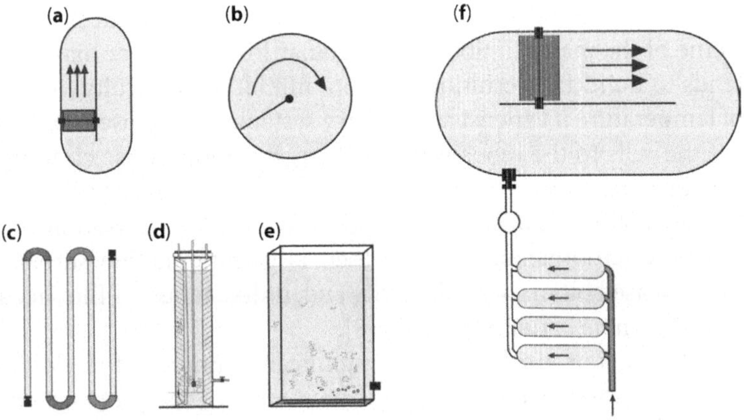

Figure 16.1 Microalgae culture systems. (a) Raceway pond; (b) circular pond; (c) horizontal tubular PBR; (d) airlift PBR, (e) flat panel, (f) hybrid system, raceway pond coupled to parallel horizontal tubes.

the inhibition of photosynthesis that frequently occurs in other culture systems [62]. The gently agitation of the medium by the gas bubbles shows very little shear stress compared to impellers and pumps, diminishing cell damage compared to other PBR systems, except at extreme superficial gas velocities [61, 63].

Internal-loop airlift PBRs typically comprise a transparent column, an internal column, and an air sparger [64, 65]. Air or CO_2-enriched air is introduced inside the internal column at the bottom; and degassing occurs in the freeboard regime, which locates at the top of the internal column. Since the gas holdup inside the internal column is much larger than in the degassed liquid outside of the internal column, an upward flow of the liquid/gas mixture will be created inside the internal column while a downward flow of degassed liquid is generated outside of it. This excellent mixing is the main advantage of the airlift PBR since it allows good exposure of cells to light radiation even with a relatively large diameter of column and high cell density. Additionally, is simple and fairly easy to maintain [61].

Tubular PBRs are one of the most suitable types for outdoor mass culture since they have large illumination surface area. Most outdoor tubular PBRs are usually constructed with either a glass or plastic tube of 0.1 m or less in diameter, to ensure light penetration of dense cultures. Using these settings, sunlight interception can be optimized by changing the orientation according to the position of the sun. The most common tubular arrange is the horizontal/serpentine type. Aeration and mixing of the cultures in tubular PBRs are usually accomplished by air-pump or airlift systems [32, 58]. Some of the major limitations of tubular PBRs are poor mass transfer (O_2 tends to build-up) resulting in photo inhibition, difficulty in the control of temperature if expensive accessory systems are not used, and adherence of the cells to the tube's walls, that largely compromise cleaning and maintenance [55, 58].

Flat-panel PBRs have received much attention for cultivation of photosynthetic microorganisms due to their large illumination surface area toward increased biomass productivity and, unlike tubular PBRs, accumulation of O_2 can be better managed [58].

Algal biomass production can be significantly improved in closed culture systems. However, installation and maintenance costs are much higher than those for open ponds and represent the major limiting factor for commercialization of closed systems [53, 66, 67]. Thus, hybrid systems that take advantage of salient features of both open ponds and PBRs have also been developed for increased biomass productivity. Hybrid systems couple a first stage in PBRs, to maintain constant conditions that favor continuous cell division and prevent contamination of the culture, and a second stage in open ponds to expose the cells to nutrient deprivation and other environmental stresses that lead to the synthesis and accumulation of carbohydrates and/or lipids [66].

16.2.2.3 Microalgae Productivity

The main constraints to microalgal productivity are environmental factors, mostly irradiance and temperature, although the culture system and fertilization (see above) introduce some variations. When algae are cultivated photosynthetically, the efficiency of photosynthesis is a crucial determinant of their productivity, affecting growth rate and biomass production. Theoretical best-case biomass productivity values are in the range of 33–42 $g/m^2/day$ [68], a productivity that could be accomplished after optimization of both biological and production systems.

The maximum conversion efficiency of total solar energy into primary photosynthetic organic products is around 10% [68, 69]. Photosynthetically active radiation (PAR) is limited by normal diurnal and seasonal fluctuations as a function of the sun's changing zenith angle throughout the year. Thus, algae cultivation sites at lower latitudes experience less change in solar insolation and will generally have a more consistent daily availability of PAR. Even though cloud cover and storms have a significant impact on available PAR, microalgae photosynthesis still occurs at a reduced rate using available diffuse radiation [70].

Optimal growth temperature is usually strain specific [71]. Although many microalgae can tolerate temperatures 15 °C below or 2°–4 °C above their optimal temperature, photosynthetic reactions become limiting beyond the optimal temperature range and extremely suboptimal temperatures will eventually reduce cell viability [72].

Most evaluations of the economic and environmental viability of algal biofuels rely on one of the following methods to estimate biomass productivity: *i*) assumed values; *ii*) modeled values based on climatology; or *iii*) measured values based on long-term outdoor production in a scalable system [73]. Studies that use assumed values for biomass productivity project yields ranging from 50 to over 100 MT/ha/yr biomass, whether in open ponds or PBRs, respectively [66, 74, 75]. Modeling studies may incorporate site-specific climatology, but they predict a similar wide range of biomass yields for both ponds and PBRs [76, 77]. To date, there are just a few reports of direct determination of yields in open ponds at a scale larger than 10 m^2, which is an order of magnitude lower than that frequently used for assumed or modeled yields in "virtual ponds" (78; 79; 80; 81; 82). A recent demonstration of productivity under favorable year-round climatic conditions in Hawaii in a hybrid closed-open system reached 23 g/m^2 [66], considerably higher than other currently achieved values of 10–15 $g/m^2/day$, which are based on sustained values averaged over the course of a year [83, 84, 85].

16.2.2.4 Harvesting and Drying Algal Biomass

One of the main drawbacks of microalgal biomass as an alternative feedstock for biofuels is their cellular organization. Frequently, they are either single cells in the range of 2– 40 mm or groups of a few cells, making the harvesting process more complex and expensive than for lignocellulosic or macroalgae feedstocks [86, 87, 88]. Cell harvesting expenses accounts for almost 20–30% of the entire cost of producing this feedstock [87] and the methods used include mass harvesting by flotation, flocculation, gravity sedimentation and thickening by centrifugation and/or filtration, a step that requires substantial amounts of energy (Figure 16.2).

Flocculation could be done by microbes or chemicals and the standard flocculants must be cheap, nonhazardous and active in small doses [87]. Additionally, it can be achieved by an electrolytic method, where an electric field is used to charge the microalgal cells in order to move out of the solution. Hydrogen produced by the electrolysis of water gets trapped in the algal flocs and promote their flotation [89]. Another alternative is flotation by sparging small air bubbles, with a size between 10 and 100 μm.

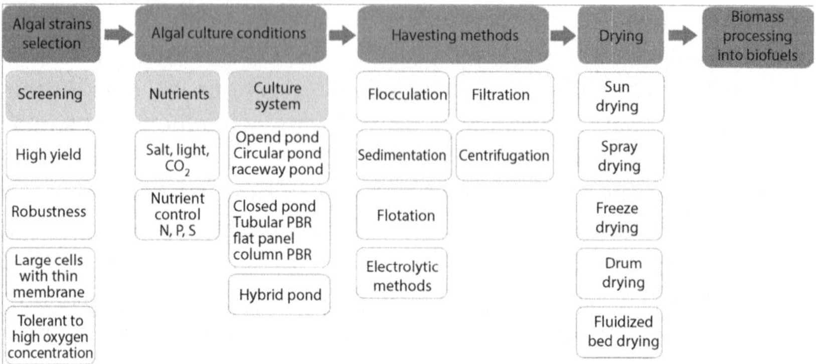

Figure 16.2 Schematic of microalgae biomass production systems.

In some occasions, this approach must be coupled with flocculants such as alum to enhance the flotation effect [90].

Filtration is usually categorized according to exclusion size of the membrane's pores: *i)* macro filtration, 10 μm; *ii)* micro-filtration, 0.1–10 μm; *iii)* reverse osmosis 0.001 μm; and *iv)* ultrafiltration, 0.02–0.2μm. The force needed to push liquid from one side to the other of the membrane rises as the membrane's aperture size drops. This approach is cost-effective for small broth volumes (less than $2m^3$/day), but filtering on a large scale (over 20 m^3/day) is less economical- and operative-effective than flotation [55].

Almost all types of microalgae can be separated reliably and without difficulty by centrifugation using a variety of models. Disc stack centrifuges are the most common industrial centrifuges and are widely used in commercial plants for high-value algal products and in algal biofuel pilot plants [87]. A disc stack centrifuge consists of a relatively shallow cylindrical bowl containing a number (stack) of closely spaced metal cones (discs) that rotate. The mixture to be separated is fed to the center of the stack of discs and the dense phase travels outwards on the underside of the discs while the lighter phase is displaced to the center. Centrifugation is both capital- and operative-expensive. However, the energy return of using centrifugation could be improved by pre-concentration using a combination of other harvesting techniques, such as the ones discussed above [87, 88].

Most times, drying the harvested biomass is necessary to increase the biomass converted into biofuels. While fresh-harvested algae paste has almost 90% of water content, dehydration to about 50% water content is essential for further manipulation [91, 92]. Drying methods may include natural sun drying or more advanced techniques like freeze drying, drum drying, oven-drying, spray drying and fluidized bed-drying. Despite sun drying being amongst the slower methods, and that it produces biomass quality deterioration in some degree, it is cost- and energy-effective when compared to the other techniques, especially important for low-value commodities such as biofuels. Freeze drying is widely used for dewatering microalgal biomass toward fine chemicals or other labile products for which capital and operation costs are justified [92].

16.2.2.5 Microalgal Biomass Conversion into Biofuels

Microalgal carbohydrates can be used to produce biofuels through several biomass conversion technologies. The most developed ones are anaerobic digestion for the production of biogas, and yeast fermentation for the production of bioethanol. However, biological biohydrogen production and hydrothermal liquefaction are also very promising alternatives for the future.

16.2.2.5.1 Anaerobic Digestion of Microalgal Biomass

Like most other organic matter (including lignocellulosic biomass, macroalgal biomass, etc.), microalgae can be subjected to anaerobic digestion to produce biogas. This process normally involves the synthrophic decomposition of the substrate by microbial communities: first, several anaerobic Bacteria hydrolyze and ferment the organic matter into acetic acid and hydrogen and afterwards, methanogenic Archaea proceed to complete the organic matter transformation into methane and carbon dioxide. The hydrolysis step is often the rate-limiting step in the anaerobic digestion and is particularly critical for most terrestrial plants' lignocellulosic biomass. Therefore, the lack of lignocellulosic or cellulosic materials in microalgae might favor the overall hydrolysis process, resulting in higher hydrolysis rates and conversion efficiencies. Additionally, high carbohydrate content, especially simple sugars like glucose, could be advantageous for anaerobic digestion. For example, *Chlamydomonas reinhrdtii* cultured in sulfur starvation showed increased carbohydrate levels, which resulted in an increased biogas production [93].

It has been shown that the optimal values of the C/N ratio are 20–30 since they tend to increase the biogas yield. Microalgae cultivated in excess of N-fertilizer accumulate higher levels of proteins and display C/N ratios of 10 and lower [42], composing a poor substrate for anaerobic digestion [94] due to the inhibition by ammonia of the methanogenic Archaea [42]. Thus, both the strategy of enriching the carbohydrates content in microalgal biomass by controlling the supply of N-fertilizer and/or co-digestion with other high-carbohydrate-containing organic matter, are convenient alternatives for transformation of microalgal biomass into biogas [42]. Also, high sodium content in most marine species can also affect the digester's performance [95].This technology could be applied in a biorefinery approach as it has been demonstrated that conversion of microalgal biomass into methane after lipid extraction can recover more energy than that from the cell lipids alone. This would be especially beneficial when the lipid content of the biomass does not exceed 40% [95].

16.2.2.5.2 Microalgal Biomass Fermentation into Bioethanol

Bioethanol production by yeast fermentation of microalgal carbohydrates is perhaps the most attractive alternative to lignocellulose-based bioethanol for the mid-term future and is the strategy that will be discussed in more detail in this chapter. Although the potential of ethanol production from microalgal biomass has been appreciated for some time now [37], research projects reporting optimization and proof-of-concept demonstrations of the tangible potential of microalgae-based bioethanol have only started to accumulate in the scientific literature in the last 5–10 years [34, 96, 91].

Microalgal biomass appears to be easier to convert into monosaccharides compared to plant lignocellulosic and macroalgal biomass, mostly because of the lack of lignin and/or other recalcitrant structures [42] (Figure 16.3). Nevertheless, microalgal carbohydrates need to be hydrolyzed into single monomers before they can be fermented into ethanol. Currently, hydrolysis and fermentation can be performed as discrete steps (separate hydrolysis and fermentation; SHF) or combined in a one-step consolidated bioprocessing (simultaneous saccharification and fermentation; SSF).

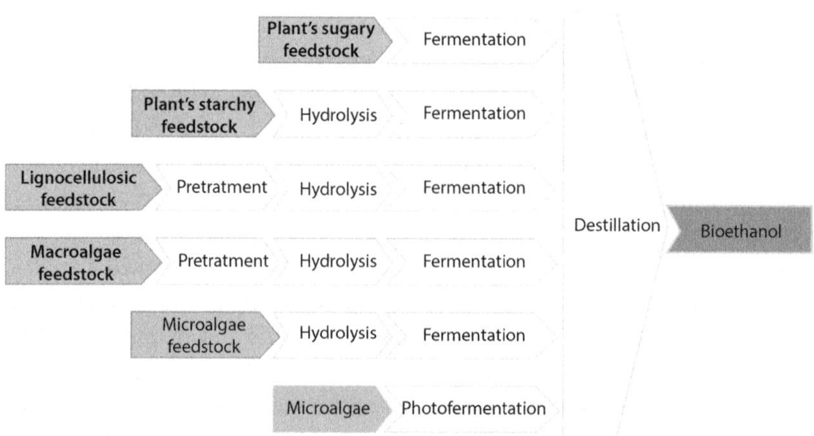

Figure 16.3 Comparison of biomass processing for bioethanol production from different raw materials.

For SHF, several successful conditions for microalgal biomass saccharification by diluted acid hydrolysis have been recently published. For a variety of carbohydrate-rich microalgae biomass, hydrolysis with either sulfuric, nitric, or hydrochloric acid at temperatures between 120 °C and 140 °C for 15–30 min resulted in over 80% saccharification [34, 96], with sulfuric acid pre-treatment being the most effective from a techno-economic perspective [97]. Also, the addition of $MgSO_4$ as a Lewis acid during saccharification resulted in a higher sugar yield than that obtained when only dilute sulfuric acid was used [98]. While the solids load more frequently used for diluted acid hydrolysis were between 10 – 50 g dry biomass/L, Sanchez Rizza and co-workers recently showed the optimized saccharification of a *Desmodesmus* sp. strain biomass containing 57% w/w carbohydrates for the release of up to 95% sugars at a solids load of 100 g dry biomass/L (close to a solids content of 15% (w/v) of the freshly collected *Desmodesmus* biomass) in the presence of 2% H_2SO_4 (v/v), at a 120 °C for 30 min. This demonstration suggests the possibility of increasing the final concentration of sugar using less acid [34]. This is a critical factor for economically-competitive bioethanol production since it has been estimated that a minimum of 40 g ethanol/L of fermentation broth would be needed to reduce distillation costs [99].

An alternative is enzymatic hydrolysis and the most commonly used enzymes are amylases, cellulases, and/or pectinases [96]. In a recent report, different pre-treatments were compared, and it was shown that the enzymatic treatment with thermostable enzymes produced the highest recovery percentage of glucose/ g total sugar [98].

Saccharified microalgal biomass can be efficiently converted into ethanol by the ethanologenic yeast *Saccharomyces cerevisiae*, although other yeasts and bacteria such as *Zymomonas mobilis* or *Escherichia coli* can be used as well (34, and references therein). The high glucose content in the saccharified microalgal biomass facilitates fermentation by common yeasts and results advantageous in comparison to C5 sugars-containing lignocellulosic or macroalgal biomass [42].

Typical ethanol yields from microalgal biomass are close to 0.25 g/ g biomass at carbohydrates contents around 50% (w/w) for nearly 85%, but quite often approaching 95% of the theoretical value of 0.51 g/g glucose (Table 16.2) [34, 96]. Previous optimistic estimates placed achievable productivities at 46,760 – 140,290 L/ha/year, largely exceeding those of more traditional plant crops [100] or macroalgae (29,658L/ha/year) [101] feedstocks. However, more recent data with realistic microalgal biomass productivities around 27 and up to 61 Tn/ha/year [102], and an achievable ethanol yield of 0.25 g/g biomass (Table 16.3) place bioethanol yields at

Table 16.2 Comparative bioethanol yields from microalgal biomass.

Strain	Hydrolysis treatment	Biomass load (g /L)	Sugar concentration (g/L)	Fermenting microorganism	Ethanol (g/L)	Ethanol (g/g biomass)	% of theoretical yield	Reference
Chlamydomonas reinhardtii	H_2SO_4	50	28.5	*Saccharomyces cerevisiae*	14.6	0.290	100	[107]
***Spirogyra* sp.**	Enzymatic	50	12.5	*S. cerevisiae*	NA	0.080	78.4	[108]
***Chlorococum* sp.**[b]	Supercritical CO_2	10	NA	*S. bayanus*	3.8	0.380	NA	[109]
Chlorococcum infusionum	NaOH (SHF)[a]	50	NA	*S. cerevisiae*	NA	0.260	NA	[110]
Chlorella vulgaris	H_2SO_4/ Enzymatic	5	5.5	*Escherichia coli*	1.7	0.400	61.0	[111]
Scenedesmus obliquus	H_2SO_4	500	63.2	*Kluyveromyces marxianus*	11.7	0.023	36.3	[112]
C. vulgaris	Enzymatic (SHF)[a]	20	7.8	*Zymomonas mobilis*	3.6	0.180	87.6	[113]
C. vulgaris	Enzymatic (SSF)[a]	20	NA	*Z. mobilis*	4.3	0.210	87.1	[113]
C. vulgaris	H_2SO_4 (SHF)[a]	50	23.6	*Z. mobilis*	11.7	0.230	96.7	[113]
***S. obliquus* CNW-N**	H_2SO_4	40	16.0	*Z. mobilis*	8.6	0.210	99.8	[113]
***S. abundans* PKUAC12**	H_2SO_4/ enzymatic	50	10.8	*S. cerevisiae*	4.7	0.100	85.5	[114]
***Mychonastes afer* PKUAC 9**	H_2SO_4/ enzymatic	50	6.0	*S. cerevisiae*	2.8	0.060	92	[114]
C. reinhardtii	H_2SO_4	NA	NA	*S. cerevisiae*	8.7	0.150	86.0	[115]
C. vulgaris	Enzymatic	10	1.2	*S. cerevisiae*	0.6	0.070	89.0	[116]
Scenedesmus bijugatus[b]	H_2SO_4 (SSF)[a]	20	5.2	*S. cerevisiae*	NA	0.158	72.5	[117]

Chlamydomonas mexicana	Enzymatic (SHF)[a]	38	22.5	*S. cerevisiae*	8.5	0.410	72.0	[118]
C. mexicana	Enzymatic (SSF)[a]	38.1	22.5	*S. cerevisiae*	10.5	0.500	88.2	[118]
C. vulgaris	hidrotermal acid (HCl)	100	10.15	*Brettanomyces custersii* H1-603	3.75	0.04	72.4	[119]
Senedesmus dimorphus[b]	Enzymatic (SSF)[a]	25	13.50	*S. cerevisiae*	7.34	0.26	95.59	[120]
Desmodesmus sp.	H_2SO_4 (SHF)[a]	100	55.3	*S. cerevisiae*	23.0	0.23	81.4	[34]
Chlorophyta sp.	H_2SO_4 (SHF)[a]	100	72.9	*S. cerevisiae*	23.6	0.24	63.7	[34]
Chlorophyta sp.[c]	H_2SO_4 (SHF)[a]	100	137.2	*S. cerevisiae*	61.2	0.31	87.4	[34]
Scenedesmus dimorphus	H_2SO_4 (SHF)[a]	18	7.7	*S. cerevisiae*	3.6	0.178	80.3	[121]
S. dimorphus	Enzymatic (SHF)[a]	18	7.6	*S. cerevisiae*	4.3	0.183	84.3	[121]
S. dimorphus	Untreated biomass (SSF)[a]	18	8.0	*S. cerevisiae*	4.5	0.181	81.2	[121]
S. dimorphus[b]	Organosolv-treated biomass (SSF)[a]	18	10.2	*S. cerevisiae*	6.8	0.266	91.3	[121]
***S. obliquus* CNW-N(outdoors)**	H_2SO_4 (SHF)[a]	40 (wet biomass)	15.9–18.1	*Z. mobilis*	8.18	0.205	94.1	[122]

NA: not available. [a]Separate hydrolysis and fermentation, (SHF); and simultaneous saccharification and fermentation (SSF). [b]Defatted microalgal biomass. [c]Concentrated hydrolysate

Table 16.3 Comparison between the major bioethanol crops and algae.

	Average world yield (kg ha^{-1}year^{-1})	Dry weight of hydrolysable carbohydrates (kg ha^{-1} year^{-1})	Potential volume of ethanol (L ha^{-1} year^{-1})	References
Corn kernel	9,900[a]	NA	3,680	[123, 124]
Corn stover	5,330[a]	NA	1,594	[123, 129]
Microalgae[b]	715,400	357,700[f]	226,679[g]	[68]
Microalgae[c]	120,450–153,300	60,225–76,650[f]	38,165–48,574[g]	[68]
Microalgae[d]	83,950	41,975 [f]	26,600 [g]	[66]
Microalgae[e]	36,500-54,750	18,250-27,375[f]	11,565-17,348[g]	[83]

[a] central Iowa, USA. [b] Productivity assumed in a theoretical case in a site on the equator, which has relatively constant solar irradiance. [c] Productivity assumed in the best case used the day with peak solar energy, and thus represents a rate that could be achieved over short periods, but not sustained, unless the site sustained a high rate of solar energy, such as those close to the equator. [d] Highest productivity demonstrated at large -scale. [e] Currently achieved values more typically observed at large -scale, based on sustained values averaged over the course of a year. [f] Assuming carbohydrate content 50%. [g] Assuming a 100% conversion efficiency. Ethanol density: 0,789 kg/L

8,500 – 19,500 L/ha/year, which would still be at the top of the list for alternative feedstocks for bioethanol [101].

SSF can be accomplished by simultaneous addition of the fermenting microorganisms and hydrolytic enzymes from different sources. For example, it has been shown that the enzymatic hydrolysis of *C. vulgaris* biomass (containing 51% carbohydrate w/w) gave a glucose yield of 90.4% and produced ethanol at 79.9% or 92.3% of theoretical yields by the SHF or SSF processes, respectively [45]. Furthermore, some developments have aimed at producing recombinant microorganisms able to produce specific hydrolases that are secreted into the fermentation broth or displayed on the cell surface. These strategies would represent a breakthrough for low-cost biomass processing due to the economic benefits of process integration and avoiding the high costs of enzymes [103]. In a representative successful study using lysozyme and a recombinant amylase-expressing yeast and the SSF strategy on carbohydrates obtained from cyanobacterium *Arthrospira platensis* (Spirulina) biomass, which accumulates large amounts of glycogen, it was achieved an ethanol production of 6.5 g/L (ethanol productivity of 1.08 g/L/day) at 86% of theoretical yield [104].

Several microalgae have themselves the capability to produce bioethanol by intracellular auto-fermentation of stored carbohydrates. Hirano and colleagues have demonstrated that almost every one of 250 studied strains displayed some capacity for the production of bioethanol through the intracellular fermentation. However, although this alternative is appealing for its simplicity and would bypass the pre-treatments and saccharification steps described above, overall yields were low. The most productive hit corresponded to a *Chlamydomonas reinhardtii* strain (UTEX2247) at conversion yields about 30–40% of the theoretical value. [105].

Unfortunately, despite its potential, current cost for microalgal biomass production has the greatest impact on bioethanol's production costs [106] and should be reduced in order to achieve large scale production and commercialization [102].

16.2.2.5.3 Genetically Modified Cyanobacteria and the Photanol Concept

Cyanobacteria *Synechocystis* sp. PCC 6803, *Synechococcus elongatus* sp. PCC 7992 and *Anabaena* sp. PCC 7120 have been genetically modified by introducing genes encoding for pyruvate decarboxylase (PDC) and ADH from *Z. mobilis* under the control of the promoter of the *rbcLS* operon encoding the cyanobacterial ribulose-1,5-biphosphate carboxylase/oxygenase with the aim to increase PDC and ADH activities in the cyanobacterial hosts. These engineered strains produced bioethanol directly from CO_2 and light according to an "artificial photofermentative" metabolic pathway that has been given the name of Photanol [126]. These modifications resulted in ethanol productivities of 0.025 mg ethanol/ L after 6 days for *Synechococcus* sp. PCC 7942 [127] and 0.55 g/Lafter 6.25 days for *Synechocystis* sp. PCC 6803 [128]. Although very promising as a proof-of-principle, these values remain low in comparison to those obtained by saccharification of microalgal biomass and fermentation with ethanologenic microorganisms.

16.2.2.5.4 Hydrogen Production from Microalgal Biomass

Biohydrogen production can be performed either by the microalgae itself, or by using microalgal biomass as feedstock for microbial processes. In general, biohydrogen can be produced either photobiologically or by fermentation. Eukaryotic microalgae and cyanobacteria can produce hydrogen by means of two kinds of enzymes: *i*) hydrogenases and/or *ii*) nitrogenases [129].

Biohydrogen production by microalgae can occur either by direct or indirect biophotolysis. Direct biophotolysis takes place briefly at the beginning of the light period or under special growth conditions such as sulfur deprivation that partially inhibits O_2 evolution activity from PSII, leading to anaerobiosis in a sealed culture. The captured solar energy by the photosynthetic apparatus is used to split water into protons, electrons, and oxygen under anoxic conditions. Then, the electrons are transferred to reduce ferredoxin which reduces the hydrogenase enzyme causing the generation of H_2. When O_2 builds up as a consequence of PSII activity, hydrogenase activity drops sharply [130]. Indirect biophotolysis is an alternative pathway through which microalgae produce biomass in the light and, when transferred into anaerobic conditions, hydrogen is produced from electrons that are extracted from the accumulated carbohydrates [130].

In the darkness, the auto-fermentation of the carbohydrates can also result in the production of H_2 by algae [42]. In addition, algal biomass can be utilized as feedstock for biohydrogen production by dark fermentation by bacteria. Some anaerobic bacteria (mainly from the genus *Clostridium*) can produce biohydrogen from fermentable organic materials. The hydrogen production rates by dark fermentation are much higher than those obtained via phototrophic hydrogen production. Under anaerobic conditions, biohydrogen is a by-product of the conversion of organic substrates into organic acids, which are naturally consumed by other microorganisms to produce methane and carbon dioxide as end-products during biomass anaerobic digestion. Inhibition or abatement of hydrogen consumer of an anaerobic digestion sludge or fermentation with axenic fermenting microorganisms results in H_2 production. As in other microbial dark fermentations, carbohydrates are considered to be the most favorable substrate for biohydrogen production [131]. For example, it has been shown that *Tetraselmis subcordiformis* increased about four-fold its carbohydrates content under nitrogen starvation and, consequently, more than 5-fold increase in H_2 levels could be obtained by fermentation of its biomass [132]. Also, optimization of S and N and their interaction with the C source, enabled a 150-fold increase in H_2 production by *Synenchocystis* sp. PCC 6803 [43].

16.2.2.5.5 Thermochemical Conversion of Microlgal Biomass into Bio-oil

As with biomass from other sources, microalgal biomass can be converted into bio-oil (a heavy and tarry oil), with an approximate composition of 73% C, 9% H, 5% N, and 13% O through different technological pathways.

Pyrolysis is performed at temperatures in the range of 350–700 °C in the absence of air. It has been shown that bio-oil derived from microalgae was of a higher quality and of a higher stability than the bio-oil derived from lignocellulosic materials [133]. Gasification takes place at temperatures of 800 to 1,000 °C, and produces syngas- a synthetic gas enriched in CO and H_2 with low calorific value. Syngas can be further converted into liquid synthetic fuels via the Fischer-Tropsch synthesis [134]. The hydrothermal liquefaction proceeds at 300–350 °C and pressure of 5–20 MPa and can be performed from wet algal biomass, bypassing the need of drying the biomass and considerably reducing operation and capital costs. The final product is bio-oil and it has been shown that proteins and lipids are converted to bio-oil most efficiently without the use of catalysts while carbohydrates are best processed using sodium carbonate [135]. The frequently high protein content of microalgae grown under N-sufficiency is problematic for high quality bio-oil, since it tends to increase NOx emissions after its combustion and/or increase expenses for its purification [136, 137]. However, as discussed in a previous section, the biomass composition can be conveniently tuned by optimization of fertilization among other culturing aspects [40].

16.3 Environmental Sustainability of Microlgal-Based Biofuels

Although much work has been done in recent years, second and third generation biofuels are still under development and thus, there is no accurate data of the impact their production has. In general, the main sustainability concerns on biofuels are climate change, energy efficiency and land occupation. Life cycle analysis (LCA) has become an essential tool in assessing the sustainability of biofuel systems [138], using three indicators to address these concerns: 1) Global Warming Potential (GWP), calculated through equivalent carbon dioxide emission per energy unit (g CO_2 eq./MJ); 2) Energy Ratio Output/Input (ER), calculated through a ratio of the biofuel's energy content (MJ of bioenergy output) over the total energy required to produce the biofuel (MJ input); 3) Land Use (LU): measures biofuels potential for area occupation in an annual basis per biofuel energy unit (dm^2.a /MJ) [139].

Well-to-wheel analyses for second generation bioethanol (including agricultural and municipal waste, energetic crops and macroalgae) have shown lower GWP average-values than third generation bioethanol [139]. The poor performance and great variability of third generation biofuels is due to the immaturity of the technology and the lack of direct large-scale determinations. However, open raceway pond technology seems to be preferable to PBRs as it looks less GHG intensive [140].

For the second parameter, ER, there is great variability in the results obtained from LCAs for either second or third generation biofuels, but the average ER values are higher for second generation than for third generation biofuels indicating that the energy required to produce the biofuel is greater when microalgae are used as feedstock [139]. Parvatker identified the harvesting step as the most energy intensive in algal culture due to their low diameter, consuming over 40% of the total energy required by the process [141]. Additionally, LCAs have been unable to point out the best cultivation method energy-wise because of the high uncertainties of the results due to the immaturity of the technology.

Lastly, third generation feedstocks occupy in average a smaller area than second generation, which has LU values comparable to those of first generation sugarcane ethanol. This is as expected since third generation biofuels do not require terrestrial areas for biomass cultivation and avoids competition with food crops. Regardless of the complexity of land use change (LUC) quantification and due to the non-existence of methodological standards, there is a clear trend of reduction in LU when comparing third generation biofuels to second generation [139]. Furthermore, PBRs show higher LU than open raceway ponds because of the PBRs' requirement of facility infrastructure [142].

Due to the diversity of approaches used, comparing LCA's results is not a straight forward task. There is a need for LCA to be harmonized (in particular for co-products credits approach and the inclusion of the indirect effects) to make the interpretation of the results easier, especially for policy makers and investors, although fixed approaches may lead to inaccurate conclusions. Additionally, the developmental stage of third generation biofuels technology and the several pathway options increase the variability and uncertainty of the LCA results. Nevertheless, data show great potential for third generation feedstocks, especially regarding GWP reduction and LUC, even if the energy efficiency is not as promising [139]. A biorefinery-based strategy could be the right approach to optimize the use of energy and lower the production costs. By converting all the available compounds in the biomass in marketable products-particularly high-added value ones- and producing no waste, it might be possible to improve its economic viability, societal acceptance and sustainability [95, 143, 144].

16.4 Prospects for Commercialization of Microalgal-Based Bioethanol

During the past last years, the private sector made impressive technological breakthroughs that placed genetically-modified-cyanobacterial-based bioethanol very close to commercialization.

In 2011, the Company Joule Unlimited (Bedford, MA, USA) cast a project that would use an engineered cyanobacterium to produce bioethanol directly from light and CO_2. In 2014, authorization from the Environmental Protection Agency (EPA) of the United States was granted for large scale cultivation of the engineered cyanobacteria in proprietary PBRs called Solar Converter®. The Company expected a huge ethanol productivity of over 230,000 L/ha/year with a production costs of \$0.16/L - \$0.32/L of ethanol, that would be ready for commercialization in 2017 [96]. However, on July 2017, during the DOE Bioeconomy 2017 conference in Washington DC, it was announced the confirmation of the closure of Joule Unlimited venture due to discontinuation in the investment needed to effectively start commercialization of cyanobacterial-based bioethanol (www.biofuelsdigest.com/bdigest/2017/07 /18).

Algenol (Fort Myers, FL, USA) is a global, industrial biotechnology company established in 2006 that is commercializing its patented algae technology platform for production of ethanol and other biofuels, chemicals and bio-based materials. They use this technology for the production of biofuels using proprietary algae, sunlight, CO_2, and saltwater, on non-arable land. The company's technology is a unique two-step, sustainable process that first produces ethanol directly by enhancing the natural ability found in several strains of cyanobacteria to produce ethanol in proprietary flexible plastic film PBRs that facilitate algae growth and product collection. Then, ethanol is purified through proprietary vapor compression steam stripping units as an energy-efficient technology, followed by standard distillation techniques. Finally, they convert the spent algae biomass into Green Crude. The Company claims that they have the only renewable fuel production process that can convert more than 85% of its CO_2 feedstock into fuel, with a carbon footprint that is 80% less than that of gasoline (http://algenol.com/).

Other companies from the biofuels, petroleum, and agribusiness industries, mostly in the US, have claimed the potential to produce algal biomass for biofuels (bioethanol) on sustainable basis. The major barriers to commercialization of algal bioethanol is the huge capital cost of facilities and other operational costs [91].

16.5 Conclusions and Perspectives

During the last two decades, the microalgae-to-biofuels discipline has produced a remarkable amount of very relevant scientific and technological knowledge. Thus, this chapter aimed at discussing the most relevant issues concerning the use of microalgae biomass as an alternative to lignocellulosic matter as a feedstock for bioethanol. Other excellent journal reviews and book chapters more directly focused on specific aspects have been recently published and are strongly recommended.

In general terms, due to microalgae productivity, the possibility of culturing them at the expense of industrial and/or domestic wastewater and in marginal lands, and its biomass composition, microalgae have very high potential as an alternative feedstock for the production of bioethanol and other biofuels. Even though every culturing systems has advantages and disadvantages, higher productivities have been shown in hybrid systems combining the strengths of both close systems (PBRs) and open systems (raceway ponds). Optimization of fertilization allowed for tuning the biomass composition and, for example, to increase the carbohydrates to protein ratio. The simpler structure of microalgal carbohydrates enables an important reduction in pre-treatment and hydrolysis efforts and the very high glucose content of microalgal carbohydrates largely facilitates fermentation with yeasts or other microorganisms. The implementation of genetic engineering for the development of genetically modified organism (GMO) strains that directly convert CO_2 into ethanol is extremely promising at the point that private companies are already envisioning commercialization.

Nevertheless, this is still a young discipline and much more pilot or pre-industrial scale demonstration plants are to be build and operated year-round for more realistic techno-economic and environmental analysis and to uncover potential new constraints. It appears that investment problems could delay progress in that direction.

References

1. Searchinger, E.T., *et al.* Use of US croplands for biofuels increases greenhouse gases through emissions from land-use change. *Science.*, 319, 1238–40, 2008.
2. Kumar P., *et al.* Methods for pretreatment of lignocellulosic biomass for efficient hydrolysis and biofuel production. *Ind. Eng. Chem. Res.*, 48[8], 3713-3729, 2009.
3. Palmqvist E., Hahn-Hagerdal B. Fermentation of lignocellulosic hydrolysates. I: inhibition and detoxification. *Bioresource Technol.*, 74, 17–24, 2000.
4. Olofsson J., *et al.* Integrating enzyme fermentation in lignocellulosic ethanol production: life-cycle assessment and techno-economic análisis. *Biotechnol. Biofuels.*, 10[1], 51, 2017.
5. Sticklen, M. B. Plant genetic engineering for biofuel production: towards affordable cellulosic ethanol. *Nat. Rev. Genet.*, 9[6], 433–43, 2008.
6. Pilate, G., *et al.* Field and pulping performances of transgenic trees with altered lignification. *Nature Biotechnol.*, 20, 607–612, 2002.

7. Hu, W.J., *et al.* Repression of lignin biosynthesis promotes cellulose accumulation and growth in transgenic trees. *Nature Biotechnol.*, 17, 808–812, 1999.
8. Li L., *et al.* Combinatorial modification of multiple lignin traits in trees through multigene co-transformation. *PNAS*, 100[8], 4939–4944, 2003.
9. Chen, F., Dixon, R.A. Lignin modification improves fermentable sugar yields for biofuel production. *Nature Biotechnol.*, 25, 759–761, 2007.
10. Wei, N., Quarterman, J., Jin, Y.S. Marine macroalgae: an untapped resource for producing fuels and chemicals. *Trends Biotechnol.*, 31[2], 70–77, 2013.
11. Adams, J., *et al.* Fermentation study on *Saccharina latissimi* for bioethanol production considering variable pre-treatments. *J. Appl. Phycol.*, 21, 569–574, 2009.
12. Horn, S.J., Aasen I.M., Østgaard K. Production of ethanol from mannitol by *Zymobacter palmae. J. Ind. Microbiol.Biotechnol.*, 24[1], 51–57, 2000.
13. Jensen, A. Present and future needs for algae and algal products. *Hydrobiologia.*, 260[1], 15–23, 1993.
14. Roesijadi, G., *et al.* Macroalgae as a Biomass Feedstock: A Preliminary Analysis (No. PNNL-19944). U.S. Department of Energy under contract DE-AC05-76RL01830 by Pacific Northwest National Laboratory (PNNL), Richland, WA (US), 2010.
15. Murdock, J.N., Wetzel, D.L. FT-IR microspectroscopy enhances biological and ecological analysis of algae. *Appl. Spectrosc. Rev.*, 44, 335–361, 2009.
16. Reith, J.H., *et al.* Seaweed potential in the Netherlands. In Macroalgae – Bioenergy Research Forum, Plymouth, UK. 2009.
17. McHugh, D.A. *Guide to the Seaweed Industry* (Technical Paper 441), FAO Fisheries *Technical Paper* 441. 2003.
18. Kraan S. Mass-cultivation of carbohydrate rich macroalgae, a possible solution for sustainable biofuel production. *Mitig. Adapt. Strateg. Glob. Change.*, 18[1], 27–46, 2013.
19. Tompkins, A. Marine biomass program. Annual Report for 1979. (https://arpae.energy.gov/sites/default/files/1979%20GRI%20Marine%20Biomass%20Program%20Annual%20Report.pdf).1979.
20. Chynoweth D.P. Review of biomethane from Marine Biomass. Review of history, results and conclusions of the "US Marine Biomass Energy Program" (1968–1990). 2002, University of Florida, 1-207.
21. Wargacki, A.J., *et al.* An engineered microbial platform for direct biofuel production from brown macroalgae. *Science.*, 335[6066], 308–313, 2012.
22. Kim, N.J., *et al.* Ethanol production from marine algal hydrolysates using *Escherichia coli* KO11. *Bioresour. Technol.*, 102, 7466–7469, 2011.
23. Saber, M., Nakhshiniev, B., Yoshikawa K. A review of production and upgrading of algal bio-oil. *Renew. Sust. Energ. Rev.*, 58, 918–930, 2016.
24. Bae, Y.J., *et al.* The characteristics of bio-oil produced from the pyrolysis of three marine macroalgae. *Bioresource Technol.*, 102, 3512–3520, 2011.
25. Díaz-Vázquez, L.M., *et al.* Demineralization of *Sargassum* macroalgae biomass: selective hydrothermal liquefaction process for bio-oil production. *Front. Energy Res.*, 3[6] 57–67, 2015.

26. Nigam, P.S., Singh, A. Production of liquid biofuels from renewable resources. *Prog. Energy Combust. Sci.* 37, 52–68, 2010
27. Meng X., *et al.* Biodiesel production from oleaginous microorganisms. *Renew. Energ.*, 34[1], 1–5, 2009.
28. Demirbas, A. Production economics of high-quality microalgae. *Energ. Source.* part B: Economics, planning, and policy, 12[5], 395–401, 2017.
29. Gonzalez-Fernandez, C., Ballesteros, M. Linking microalgae and cyanobacteria culture conditions and key-enzymes for carbohydrate accumulation. *Biotechnol. Adv.*, 30[6], 1655–1661, 2012.
30. Salerno, G.L., Curatti, L. Origin of sucrose metabolism in higher plants: when, how and why?. *Trends Plant. Sci.*, 8[2], 63–69, 2003.
31. Peccia J., *et al.* Nitrogen supply is an important driver of sustainable microalgae biofuel production. *Trends Biotechnol.*, 31[3], 134–138, 2013.
32. Chisti, Y. Biodiesel from microalgae. *Biotechnol. Adv.*, 25[3], 294–306, 2007.
33. Moheimani, N.R., Borowitzka, M.A. The long-term culture of the coccolithophore *Pleurochrysis carterae* (Haptophyta) in outdoor raceway ponds. *J. Appl. Phycol.*, 18[6], 703–712, 2006.
34. Sánchez Rizza L., *et al.* Bioprospecting for native microalgae as an alternative source of sugars for the production of bioethanol. *Algal. Res.*, 22, 140–147, 2017.
35. Wijffels, R.H.,Barbosa, M.J. An outlook on microalgal biofuels. *Science*, 329[5993], 796–799, 2010.
36. Chaudhary, L., *et al.* Algae as a feedstock for bioethanol production: new entrance in biofuel world. *Int. J. Chem. Technol. Res.*, 6, 1381–1389, 2014.
37. Sheehan, J., *et al.* A look back at the U.S. Department of Energy's Aquatic Species Program: Biodiesel from algae, Golden, Colorado. National Renewable Energy Lab. NREL/TP-580–24190. 1998.
38. Wijffels, R.H., Kruse, O., Hellingwerf, K. J. Potential of industrial biotechnology with cyanobacteria and eukaryotic microalgae. *Curr. Opin. Biotechnol.*, 24, 405–413, 2013.
39. John R.P., *et al.* Micro and macroalgal biomass: a renewable source for bioethanol. *Bioresource Technol.*, 102[1], 186–193, 2011.
40. Dragone, G., *et al.* Nutrient limitation as a strategy for increasing starch accumulation in microalgae. *Appl. Energ.*, 88, 3331–3335, 2011.
41. Domozych, D. S., *et al.* The cell walls of green algae: a journey through evolution and diversity. *Front. Plant Sci.*, 3, 82, 2012.
42. Markou G., Angelidaki I., Georgakakis D. Microalgal carbohydrates: an overview of the factors influencing carbohydrates production, and of main bioconversion technologies for production of biofuels. *Appl. Microbiol. Biotechnol.*, 96, 631–645, 2012.
43. Burrows, E.H., *et al.* Optimization of media nutrient composition for increased photofermentative hydrogen production by *Synechocystis* sp. PCC 6803. *Int. J. Hydrog. Energ.*, 33[21], 6092–6099, 2008.
44. Olguín, E.J. Dual purpose microalgae–bacteria-based systems that treat wastewater and produce biodiesel and chemical products within a Biorefinery. *Biotechnol. Adv.*, 30[5], 1031–1046, 2012.

45. Ho S.H., *et al.* Bioethanol production using carbohydrate-rich microalgae biomass as feedstock, *Bioresource Technol.*, 135, 191–198, 2013.
46. Brown, M.R. The amino-acid and sugar composition of 16 species of microalgae used in mariculture. *J. Exp. Mar. Biol. Ecol.*, 145[1], 79–99, 1991.
47. Huesemann M., *et al.* The laboratory environmental algae pond simulator (LEAPS) photobioreactor: Validation using outdoor pond cultures of *Chlorella sorokiniana* and *Nannochloropsis salina*. *Algal. Res.*, 26, 39–46, 2017.
48. Demirbas, A. Use of algae as biofuel sources. *Energ. Convers. Manage.*, 51, 2738–2749, 2010.
49. Trivedin, J., *et al.* Algae based biorefinery—How to make sense?. *Renew. Sust. Energ. Rev.*, 47, 295–307, 2015.
50. Huber G.W., Iborra S., Corma A. Synthesis of transportation fuels from biomass: chemistry, catalysts, and engineering. *Chem Rev.*, 106, 4044–4098, 2006.
51. Xu, L., *et al.* Assessment of a dry and a wet route for the production of biofuels from microalgae: Energy balance analysis, *Bioresource Technol.*, 102 [8] 5113–5122, 2011.
52. Carvalho, A.P., *et al.* Microalgae reactors: a review of enclosed system designs and performances. *Biotechnol. Prog.*, 22, 1490–1506, 2006.
53. Cai, T., *et al.* Nutrient recovery from wastewater streams by microalgae: status and prospects. *Renew. Sust. Energ. Rev.*, 19, 360–369, 2013.
54. Mata T.M., Martins A.A., Caetano N.S. Microalgae for biodiesel production and other applications: a review. *Renew. Sustain. Energy Rev.*, 14, 217–32, 2010.
55. Brennan, L., Owende, P. Biofuels from microalgae—a review of technologies for production, processing, and extractions of biofuels and co-products. *Renew. Sust. Energ. Rev.*, 14[2], 557–577, 2010.
56. Rodolfi, L., *et al.* Microalgae for oil: strain selection, induction of lipid synthesis and outdoor mass cultivation in a low-cost photobioreactor. *Biotechnol. Bioeng.*, 102 [1], 100–112, 2009.
57. Richmond, A. Outdoor mass cultures of microalgae. Handbook of microalgal mass culture. 285–330, 1986.
58. Ugwu, C. U., Aoyagi, H., Uchiyama, H. Photobioreactors for mass cultivation of algae. *Bioresource Technol.*, 99[10], 4021–4028, 2008.
59. Borowitzka, M.A. Commercial production of microalgae: ponds, tanks, tubes and fermenters. *J. Biotechnol.*, 70[1], 313–321, 1999.
60. Terry, K. L., Raymond, L.P. System design for the autotrophic production of microalgae. *Enzyme Microb. Tech.*, 7[10], 474–487, 1985.
61. Wang, B., Lan, C.Q., Horsman, M. Closed photobioreactors for production of microalgal biomasses. *Biotechnol. Adv.*, 30[4], 904–912, 2012.
62. Sanchez Mirón, A.S., *et al.* Growth and biochemical characterization of microalgal biomass produced in bubble column and airlift photobioreactors: studies in fed-batch culture. *Enzyme Microb. Tech.*, 31[7], 1015–1023, 2002.
63. Vega-Estrada, J., *et al. Haematococcus pluvialis* cultivation in split-cylinder internal-loop airlift photobioreactor under aeration conditions avoiding cell damage. *Appl. Microbiol. Biot.*, 68[1], 31–35, 2005.

64. Camacho, F.G., *et al.* Use of concentric-tube airlift photobioreactors for microalgal outdoor mass cultures. *Enzyme Microb. Tech.*, 24[3], 164–172, 1999.
65. Krichnavaruk, S., Powtongsook, S., Pavasant, P. Enhanced productivity of Chaetoceros calcitrans in airlift photobioreactors. *Bioresource Technol.*, 98[11], 2123–2130, 2007.
66. Huntley, M.E., *et al.* Demonstrated large-scale production of marine microalgae for fuels and feed. *Algal Res.*, 10, 249–265, 2015.
67. Schenk, P. M., *et al.* Second generation biofuels: high-efficiency microalgae for biodiesel production. *Bioenerg. Res.*, 1[1], 20–43, 2008.
68. Weyer, K.M., *et al.* Theoretical maximum algal oil production. *Bioenerg. Res.*, 3[2], 204–213, 2010.
69. Williams, P.J.L.B., Laurens, L.M. Microalgae as biodiesel & biomass feedstocks: review & analysis of the biochemistry, energetics & economics. *Energ. Environ. Sci.*, 3[5], 554–590, 2010.
70. Churkina, G., Running, S.W. Contrasting climatic controls on the estimated productivity of global terrestrial biomes. *Ecosystems*, 1[2], 206–215, 1998.
71. Pate, R.C. Resource requirements for the large-scale production of algal biofuels. *Biofuels.*, 4[4], 409–435, 2013.
72. Langholtz, M.H., Stokes, B.J., Eaton, L.M. 2016 Billion-Ton Report. ORNL (Oak Ridge National Laboratory (ORNL), Oak Ridge, TN (United States), managed by UT-Battelle, LLC for the US DEPARTMENT OF ENERGY, 2016, 1–411 (http://energy.gov/eere/bioenergy/2016-billion-ton-report).
73. Beal, C.M., *et al.* Algal biofuel production for fuels and feed in a 100-ha facility: a comprehensive techno-economic analysis and life cycle assessment. *Algal. Res.*, 10, 266–279, 2015.
74. Davis, R., *et al.* Techno-economic analysis of autotrophic microalgae for fuel production. *Appl. Energ.*, 88[10], 3524–3531, 2011.
75. Stephenson, A.L., *et al.* Life-cycle assessment of potential algal biodiesel production in the United Kingdom: a comparison of raceways and air-lift tubular bioreactors. *Energ. Fuel.*, 24[7], 4062–4077, 2010.
76. Davis, R.E., *et al.* Integrated evaluation of cost, emissions, and resource potential for algal biofuels at the national scale. *Environ. Sci. Technol.*, 48[10], 6035–6042, 2014.
77. Jonker, J.G.G., Faaij, A.P.C., Techno-economic assessment of micro-algae as feedstock for renewable bio-energy production. *Applied Energy.*, 102, 461–475, 2013.
78. Liu, X., *et al.* Pilot-scale data provide enhanced estimates of the life cycle energy and emissions profile of algae biofuels produced via hydrothermal liquefaction. *Bioresource Technol.*, 148, 163–171, 2013.
79. Passell, H., *et al.* Algae biodiesel life cycle assessment using current commercial data. *J. Environ. Manage.*, 129, 103–111, 2013.
80. Craggs, R., *et al.* Hectare-scale demonstration of high rate algal ponds for enhanced wastewater treatment and biofuel production. *J. App. Phycol.*, 24[3], 329–337, 2012.

81. Jiménez, C., *et al.* The feasibility of industrial production of Spirulina (Arthrospira) in Southern Spain. *Aquaculture.*, 217[1], 179–190, 2003.
82. Park, J.B.K., Craggs, R.J., Shilton, A.N. Enhancing biomass energy yield from pilot-scale high rate algal ponds with recycling. *Water res.*, 47[13], 4422–4432, 2013.
83. Dirks, G. Paper read at 2015 BETO Algae Peer Review. www.energy.gov/sites/prod/files/2015/04/f21/algae_dirks_135100.pdf
84. Ogden, K. Paper read at 2015 BETO Algae Peer Review. www.energy.gov/sites/prod/files/2015/04/f21/algae_ogden_135111.pdf
85. NAABB, NAABB Full Final Report. DE-EE0003046 U.S. DOE, Bioenergy Technologies Office, 2014.
86. Li, Y., *et al.* Biofuels from microalgae. *Biotechnol.Progr.*, 24[4], 815–820, 2008.
87. Molina Grima, E.M., *et al.* Recovery of microalgal biomass and metabolites: process options and economics. *Biotechnol. adv.*, 20[7], 491–515, 2003.
88. Milledge, J.J., Heaven, S. A review of the harvesting of micro-algae for biofuel production. *Rev. Environ. Sci. Bio.*, 12[2], 165–178, 2013.
89. Mollah, M.Y., *et al.* Fundamentals, present and future perspectives of electrocoagulation. *J. Hazard. Matter.*, 114[1], 199–210, 2004.
90. Hanotu, J., *et al.* Microalgae recovery by microflotation for biofuel production using metallic coagulants. *Biofuels.*, 4[4], 363–369, 2013.
91. Bibi, R., *et al.* Algal bioethanol production technology: A trend towards sustainable development. *Renew. Sust. Energ. Rev.*, 71, 976–985, 2017.
92. Guldhe, A., *et al.* Efficacy of drying and cell disruption techniques on lipid recovery from microalgae for biodiesel production. *Fuel.*, 128, 46–52, 2014.
93. Mussgnug, J.H., *et al.* Microalgae as substrates for fermentative biogas production in a combined biorefinery concept. *J. Biotechnol.*, 150[1], 51–56, 2010.
94. Yen, H. W., Brune, D. E. Anaerobic co-digestion of algal sludge and waste paper to produce methane. *Bioresour. Technol.*, 98[1], 130–134, 2007.
95. Sialve, B., Bernet, N., Bernard, O. Anaerobic digestion of microalgae as a necessary step to make microalgal biodiesel sustainable. *Biotechnol. Adv.*, 27[4], 409–416, 2009.
96. de Farias Silva, C.E., Bertucco, A. Bioethanol from microalgae and cyanobacteria: A review and technological outlook. *Process Biochem.*, 51, 1833–1842, 2016.
97. Li K., Liu S., Liu X., An overview of algae bioethanol production. *Int. J. Energy Res.*, 38, 965–977, 2014.
98. Shokrkar, H., Ebrahimi, S., Zamani, M., Bioethanol production from acidic and enzymatic hydrolysates of mixed microalgae culture. *Fuel.*, 200, 380–386, 2017.
99. Möllers, K.B., *et al.* Cyanobacterial biomass as carbohydrate and nutrient feedstock for bioethanol production by yeast fermentation. *Biotechnol. biofuels.*, 7[1], 64, 2014.

100. Mussatto S. I., *et al.* Technological trends, global market, and challenges of bio-ethanol production. *Biotechnol. Adv.*, 28[6], 817–830, 2010.
101. Noraini M.Y., *et al.* A review on potential enzymatic reaction for biofuel production from algae. *Renew. Sust. Energ. Rev.*, 39, 24–34, 2014.
102. Ruiz, J., *et al.* Towards industrial products from microalgae. *Energ. Environ. Sci.*, 9, 3036–3043, 2016.
103. den Haan, R., *et* al. Engineering *Saccharomyces cerevisiae* for next generation ethanol production. *J. Chem. Technol. Biot.*, 88[6], 983–991, 2013.
104. Aikawa, S., *et. al.* Direct conversion of Spirulina to ethanol without pretreatment or enzymatic hydrolysis processes. *Energ. Environ. Sci.*, 6[6], 1844–1849, 2013.
105. Hirano A., *et al.* CO_2 fixation and ethanol production with microalgal photosynthesis and intracellular anaerobic fermentation. *Energ.*, 22(2–3), 137–142, 1997.
106. Kwiatkowski J.R., *et al.* Modeling the process and costs of fuel ethanol production by the corn dry-grind process. *Ind. Crops Prod.*, 23, 288–296, 2006.
107. Nguyen, M.T., *et al.* Hydrothermal acid pretreatment of *Chlamydomonas reinhardtii* biomass for ethanol production. *J. Microbiol. Biotechnol.*, 19[2], 161–166, 2009.
108. Eshaq, F.S., *et al.* Spirogyra biomass a renewable source for biofuel (bioethanol) production. *Eng. Sci. Technol.*, 2[12], 7045–7054, 2010.
109. Harun, R., Danquah, M.K., Forde, G.M., Microalgal biomass as a fermentation feedstock for bioethanol production. *J. Chem. Technol. Biotechnol.*, 85[2], 199–203, 2010.
110. Harun, R., *et al.* Exploring alkaline pre-treatment of microalgal biomass for bioethanol production. *Appl. Energ.*, 88[10], 3464–3467, 2011.
111. Lee S., *et al.* Converting Carbohydrates Extracted from Marine Algae into Ethanol Using Various Ethanolic *Escherichia coli* strains. *Appl. Biochem. Biotechnol.*, 164, 878–888, 2011.
112. Miranda, J.R., Passarinho P.C., Gouveia L. Bioethanol production from *Scenedesmus obliquus* sugars: the influence of photobioreactors and culture conditions on biomass production. *Appl. Microbiol. Biot.*, 96[2], 555–564, 2012.
113. Ho S.H., *et al.* Bioethanol production using carbohydrate-rich microalgae biomass as feedstock, *Bioresource Technol.*, 135, 191–198, 2013.
114. Guo, H., *et al.* Biochemical features and bioethanol production of microalgae from coastal waters of Pearl River Delta. *Bioresource Technol.*, 127, 422–428, 2013.
115. Scholz, M.J., Riley, M.R., Cuello, J.L. Acid hydrolysis and fermentation of microalgal starches to ethanol by the yeast *Saccharomyces cerevisiae. Biomass Bioenerg.*, 48, 59–65, 2013.
116. Kim K.H., *et al.* Bioethanol production from the nutrient stress-induced microalga *Chlorella vulgaris* by enzymatic hydrolysis and immobilized yeast fermentation. *Bioresource Technology.*, 153, 47–54, 2014.

117. Ashokkumar, V., *et al.* An integrated approach for biodiesel and bioethanol production from *Scenedesmus bijugatus* cultivated in a vertical tubular photobioreactor. *Energ. Convers. Manage.*, 101, 778–786, 2015.
118. El-Dalatony, M.M., *et al.* Long-term production of bioethanol in repeated-batch fermentation of microalgal biomass using immobilized *Saccharomyces cerevisiae. Bioresource Technol.*, 219, 98–105, 2016.
119. Park C., *et al.* Enhancement of hydrolysis of *Chlorella vulgaris* by hydrochloric acid. *Bioproc. Biosyst. Eng.*, 39[6], 1015–1021, 2016.
120. Chng, L.M., *et al.* Sustainable production of bioethanol using lipid-extracted biomass from *Scenedesmus dimorphus. J. Clean. Prod.*, 130, 68–73, 2016.
121. Chng, L.M., *et al.* Synergistic effect of pretreatment and fermentation process on carbohydrate-rich *Scenedesmus dimorphus* for bioethanol production. *Energ. Convers. Manage.*, *141*, 410–419, 2017.
122. Ho S-H., *et al.* Feasibility of CO_2 mitigation and carbohydrate production by microalga *Scenedesmus obliquus* CNW-N used for bioethanol fermentation under outdoor conditions: effects of seasonal changes. *Biotechnol. Biofuels.*, 10:27, 2017.
123. Karlen, D. L., *et al.* A five-year assessment of corn stover harvest in central Iowa, USA. *Soil Tillage Res.* 115, 47–55, 2011.
124. Pimentel, D., Patzek, T. W. Ethanol production using corn, switchgrass, and wood; biodiesel production using soybean and sunflower. *Nat. Resources Res.* 14, 65–76, 2005.
125. Humbird, D., *et al.* Process design and economics for biochemical conversion of lignocellulosic biomass to ethanol: Dilute-acid pretreatment and enzymatic hydrolysis of corn stover. Technical Report NREL/TP-5100- 47764, National Renewable Energy Laboratory, Golden, CO. 2011.
126. Savakis, P., Hellingwerf, K.J. Engineering cyanobacteria for direct biofuel production from CO_2. *Curr. Opin. Biotechnol.*, 33, 8–14, 2015.
127. Deng, M., Coleman, J.R. Ethanol synthesis by genetic engineering in cyanobacteria. *App. Environ. Microb.*, 65[2], 523–528, 1999.
128. Dexter, J., Fu, P. Metabolic engineering of cyanobacteria for ethanol production. *Energy Environ. Sci.*, 2[8], 857–864, 2009.
129. Ghirardi, M.L., *et al.* Photobiological hydrogen-producing systems. *Chem. Soc. Rev.*, 38[1], 52–61, 2009.
130. Melis, A. Photosynthetic H_2 metabolism in *Chlamydomonas reinhardtii* (unicellular green algae). *Plant.*, 226[5], 1075–1086, 2007.
131. Kim M-S., *et al.* Hydrogen production from *Chlamydomonas reinhardtii* biomass using a two-step conversion process: anaerobic conversion and photosynthetic fermentation. *Int. J. Hydrog. Ener.*, 31[6], 812–816, 2006.
132. Ji, C.F., *et al.* Effects of nutrient deprivation on biochemical compositions and photo-hydrogen production of *Tetraselmis subcordiformis. Int. J. Hydrogen Energ.*, 36[10], 5817–5821, 2011.
133. Amin, S. Review on biofuel oil and gas production processes from microalgae. *Energ. Convers. Manage.*, 50[7], 1834–1840, 2009.

134. Posten, C., Schaub, G. Microalgae and terrestrial biomass as source for fuels—a process view. *J. Biotechnol.*, 142[1], 64–69, 2009.
135. Biller, P., Ross, A.B. Potential yields and properties of oil from the hydrothermal liquefaction of microalgae with different biochemical content. *Bioresource Technol.*, 102[1], 215–225, 2011.
136. Dote, Y., *et al.* Studies on the direct liquefaction of protein-contained biomass: the distribution of nitrogen in the products. *Biomass Bioenerg.*, 11[6], 491–498, 1996.
137. Du, Z., *et al.* Hydrothermal pretreatment of microalgae for production of pyrolytic bio-oil with a low nitrogen content. *Bioresource Technol.*, 120, 13–18, 2012.
138. Thamsiriroj, T.,Murphy, J.D. Can rape seed biodiesel meet the European Union sustainability criteria for biofuels? *Energ. Fuel., 24*, 1720–1730, 2010.
139. Carneiro, M.L.N.M., *et al.* Potential of biofuels from algae: Comparison with fossil fuels, ethanol and biodiesel in Europe and Brazil through life cycle assessment (LCA). *Renew. Sust. Energ. Rev.*, 73, 632–653, 2017.
140. Weinberg, J., Kaltschmitt, M., Wilhelm, C. Analysis of greenhouse gas emissions from microalgae-based biofuels. *Biomass Conv. Bioref.*, 2, 179–94, 2012.
141. Parvatker A.G. Biodiesel From microalgae – A sustainability analysis using life cycle assessment. *Int. J. Chem. Phys. Sci.*, 2, 159–69, 2013.
142. Lardon L., *et al.* Life-cycle assessment of biodiesel production from microalgae. *Environ. Sci. Technol.*, 43[17], 6475–6481, 2009.
143. Adesanya, *et al.* Life cycle assessment on microalgal biodiesel production using a hybrid cultivation system. *Bioresour. Technol.*, 163, 343–355, 2014.
144. Parajuli, R., *et al.* Biorefining in the prevailing energy and materials crisis: a review of sustainable pathways for biorefinery value chains and sustainability assessment methodologies. *Renew. Sustain. Energy Rev.*, 43, 244–263, 2015.

17

A Sustainable Process for Nutrient Enriched Fruit Juice Processing: An Enzymatic Venture

Debajyoti Kundu, Jagriti Singh, Mohan Das, Akanksha Rastogi and Rintu Banerjee*

Agricultural and Food Engineering Department, Indian Institute of Technology, Kharagpur, India

Abstract

Fruits and vegetables hold a superior place in our routine life not only from economic aspects but also from health benefit for curing several illness. India is the second largest producer of fruits all over the world. But improper storage and exposure to the external factors accelerate the loss of quality attributes. The gap between seasonal supply of fruits and its demand throughout the year enforced us for proper storage. Preservation through juice production ensures availability and supply around the year. Like fresh fruits, fruit juice is also gaining more impetus throughout the world for its refreshing nature, therapeutic value and health benefits. Modern life style led us to take fruit juice rather than fresh fruit and becoming more popular with time span.

Juice processing have a major influence on phytochemicals property, antioxidant property, texture, and shelf life. Conventional processing are reported for hampering the original properties of juice. In this consequence enzyme mediated juice processing offer tremendous advantages over traditional methods. It is an integral component of juice processing and highly acceptable from industrial point of view. The main purpose of paradigm shifting from traditional to alternative processing is concentrated on the high yield, process efficiency, cost effective production, upliftment of nutritional properties, soothing appearance, and extent shelf life.

Keywords: Juice processing, traditional methods, enzymatic processing, nutritional property, shelf life

**Corresponding author*: rb@iitkgp.ac.in

Arindam Kuila and Vinay Sharma (eds.) Principles and Applications of Fermentation Technology, (387–400) © 2018 Scrivener Publishing LLC

17.1 Introduction

In recent time, fruit juice based diet is gaining more popularity all over the world, which provide essential nutrients to human health. As fruit juices are very rich in bio-active components, therefore, it contributes in health promotion and diminish the risk of chronic diseases [1]. Several diseases like obesity, diabetes, coronary heart disease, cancer, bowel disease, skin disease, rheumatoid arthritis, pulmonary disease, asthma, osteoporosis, eye diseases, dementia, stroke etc. can be prevented through consumption of fruits [2].

As all fruits are not available throughout the year, thus, there is an urge to preserve the fruit materials to meet the demand in the off season. There is a limitation for all fruit storage, after post-harvest due to its perishable nature. According to the third estimate of Indian Horticulture Database, total fruit production is 93,707,000 MT in the year 2017 [3]. Fresh fruits are consumed directly, but a large portion of the annual harvesting is processed as consumable goods in the form of fruit juice and canned fruits. In our country, >20–25% fruits get spoiled before consumption due to improper storage and only 1.5% of total are being processed [4]. Fruit juice preparation ensures the availability of the product in all seasons and occupies a good position from a commercial point of view. The major challenges in the juice processing industry lie in the retainment of nutritional property and improvement of organoleptic characteristics.

In industry, several techniques like use of mechanical device for peeling (abrasive devices, drums, rollers, knives, and milling cutters), juice extraction (traditional rack and cloth press, screw presses, horizontal press, and the belt press), use of chemicals for clarification and debittering, thermal processing (blanching, pasteurization, heat sterilization, radio frequency heating, microwave heating, ohmic, and dielectric heating) has been practiced extensively [5, 6].

Hindrances in traditional methods of juice processing drive the industry towards biotechnological approaches. Enzymatic juice processing facilitate the improvement of process efficiency and product quality. Enzyme application is a best alternative of chemical methods as it is environmentally benign. Different factors like high activity, rapid and efficient action in low amount, substrate specificity, turn over number, reuse of the enzymes, no impact on nutritional and organoleptic property and mild reaction condition make enzymes more preferable and acceptable [7]. Thus, there is an outstanding growth in enzyme market for fruit processing industry. It is estimated that in 2020 market value of the industrial enzyme will be 6.2 billion USD with a compound annual growth rate of 7% from 2015 and the demand of industrial enzymes in food and beverage sector will be 2.0 billion USD [8].

This chapter focus on the limitation of conventional juice processing and the application of various enzymes in fruit processing that makes the process nutritionally enriched fruit juice production along with its eco-friendly and cost effectiveness.

17.2 Conventional Methods for Juice Processing and Their Drawbacks

Juice processing involves different set of operational activities like peeling, extraction, yield recovery, clarification, bitterness removal, etc. Basically, these process are required to make attractive, to increase the acceptability and to retain or increase the nutritional property of the juice. Palatability of juice is highly dependent on the appearance, flavour, texture, and nutrition of the juice.

The conventional peeling methods are categorized into mechanical, chemical, and thermal process, where mechanical peeling includes abrasive devices, drums, rollers, knives and milling cutters, etc. [9]. Chemical treatments process include application of chemicals such as NaOH solution, which is most preferred because of its easy penetration in the pores that facilitate in the degradation of the pulp layer, by acting on the thin waxy layer of the skin [5]. This chemical process is also known as lye peeling. Thermal peeling can be performed by wet heat or dry heat. In steam peeling process, initially internal pressure has been build, resulted in mechanical failure of the cell. After that it affects the tissue and disorganize the cell wall components, such as pectin and polysaccharides. In flame or dry heat peeling 1000 °C temperature has been used for peel removal. Infrared radiation (IR) separate peels by rapid surface heating, which permit effective heating of shallow layer of the fruit or vegetable surface [10].

Clarification can be performed by traditional physico-chemical or mechanical methods or combination of these. Centrifugation and filtration are used as a mechanical tool for removal of suspended from cloudy juice. In physico-chemical methods clarifying agents like minerals, natural organic or organic polymers are often used [11].

Membrane ultrafiltration (UF) is also a very promising tool for juice clarification because it facilitate in the permeability of small molecules across the membrane, thus water and small molecules across the membrane while other big molecules such as protein and colloids molecules can retain in the juice. Due to large membrane pore size, microfiltration (MF) is also another promising membrane process which is used for clarification of juice [12].

Concentration is another processing step in juice industry for longer shelf life by decreasing the water content and increasing the total soluble solid (TSS) content in juice. Common method of juice concentration is evaporation, freezing, and membrane separation processes [11]. Reverse osmosis (RO) is used for the separation of water to concentrate the juice [12]. For the bitterness removal different physical separation using adsorbent resins, polyamide, polystyrene, and ion exchange has been used for limonin and naringin separation from citrus juices. These resins can remove up to 85% limonin from grapefruit juice while polystyrene divinyl benzene XAD-16 and anionic resin IRA-93 are used for limonin removal respectively from sour orange juice. Low molecular weight polyvinyl chloride beads for navel orange juice and Amberlite XAD-16HP and Dowex-L285 reduce bitterness to acceptable levels in Washington navel orange juices. For kinnow orange juice debittering Amberlite XAD-16 packed in glass column was used as adsorbent [13].

As nutrients composition and phytochemicals property are very much important in any food product for health benefit thus detainment of this composition in processed food is necessary. Above mentioned conventional methods changes several physical, biochemical, and nutritional parameters of the food product [14]. Conventional juice extraction process consume more energy, time and low yield. Whereas, thermal process are responsible for enzyme deactivation, degradation of nutrients and vitamins, changes in appearance, etc. [15]. Ohomic heat treatment, ultrasound, radiation, evaporation, and filtration techniques are reported for the loss of bioactive compounds, phytochemicals, vitamins, anti-oxidant property, and phenolic contents [16–20]. In mechanical, thermal, and chemical peeling of fruits resulted into high cost and loss of mash [21], while during bitterness removal the conventional practices of exchange resin resulted in the anti-oxidant, phytochemicals and DNA damage [22].

17.3 Enzyme Technology in Different Step of Juice Processing

Juice processing techniques involve in different operational set viz. peeling, extraction, yield recovery, clarification, concentration, bitterness removal, etc. (Figure 17.1). Generally, processing facilitate in the improvisation of shelf life, organoleptic properties, and nutritional property of the juices. Drawbacks in conventional processing hasten the application of

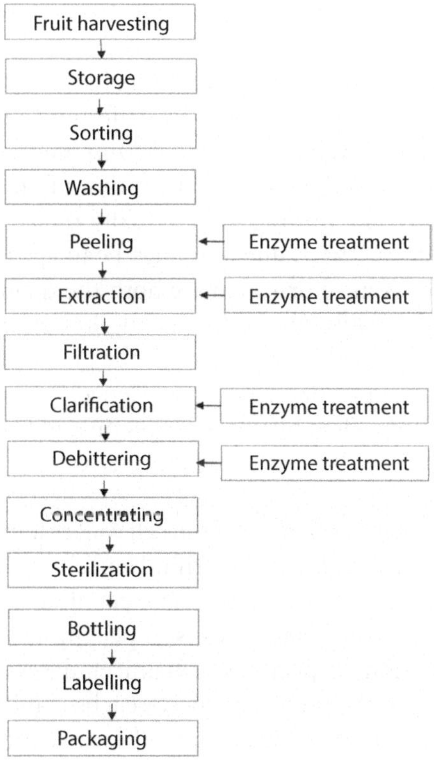

Figure 17.1 General steps involved in juice processing.

biocatalysis in juice processing. Broad range of application makes enzyme more suitable for bioconversion process. Application of enzymes in juice processing lies on peeling, extraction, clarification, and bitterness removal, where different types of enzymes like pectinase, cellulase, laccase, tannase, and naringinase play a crucial role. A brief description of application of suitable enzymes in different steps of juice process are discussed below:

17.3.1 Peeling and Extraction

Before extraction of juice from fruits, raw fruits undergo various processing, that is, washing, sorting, and crushing. After proper washing, peeling is the first step for juice extraction, which broadly depends on the variety of the fruit. Generally, citrus and tropical fruits such as bananas, avocados, oranges, beets, and pumpkins require removal of the peel and need to be sliced before juice extraction. Traditional peeling process consist of manual or mechanical removal of peel followed by chemical treatment which is labour intensive and high cost and also pollute environment due to discharge of caustic agent as effluent. Furthermore aggressive chemical treatment alters the original taste of the fruits [23].

Thus, this sector of processing is badly suffering from the lack of alternative technology which can give the relief from the traditional method as well as from economic and environmental point of view. In this scenario, enzymatic processing for peeling are gradually replacing chemical, thermal, and mechanical methods due to reduced damage to the pulp of fruits and vegetables, which in return increase both the quality and quantity of juice production. Basically, enzymatic peeling lies on the digestion of the cell wall. Sticking of peel with fruit is due to the presence of polysaccharide viz. pectin, cellulose, and hemicellulose. Hence, cellulase and pectinase plays a crucial role in peeling by releasing pectin and hydrolysing the polysaccharide. On the other hand polygalacturonase facilitate in the degradation of albedo [23].

Alteration of the polysaccharide of the cell wall resulted into the changes in smell, texture, and shelf life of the fruit. Hence, a concoction of polygalacturonase, pectinlyase, pectinestarase, and cellulase is required to restrict the changes occurred from other means of peeling [23]. Enzymatic treatment using concoction of pectinases, hemicellulases, and cellulases for peeling of apricots, nectarines, mangoes, peaches, and other stone fruits showed considerably good results. Among these peeling enzymes, mixture of cellulases, hemicellulases, arabinases, and pectinase showed better peeling activity on lemon, orange, and grape fruits [24, 25]. Enzymatic peeling is very suitable for citrus fruit because the mesocarp or albedo is very porous in nature and enzyme can easily occupy this intercellular place by replacing air. As morphology of albedo of all fruits are not porous hence pretreatment (like scalding, cut in the flavedo, vaccum pressure, etc.) prior to enzymatic peeling showed effective result [23]. Thus, enzymatic treatment offers most effective and convenient ways to get high quality juices from wide varieties of fruit and vegetable which come directly from cold storage [26].

A concoction of enzymes (polygalacturonase, hemicellulase, and cellulase) were employed for the peeling of peach (*Prunuspersica*), nectarine (*Prunuspersica var. nucipersica*), and apricot (*Prunusarmeniaca*) at the temperature around 45 °C [27]. As Cellulase is a complex of three different enzymes (endoglucanase, exoglucanase, and β-glucosidase), here endoglucanase and exoglucanase alter cellulose into cello-oligosaccharides and β-glucosidase hydrolyzes the cello-oligosaccharides into glucose [28]. Persimmon (*Diospyros kaki*) fruits was successfully peeled using polygalacturonase followed by heat treatment [29]. Heat treatment forms cracks in the cuticle, through which enzyme can easily penetrate into it and degrade within several hour. Furthermore, heat treatment also inactivates the polygalacturonase-inhibiting proteins. Ni *et al.* [30] used pectinase for higher yield of pomelo (*Citrus grandis*) juice. After combined peeling and enzymatic hydrolyses using this enzyme yield was obtained about 43%. Simultaneous application of Pectinex® smash XXL or Cellubrix® L resulted into 46% increment of extracted sugar [31]. Another application of pectinase and cellulase showed two times more soluble solid extraction in date syrup producton [32]. Increased juice yield (up to 30%) was found in grape juice by pectinase treatment [33].

17.3.2 Clarification

Due to the presence of polysaccharides, extracted juice become opaque and viscous. As clear juice is a determinant factor for consumer acceptability, clarification of juice play an important role. This process facilitate in the breakdown of colloidal plant carbohydrates which is insoluble cloud material [6]. Though there are two method of clarification viz. enzymatic and non-enzymatic (screening, sedimentation, filtration, heat treatment, addition of gelatin and casein, etc.) but enzymatic process is more advantageous than non-enzymatic process because it help in the increment of the juice yield, sugar moieties, soluble dry matter, phytochemicals and anti-oxidant property of the products. Enzymatic clarification resulted in viscosity reduction and cluster formation and thus gives high clarity, concentrated color and aroma, and haze-free clear juice [34].

Pectinases, cellulases, hemicellulases, α-amylase, and amyloglucosidase can be effectively use for juice clarification where α-amylase and amyloglucosidase broadly applicable for starchy based fruit. Enzymatic clarification of the juice depend upon the several factors like nature of enzyme, reaction time, temperature, amount of enzyme, proper mixing, pH of the juice, and enzyme combinations [6].

For homogeneity increment and turbidity decrease Okoth *et al.* [35] used amylase before pasteurization for passion fruit juice. Application of polygalacturonase (from *Aspergillus niger*) with addition of gelatin possess 85% clarification of apple juice, 35% viscosity reduction and no further haze during storage [36]. As storage of juice is important then prevention of haze formation is also a concern. In general instant turbidity caused by pectin while protein-phenol interaction is blamed for haze formation during cold storage. But in the case of cherry juice it is opposite. Application of pectinase (Pectinex Smash) and protease (Enzeco) showed effective turbidity removal in immediate pressed and cold storage juice [37]. Apple juice clarification was carried by Carrin *et al.* [38] using α-Amylases at different degree of ripeness. Due to the presence of calcium in enzyme along with treatment facilitates the enrichment of calcium in juice [39]. A combination of pectinex Ultra SP-L and amylase AG XXL was showed increased juice yield (23%) and turbidity (31.63%) with reduction in viscosity (1.84cP) [40]. Pectinase derived from *Aspergillus niger* T0005007-2 and *Aspergillus oryzae* IPT 301, were used for clear and dark juice extracted from apple, butia palm and blueberry, and grape fruits respectively. Excellent clarification was obtained using fungal enzyme compared to commercial one (Pectinex Clear and Pectinex BE Colour) [41]. Though applied enzyme activity for produced and commercial was same but broad range of pH tolerance of produced enzyme facilitate in the better clarification than commercial one. Normally syrup production from date palm (*Phoenix dactylifera* L) is a low-quality product due to unpleasant texture. Addition of pectinases and cellulases reduced the turbidity and increase the extraction of soluble solids of interest [42]. Cellulase is used to hydrolyze anthocyanins and thus prevents discoloration caused by anthocyanins during juice pasteurization [43]. Will *et al.* [44] used combination of pectinases and cellulases in hot water extraction of apple pomace and observed 37% increase in juice yield. In addition to this, the organoleptic properties of vegetable and fruit juices can be improved upon application of a concoction of pectinases and β-glucosidase during processing [45]. The yield of juices can be enhanced by adding complex of macerating enzymes including cellulases, xylanases, and pectinases; although, cloud stability, viscosity, and texture of juices can be enhanced by adding macerating enzymes [45]. Generally, fruits and vegetables are rich in pectin but certain fruits like pineapple, apple, and vegetables like tomatoes contain considerably high amounts of xylan, which is mainly attributed due to their high hemicelluloses content. Therefore, treatment of such vegetables and fruits with xylanases improves the transparency of pineapple, apple, and tomato juices by 22.20%, 14.30%, and 19.80% and their corresponding yield by 23.53%, 20.78%, and 10.78%, respectively [46]. For apple juice clarification, Kothari *et al.* [47] used concoction of Pectinase, Cellulase, and Amylase, where 50% of clarified juice was obtained after an incubation of 4 h. Macerating enzymes are mostly added during crushing and extraction of juice. The role of these enzymes is to hydrolyze pectins and cell walls, so as to render reduced viscosity and improved retention of nutritional components [48]. Application of chitosan with Citrozym-Ultra L (a commercial enzyme) showed improved results in by reducing insoluble solids, lipids, and other substances, mainly responsible for causing haziness in juices [49]. Clarification of apple, tomato, and pineapple juices by infusion of pectinase, amylase, xylanases, and carboxymethyl cellulase has also been used. The clarity of apple, orange, and grape increased to 17.85%, 18.36%, and 19.19%, respectively upon application of xylanases [50].

17.3.3 Debittering

Juice processing and commercial utilization of citrus fruits have faced a great hindrance due to the development of bitterness [51–53]. Mainly bitterness is caused for the presence of three bittering components – limonin (a limonoid), naringin (a flavanoid), and tannin therefore affecting its consumer acceptability [51, 52, 54]. Tannin is a compound which contributes to the bitterness by its astringent nature. Biochemically, tannin is a plant polyphenolic compound that either binds and precipitates or shrinks proteins and various other organic compounds including amino acids and alkaloids. The astringency from the tannins is responsible for the dry and puckery feeling in the mouth followed by the consumption of unripened fruit. Limonin is responsible for delayed bitterness.

Tannin acyl hydrolase (E.C. 3.1.1.20) is basically known as tannase involve in the hydrolysis of hydrolysable tannins. Tannase act on the ester and depside bonds in tannic acid present in citrus juice thus help in tannin hydrolysis [55]. In the pomegranate juice debittering process, tannase reduce 25% tannin content in the juice without hampering its nutritional profile [56]. 57% of tannin was removed by applying 4 ml enzyme with 8 ml juice in pomegranate juice [58]. In Indian gooseberry (*Phyllanthusemblica*) juice debittering process, tannase facilitate in 73.6% bitterness removal in terms of tannin [59]. For 10 ml of grape juice debittering 2 ml tannase application reduce 46% tannin after 120 min of incubation period [58]. Detannification of guava juice (*Psidiumguajava*) using 2% tannase resulting into 59.23% of tannin removal after 60 min of enzymatic reaction [6].

Unlike tannin another bitterness causing compound is naringin which is intensely bitter compound slightly soluble in cold water but moderately soluble in warm water and present in all part especially in peel albedo core and rag of the fruit. Application of naringinase (EC3.2.1.40) can act as promising tool for naringin hydrolysis to form tasteless component, naringenin in the juice. 74% hydrolysis of naringin was carried out by using 1.0g/L naringinase in citrus fruit juice after 4 h of treatment at 40 °C [60]. Immobilized naringinase was successfully employed for the conversion of 22.72% naringin to prulin, and removed 60.71% limonin [61].

Limonate dehydrogenase, another hydrolysable enzyme dehydrogenate the hydroxyl group at C-17 of limonoids and make 17-dehydrolimonoids which is non bitter [62]. *Pseudomonas putida* producing limonoate dehydrogenase is used for 9 folds reduction of limonin content in citrus juice [63].

17.4 Conclusion

Though India holds second place in fruit production yet poor processing techniques, lack of knowledge, traditional techniques resulted into low yield, nutrition, sensory attributes, and shelf life. Consumption of fruit juice is directly linked with health risk reduction and cure of diseases. As health benefit is also associated with processed juice thus minimally processing is widely acceptable rather than harshly processed juice because detainment of nutritional value is possible only in minimal processing. On the other hand enrichment of juice with nutrition and minerals is also possible through this alternative process.

Enzyme mediated juice processing is well studied and broadly accepted but opportunities are always open for new development and further improvement. More emphasis on enzyme kinetics, structure and behaviour of enzymes, source of enzyme production, and improvement of the juice processing pathway, can increase yield, storage time, and reduce cost which can put a new light in juice industry.

References

1. Henning, S.M., Yang, J., Shao, P., Lee, R-P., Huang, J., Ly, A., Hsu, M., Lu, Q-Y., Thames, G., Heber, D., Li, Z., Health benefit of vegetable/ fruit juice-based diet: Role of microbiome. *Sci. Rep.*, 7(2167), 1–9, 2017.
2. Boeing, H., Angela, B., Bub, A., Ellinger, S., Haller, D., Kroke, A., Leschik-Bonnet, E., Mu¨ller, M.J., Oberritter, H., Schulze, M., Stehle, P., Watzl, B., Critical review: Vegetables and fruit in the prevention of chronic diseases. *Eur. J. Nutr.*, 51, 637–663, 2012.
3. National Horticulture Board. F.No.8–10(3)/2017/HS, 2017. Accessed on: http://nhb.gov.in.
4. Kumar, S., Role of enzymes in fruit juice processing and its quality enhancement. *Adv. Appl. Sci. Res.*, 6(6), 114–124, 2015.
5. Caceres, L.G., Andrade, J.S., Filho, D.F.D.S., Effects of peeling methods on the quality of cubiu fruits. *Ciênc. Tecnol. Aliment.*, 32(2), 255–260, 2012.
6. Sharma, H.P., Patel, H., Sharma, S., Enzymatic extraction and clarification of juice from various fruits-A review. *Trends Post Harv. Technol.*, 2(1), 1–14, 2014.

7. Dash, A., Kundu, D., Das, M., Bose, D., Adak, S., Banerjee, R., Food biotechnology: A step towards improving nutritional quality of food for Asian countries. *Recent Pat. Biotechnol.*, 10, 43–57, 2016.
8. www.marketsandmarkets.com, Industrial enzymes market by type (carbohydrases, proteases, non-starch polysaccharides & others), application (food & beverage, cleaning agents, animal feed & others), brands & by region – Global trends and forecasts to 2020. 2017, Available from: http://www.marketsandmarkets.com/PressReleases/industrial-enzymes.asp. Accessed on: 25/12/2017.
9. Shirmohammadi, M., Yarlagadda, P.K., Kosse, V., Gu, Y., Study of mechanical deformations on tough skinned vegetables during mechanical peeling process. *GSTF J. Eng. Technol.*, 1, 31–37, 2012.
10. Tapia, M.R., Gutierrez-Pacheco, M.M., Vazquez-Armenta, F.J., González-Aguilar G.A., Ayala Zavala, J.F., Rahman, M.S.H., Siddiqui M.W., Washing, peeling and cutting of fresh-cut fruits and vegetables. In: *Minimally processed foods*, M.W. Siddiqui, M.S. Rahman (Eds.), Springer, Switzerland, 57–58, 2015.
11. Vatai, G., Separation technologies in the processing of fruit juices. In: *Separation, extraction and concentration processes in the food, beverage and nutraceutical industries*, S.S.H. Rizvi (Ed.), Woodhead Publishing Limited, Cambridge, 81–395, 2010.
12. Susmit, A., Ilame, Singh, S.V. Application of membrane separation in fruit and vegetable juice processing: A review. *Crit. Rev. Food Sci. Nutr.*, 55(7), 964–987, 2015.
13. Siddiqui, A.N., Kulkarni, D.N., Kulkarni, K.D., Mulla, M.Z., Studies on debittering of sweet orange juice. *World. J. Dairy. Food. Sci.*, 8(2), 185–189, 2013.
14. Tiwari, U., Cummins, E., Factors influencing levels of phytochemicals in selected fruit and vegetables during pre- and post-harvest food processing operations. *Food. Res. Int.*, 50, 497–506, 2013.
15. Çopur, Ö.U., Tamer, C.E., Fruit processing. In: *Food processing: Strategies for quality assessment*, A. Malik, Z. Erginkaya, S. Ahmad, H. Erten, (Eds.), Springer, New York, 9–35, 2014.
16. Vikram, V.B, Ramesh, M.N., Prapulla, S.G., Thermal degradation kinetics of nutrients in orange juice heated by electromagnetic and conventional methods. *J. Food. Eng.*, 69, 31–40, 2005.
17. Tiwari, B.K., O'Donnell, C.P., Muthukumarappan, K., Culllen, P.J., Ascorbic acid degradation kinetics of sonicated orange juice during storage and comparison with thermally pasteurized juice. *LWT Food. Sci. Technol.*, 42, 700–704, 2009.
18. Alighourchi, H., Barzegar, M., Abbasi, S., Effect of gamma irradiation on the stability of anthocyanins and shelf-life of various pomegranate juices. *Food. Chem.*, 110, 1036–1040, 2008.
19. Iborra, M., Miranda, M.I.A., Álvarez, S., Membrane processes in juice production. In: *Juice processing: Quality, safety and value-added opportunities*, V. Falguera, A. Ibarz, (Eds.), CRC Press, NY, 265–300, 2016.

20. Pap, N., Pongrácz, E., Jaakkola, M., Tolonen, T., Virtanen, V., Turkki, A., Horváth-Hovorka, Z., Vatai, G., Keiski, R.L., The effect of pre-treatment on the anthocyanin and flavonol content of black currant juice (*Ribes nigrum* L.) in concentration by reverse osmosis. *J. Food. Eng.*, 98, 429–436, 2010.
21. Bishai, M., Singh, A., Adak, S., Prakash, J., Roy, L., Banerjee, R., Enzymatic peeling of potato: A novel processing technology. Potato. Res., 58(4), 301–311, 2015.
22. Cavia-Saiz, M., Muñiz, P., Ortega, N., Busto, M.D., Effect of enzymatic debittering on antioxidant capacity and protective role against oxidative stress of grapefruit juice in comparison with adsorption on exchange resin. *Food. Chem.*, 125, 158–163, 2011.
23. Pretel, M.T., Sanchez-Bel, P., Egea, I., Romojaro, F., Enzymatic peeling of citrus fruits: Factors affecting degradation of the albedo. *Tree. For. Sci. Biotech.*, 2(Special issue 1), 52–59, 2008.
24. Pag´an, A., Conde, J., Ibarz, A., Pag´an, J., Albedo hydrolysis modelling and digestion withreused effluents in the enzymatic peeling process of grapefruits. *J. Sci. Food. Agr.*, 90, 2433–2439, 2010.
25. Sanchez-Bel, P., Egea, I., Sanchez-Ballesta, M.T., Sevillano, L., Del Carmen Bolarin, M., Flores, F.B., Proteome changes in tomato fruits prior to visible symptoms of chilling injury are linked to defensive mechanisms, uncoupling of photosynthetic processes and protein degradation machinery. *Plant. Cell. Physiol.*, 53, 470–484, 2012.
26. Lozano, J.E., Processing of fruits: ambient and low temperature processing. In: *Fruit manufacturing: scientific basis, engineering properties, and deteriorative reactions of technological importance*, G.V. Barbosa-Canovas (Ed.), Springer, US, 21–54, 2006.
27. Toker, İ., Bayindirli, A., Enzymatic peeling of apricots, nectarines and peaches. *LWT - Food. Sci. Technol.*, 36, 215–221, 2003.
28. Althuri, A., Chintagunta, A.D., Sherpa, K.C., Rajak, R.C., Kundu, D., Singh, J., Rastogi, A., Banerjee, R., Microbial enzymes and lignocellulosic fuel production. In: *Lignocellulosic biomass production and industrial applications*, A. Kuila, V. Sharma, (Eds.), Scrivener Publishing LLC, USA, 135–170, 2017.
29. Noguchi, M., Ozaki, Y., Azuma, J.-I., Recent progress in technologies for enzymatic peeling of fruit. *Jpn. Agric. Res. Q,* 49(4), 313–318, 2015.
30. Ni, H., Yang, Y., Fan, Chen, F., Ji, H.F., Ling, H.Y., Wu, L., Cai, H.N., Pectinase and naringinase help to improve juice production and quality from pummelo (*Citrus grandis*) fruit. *Food. Sci. Biotechnol.*, 23(3), 739–746, 2014.
31. Bahramian, S., Azin, M., Chamani, M., Gerami, A., Optimization of enzymatic extraction of sugars from kabkab date fruit. *Middle-East J. Sci.*, 7(2), 211–216, 2011.
32. Al-Hooti, S.N., Sidhu, J.S., Al-Saqer, J.M., Al-Othman, A., Chemical composition and quality of date syrup as affected by enzyme treatment. *Food. Chem.*, 79, 215–220, 2002.
33. Villettaz, J.C., *Wine: Enzymes in food processing*, 3rd ed. Academic Press Inc., 1993.

34. Abdullah, A.G.L., Sulaiman, N.M., Aroua, M.K., Megat, M.N.M.J., Response surface optimization of conditions for clarification of carambola fruit juice using a commercial enzyme. *J. Food. Eng.*, 8, 65–71, 2007.
35. Okoth, M.W., Kaahwa, A.R., Imungi, J.K., The effect of homogenisation, stabiliser and amylase on cloudiness of passion fruit juice. *Food Control*, 11, 305–311, 2000.
36. Singh, S., Gupta, R., Apple juice clarification using fungal pectionolytic enzyme and gelatin. *Indian J. Biotechnol.*, 3, 573–576, 2004.
37. Pinelo M., Zeuner, B., Meyer, A.S., Juice clarification by protease and pectinase treatments indicates new roles of pectin and protein in cherry juice turbidity. *Food Bioprod. Process.*, 88, 259–265, 2010.
38. Carrin, M.E., Ceci, L.N., Lozano, J.E., Characterization of starch in apple juice and its degradation by amylases. *Food Chem.*, 87, 173–178, 2004.
39. Dey, T.B., Adak, S., Bhattacharya, P., Banerjee, R., Purification of polygalacturonase from *Aspergillus awamori Nakazawa* MTCC 6652 and its application in apple juice clarification. *Food Sci. Technol.*, 59(1), 591–595, 2014.
40. Telesphore, M., He, Q., Optimization of processing parameters for cloudy passion fruit juice processing using pectolytic and amylolytic enzymes. *Pakistan J. Nutr.*, 8(11), 1806–1813, 2009.
41. Sandri, I.G., Fontana, R.C., Barfknecht, D.M., da Silveira, M.M., Clarification of fruit juices by fungal pectinases. *LWT - Food Sci. Technol.*, 44(10), 2217–2222, 2011.
42. Abbès, F., Bouaziz, M.A., Blecker, C., Masmoudi, M., Attia, H., Besbes, S., Date syrup: Effect of hydrolytic enzymes (pectinase/cellulase) on physicochemical characteristics, sensory and functional properties. *LWT - Food. Sci. Technol.*, 44, 1827–1834, 2011.
43. Fleuri, L.F., Delgado, C.H.O., Novelli, P.K., Pivetta, M.R., Prado, D.Z., Simon, J.W., Enzymes in fruit juice and vegetable processing. In: *Enzymes in food and beverage processing*, C. Muthusamy, (Ed.), Taylor & Francis Group, NW, 255–279, 2015.
44. Will, F., Bauckhage, K., Dietrich, H., Apple pomace liquefacation with pectinases and hemicellulases: Analytical data of the corresponding juice. *Eur. Food. Res. Technol.*, 211, 291–297, 2000.
45. Kuhad, R.C., Gupta, R., Singh, A., Microbial cellulases and their industrial applications. *Enzyme Res.*, 2011, 1–10, 2011.
46. Nagar, S., Mittal, A., Gupta, V.K., Enzymatic clarification of fruit juices (apple, pineapple, and tomato) using purified *Bacillus pumilus* SV-85S xylanase. *Biotechnol. Bioproc. E*, 17, 1165–75, 2012.
47. Kothari, M.N., Kulkarni, J.A., Maid, P.M., Baig M.M.V., Clarification of apple juice by using enzymes and their mixture. *World Res. J. Biotechnol.*, 1(2):29–31, 2017
48. Sharada, R., Venkateswarlu, G., Venkateswar, S., Anand Rao, M., Applications of cellulases – Review. *Int. J. Pharm. Chem. Biol. Sci.*, 4, 424–37, 2014.
49. Cesar, L.T., de Freitas, C.M., Maia, G.A., de Figueiredo, R.W., de Miranda, M.R.A., de Sousa, P.H.M., Brasil, I.M., Gomes, C.L., Effects of clarification on

physicochemical characteristics, antioxidant capacity and quality attributes of ac¸a´ı (Euterpe oleracea Mart.) juice. *J. Food. Sci. Tech.*, 51, 3293–300, 2014.

50. Adig¨uzel, A.O., Tuncer, M., Production, characterization and application of a xylanase from Streptomyces sp. AOA40 in fruit juice and bakery industries. *Food Biotechnol.*, 30, 189–218, 2016.
51. Premi, B.R., Lal, B.B., Joshi, V.K., Distribution pattern of bittering principles in kinnow fruit. *J. Food Sci.*, 31, 140–141, 1995.
52. Puri, J.S., Kothari, R.M., Kennedy, J.F., Biochemical basis of bitterness in citrus fruits and biotech approaches for debittering. *Crit. Rev. Biotechnol.*, 16, 145–155, 1996.
53. Sandhu, K.S., Bhatia, B.S., Shukla, F.C., Physiochemical changes during storage of kinnow mandarin oranges and pineapples juice concentrates. *J. Food Sci. Technol.*, 22, 342–345, 1985.
54. Kamaljeet, K., Application of novel juice extraction methods and bacterial utilization of limonin for control of bitterness in kinnow juice. M.Sc Thesis, Thapar Institute of Engineering and Technology, Patiala, Punjab, India, 2002.
55. Mahapatra, K., Nanda, R.K., Bag, S.S., Banerjee, R., Pandey, A., Szakacs, G., Purification, characterization and some studies on secondary structure of tannase from *Aspergillus awamori Nakazawa. Process. Biochem.*, 40(10), 3251–3254, 2005.
56. Rout, S., Banerjee, R., Production of tannase under mSSF and its application in fruit juice debittering. *Indian J. Biotechnol.*, 5(3), 346–350, 2006.
57. Kapoor, A., Iqbal, H., Efficiency of tannase produced by *Trichoderma Harzianum* MTCC 10841 in Pomegranate juice clarification and natural tannin degradation. *Int. J. Biotechnol. Bioeng. Res.*, 4(6), 641–650, 2013.
58. de Lima, J.S., Cruz, R., Fonseca, J.C., de Medeiros, E.V., Maciel, M.H.C., Moreira, K.A., Motta, C.M.S., Production, characterization of tannase from *Penicillium montanense* URM 6286 under SSF using agroindustrial wastes, and application in the clarification of grape juice (*Vitis vinifera* L.). *Sci. World J.*, 2014, 1–9, 2014.
59. Srivastava, A., Kar, R., Application of immobilized tannase from *Aspergillus niger* for the removal of tannin from myrobalan juice. *Indian. J. Microbiol.*, 50(1), 46–51, 2010.
60. Patil, M.B., Dhake, A.B., Debittering of citrus fruit juice by naringinase of penicillium purpurogenum. *Int. J. Engg. Res. Sci. Tech.*, 3(2), 266–270, 2014.
61. Huang, W., Zhan, Y., Shi, X., Chen, J., Deng, H., Du, Y., Controllable immobilization of naringinase on electrospun cellulose acetate nanofibers and their application to juice debittering. *Int. J. Biol. Macromol.*, 98, 630–636, 2017.
62. Munish, P., Lakhwinder, K., Singh, M.S., Partial purification and characterization of limonoate dehydrogenase from *Rhodococcus fascians* for the degradation of limonin. *J. Microbiol. Biotechnol.*, 12(4), 669–673, 2002.
63. Verma, J.P., Singh, S., Ghosh, M., Srivastava, P.K., Identification and characterization of cellular locus of limonin biotransforming enzyme in *Pseudomonas putida. Int. J. Food. Sci. Tech.*, 45, 319–326, 2010.

18

Biotechnological Exploitation of Poly-Lactide Produced from Cost Effective Lactic Acid

Mohan Das, Debajyoti Kundu, Akanksha Rastogi, Jagriti Singh and Rintu Banerjee*

Agricultural and Food Engineering Department, Indian Institute of Technology, Kharagpur, India

Abstract

Lactic acid, a green chemical is found in many naturally processed products. Although, the organic acid can be manufactured by chemical method but cost effective utilization of agro-residues by carbohydrate fermentation is mostly preferred. This integrated approach not only minimizes the entire production cost but also reduces global waste burden. Furthermore, polymerization of monomers of lactic acid by poly-condensation leads to the formation of Poly-lactic acid (PLA). Biodegradability and biocompatibility of PLA, makes the polymer safe and ideal for its application in a wide range of industries. Unique surface modifiable characteristics of PLA also increase its versatility. Growing global awareness, public demand and governmental policies promotes the development of such technologies for making a sustainable environment. Although, rigorous research is required to make the polymer acceptable and commercially viable. In this chapter, an emphasis has been given to explain the possible ways of lactic acid production, polymerization and applications in a concise manner.

Keywords: Lactic acid, poly-lactic acid, polymerization, agro-residues, biopolymer

**Corresponding author*: rb@iitkgp.ac.in

Arindam Kuila and Vinay Sharma (eds.) Principles and Applications of Fermentation Technology, (401–416) © 2018 Scrivener Publishing LLC

18.1 Introduction

Lactic acid or 2-hydroxyl propanoic acid is one of the abundantly existing hydroxy form of carboxylic acid. In 1780, C.W. Scheele, a Swedish Chemist first discovered the chemical. This organic acid can be produced both by chemical and microbial means. It is found as one of the principal intermediate metabolite of living organisms ranging from prokaryotes to eukaryotes. Although, it is produced as an intermediate by-product of different bioprocessing industries but was found to be insufficient to meet the demands. During 1990s, an USA based company named Sterling Chemicals, Texas and CCA Biochemicals, Netherlands with 2 subsidiaries, each in Spain and Brazil were found to be the primary manufacturers [1].

Sterling started the manufacturing unit using chemical technology, whereas CCA Biochemicals proceeded by fermentation of carbohydrate rich feedstock. During earlier times, the chemical was used in plastic industry or in the synthesis of oxygenated chemicals. By 2003, Cargill Dow and Archers Daniel Midland also entered in the business of Lactic acid. Both the companies, started production by carbohydrate fermentation technology. Musashino Chemical Co., Japan, also started producing lactic acid through lactonitrile method followed by Sterling. Lactic acid, gained huge acclaim after its consideration as GRAS chemical by FDA, USA [2].

In recent years, escalating environmental concern and global energy demands urged to develop methods that produces biochemicals through green and sustainable means. During that time, food and its related-industries are the main consumers of lactic acid. However, with advancement in science and technology, they are used in different sectors, starting from textile and cosmetic industry to biomedical companies. Lactic acid is mildly acidic in nature and therefore acts as a buffering agent. It acts as a bacterial inhibitor and so used for preservation of a range of food products including jams, jellies, soups, soft drinks, and bakery as well as dairy products [3].

Despite of its application in multidisciplinary fields, Poly-lactide (PLA), a polymer of lactic acid gained popularity and was heartily welcomed by almost every industrial sectors and environmentalists. The reason behind this is its unique physico-chemical properties like biodegradability, biocompatibility, tensile strength, ease of molding, strength etc. There are a number of methodologies for the polymerization of monomers of lactic acid, which depends on the type of applications [4].

The authors in this chapter attempted to portray cost effective production of lactic acid, its purification, different methods of polymerization and its wide spectrum of applications in the subsequent sections.

18.2 Need for Ideal Substrates for Lactic Acid Production

Pure form of sugars or usages of edible sugars are source of substrates used for the production of lactic acid. In fact, addition of yeast extract or any other form of complex nitrogenous sources enhances the production of Lactic acid. The aforesaid system is efficient as it lowers the processing cost by eliminating pre-treatment procedures but while scaling up the entire system, usage of pure form of carbon and nitrogen sources will not make the system economically viable. Therefore, extensive studies were carried by different group of researchers throughout the geographical extent to scale up the system and make the process economically viable [5]. To overcome the above bottlenecks, the raw material should be selected on the basis of following criterias:

- availability and sustainability
- rich in carbohydrates
- high bioconversion efficiency
- low cost
- no need of pre-treatment

During the course of research, it has been observed that by-products and waste residues of food industries, agro-waste residues and different natural biomass like lignocellulosic and algal biomass are ideal substrates for lactic acid production. Although, direct fermentation of carbohydrate rich biomass leads to the production of lactic acid but pre-treatment of the substrates by physical, chemical, physico–chemical and enzymatic means may alleviate the bottlenecks of the entire production system. Apart from these, lignocellulosic biomass has already drawn interests of researchers because they are rich sources of sugar residues as well as increases food security by eliminating the usage of food crops. Sugarcane bagasse, corn cobs, corn stover and cheese whey from food processing industry; wood processing waste and pulp from paper industry; agricultural residues or leftovers are potential biomasses for the production of lactic acid. The possible substrates outsourced from different industry as wastes are depicted in Figure 18.1 [6, 7]. This kind of innovative approaches not only eliminates the problem of substrate availability but also reduces a global burden by reducing the amount of wastes [6, 7].

Whey is a by-product of cheese manufacturing industry. It is generally discharged with wastewater; therefore, treatment of the liquid biomass is important prior to its disposal. The reason behind this is its richness in

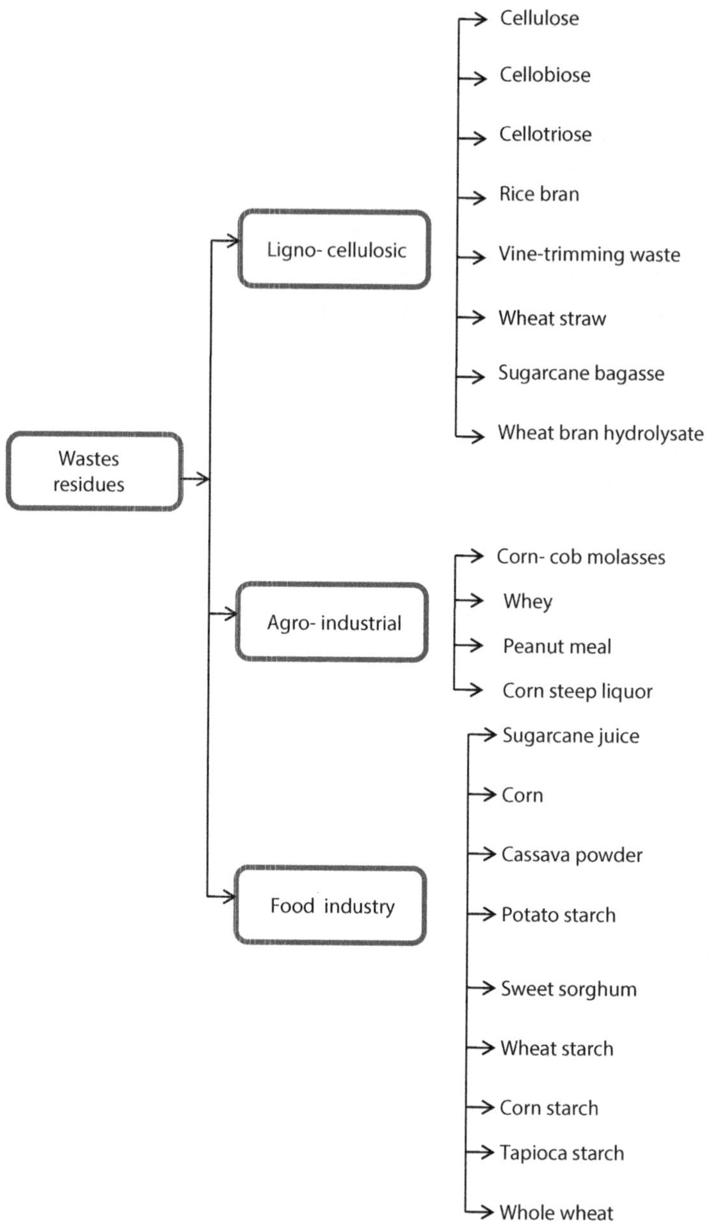

Figure 18.1 Possible substrates for Lactic acid production.

nutrients including fats, protein, lactose, mineral salts, and soluble vitamins [8]. This makes the system ideal for the growth of Lactic Acid Bacteria (LAB). Theoretically, 1 mol each of glucose and galactose is released from the breakdown of 1 mol of lactose and 4 mol of lactic acid that is produced from 1 mol of lactose [7]. Lactic acid production by continuous fermentation is preferred over batch fermentation because in batch techniques the lag phase gets extended which increases the operational costs; whereas in comparison continuous technique reduces the production cost as it eliminates the need of large fermentor for production [9].

Microalgae has been found to be an alternative for lignocellulosic biomasses. Since, removal of lignin from lignocellulosics possess a problem for the effective release of reducing sugars, therefore, use of microalga eliminates such issues. Even, microalga can be grown anywhere with minimal nutrient supply within 1 to 10 days. It has been reported that *Hydrodictyon reticulum*, a green alga contains high percentage of reducing sugar (47%), out of which glucose content is 35%. This sugar can be utilized by *Lactobacillus paracasei* LA104 and *Lactobacillus coryniformis* sp. *torquenes* for the production of lactic acid. Further, research need to be carried out to develop technologies for the pre-treatment of high amount of biomass of microalgae and subsequent utilization of the biomass for lactic acid production at commercial level [10].

18.3 Role of Microbes and Biochemical Pathways in Lactic Acid Production

Lactic acid from microbial source includes a variety of bacteria, cyanobacteria, yeast and algae. Each microbial machinery has gained one or more improved characteristics over the other including broad spectrum of substrate utilization, less need of nutrients, enhanced yield and purity. Although, employment of concoction of microbes serves the purpose of utilization of complex substrates and their consequent increase in conversion and production of lactic acid. Fermentative conversion of reducing sugars (e.g., hexoses and pentoses) through channel of metabolic pathways, by a wide range of microbes leads to the production of lactic acid [9]. Lactic Acid Bacteria (LAB) and filamentous fungi are the class of microbes preferred for the production of lactic acid. However, comparatively LAB are preferred over fungi because they comes with several disadvantages as low production rate, high requirement of aeration, increased production of by-products (e.g., ethanol and fumaric acid). About 90% of lactic acid produced worldwide throughout the year are based on LAB [7].

Lactic acid production by LAB, proceeds mainly by homo-lactic and hetero-lactic fermentation. Homo-lactic fermentation proceeds through utilization of hexose via Embden–Meyerhoff–Parnas (EMP) pathway and pentose via Pentose Phosphate (PP) pathway. In homo-lactic fermentation, only lactic acid is produced as the sole secondary metabolite whereas in hetero-lactic fermentation alcohols, organic acids, aldehydes and ketones are produced as secondary metabolites along with lactic acids. Therefore, homo-lactic fermentation is preferred over hetero-lactic fermentation because the former technique serves to manufacture the desired product in high percentage and purity. LAB that predominates the lactic acid production business are *Lactobacillus amylophilus, Lactobacillus acidophilus, Lactobacillus bavaricus, Lactobacillus salivarius, Lactobacillus delbrueckii, Lactobacillus maltaromicus, Lactobacillus casei, Lactobacillus jensenii,* etc. [7, 10].

In, EMP pathway firstly glucose is converted into fructose 1,6-diphosphate (FDP) via phosphorylation of glucose. Enzymatic cleavage of FDP generates glyceraldehyde-3-phosphate and dihydroxy acetone phosphate. Thereafter, pyruvate is formed from GAP via steps of substrate level phosphorylation. Finally, lactic acid dehydrogenase reduces pyruvate to lactic acid through NADH to NAD^+ oxidation. Whereas in PP pathway, 5 mol of GAP is generated via conversion of 3 mol of xylulose-5-phosphate using 2 important enzymes: transaldolase and transketolase. Thereafter, GAP is converted to lactic acid via the pathway already discussed earlier. Theoretically, 2 mol mol^{-1} and 1 mol mol^{-1} of lactic are produced using glucose and pentose via EMP pathway and PP pathway, respectively [10, 11].

Hetero-lactic fermentation generally proceeds via phosphoketolase (PK) pathway where ribulose-5-phosphate is formed from glucose-6-phosphate. The resulting product gets converted into equal amounts of GAP and acetyl phosphate. GAP is then finally converted to lactic acid via the EMP pathway. Acetyl phosphate leads to the formation of ethanol via the formation of acetaldehyde and acetyl-coA as intermediates. As a result, 0.5 mol mol^{-1} of lactic acid is produced in hetero-lactic fermentation. From the above discussion, it can be concluded that along with EMP and PP pathway, PK pathway is equally important in terms of lactic acid production as well as reducing sugar utilization. Although, different reducing sugars can be utilized for lactic acid it depends on the agro-residue used and the microbe employed for the fermentative conversion [11]. An overview of different carbon utilizing pathways is given in Figure 18.2 [7, 10].

18.4 Purification of Lactic Acid

Downstream processing is an important and vital step for any biologicals produced under different production conditions. Unlike other products, lactic acid produced under anaerobic fermentation through LAB undergo the similar challenges. To make the process economically viable the basic

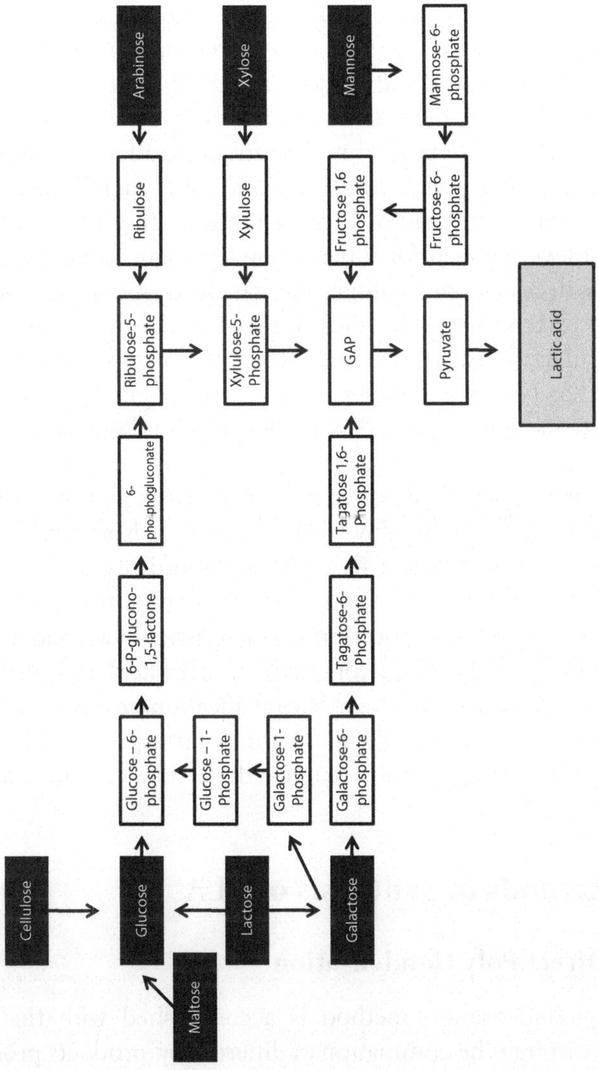

Figure 18.2 An overview of metabolic pathways showing utilization of different carbon sources for the production of Lactic acid.

objective of the research is to go for higher recovery so that the overall processing cost can be minimised. Ion exchange, liquid extraction [12], electro-dialysis [13], reverse osmosis [14], ultra-filtration [15], precipitation, adsorption [16], liquid surfactant membrane separation [17], and distillation are the methods employed to remove or reduce impurities from fermented broths. Lactic acid is less volatile, therefore distillation is very difficult and electro-dialysis is incapable of removing charged compounds like organic acids and amino acids. However, nanofiltration in combination with bipolar electro-dialysis acts as an efficient substitute to multiple step procedure for the purification of lactic acid in monomeric form [18]. Besides, the use of several membrane separations has been explored by using nano, micro, ultra and electro-dialysis membranes. Integrative approach for the utilization of such membranes in a single step although failed to achieve the goal for intensifying the manufacturing process. For the last few decades, chromatography has been one of the tools used by pharmaceutical companies for the production of fine grade chemicals [19].

In particular, ion exchange chromatography is mostly used for bio-separations. It has been successfully used to recover pure form of lactic acid. Ion exchange has been accepted as a reliable technology, since permutation and combination of different anionic and cationic exchangers have been reported over the last few decades [18]. Evangelista and Nikolov [20] used VI-15, IRA-35, MWA-1 (weakly basic polymeric adsorbents) whereas Cao *et al.* [21] reported the use of IRA-400, an anionic exchange resin for the direct recovery of lactic acid from crude fermented broth. Lee *et al.* [22] attempted to explore the adsorption characteristics of activated carbon and PVP (poly vinyl pyridine) resin for purifying lactic acid. Despite of extensive research, there remains a lacuna in the purification process. Therefore, each minute intricacy should be given attention starting from selection of the substrate to processing variables to make the system commercially viable.

18.5 Methods of Synthesis of PLA

18.5.1 Direct Poly Condensation

Direct poly-condensation method is accomplished with the linking of monomers through the elimination of different by-products produced during the time of processing. Water and alcohol are the main by-products of the direct poly-condensation. Synthesis of PLA occurs through the connection of carboxyl and hydroxyl group, which release water as a by-product. Due to incomplete removal of this by-product, produced PLA not only shows low molecular weight but also less in quantity [23].To overcome these problem another two new direction in poly-condensation are azeotropic polycondensation (AP) and solid state polymerization (SSP). An overview of different method of synthesis of PLA is depicted in Figure 18.3 [23–25, 29–32].

In AP, appropriate selection of azeotropic solvent facilitate in the efficient removal of water from the matrix in one step and thus form

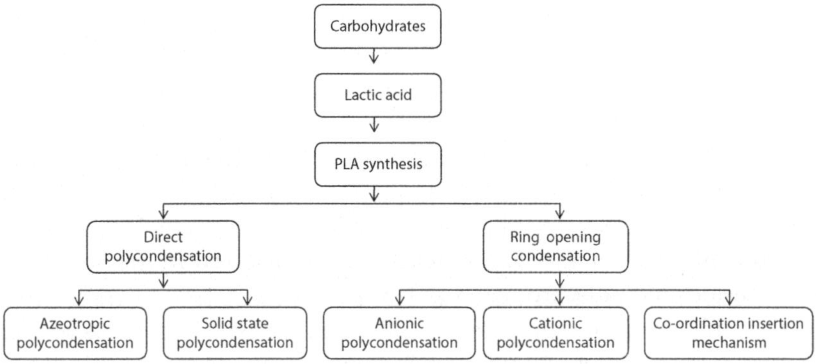

Figure 18.3 An overview of methods of synthesis of PLA.

equilibrium between monomer and polymer, producing high molecular weight polymer. Temperature and solvent plays a crucial role in this process. Implementation of low temperature, which is below polymer melting point, determines the formation of impurities. On the other hand, selection of solvent has a great impact on the property of polymer [24].

SSP occurs via two phases, in the first step monomers are directly heated at high temperature to produce polymers with low molecular weight. This phase is formally known as melting phase. The second phase is formally known as solid-state phase, where produced low molecular weight polymer are further processed by heating (temperatures between the glass transition and the onset of melting) to form high molecular weight polymer [25].

Through direct poly-condensation method, PLA is synthesized without using catalyst and initiator, having the molecular weight of 90,000 g/mol. To avoid heat consumption in this process, microwave is implemented to produce PLA (molecular weight-16,000 g/mol) in 30 mins [26].To make the process more efficient soxhlet extractors were used for the removal of water in AP method which facilitate in the production of higher molecular weight of PLA (30,000 g/mol) [27]. Furthermore, utilization of pulverized pre-polymer (diameter less than 150 μm), in the second step of SSP produces ~ 200,000 g/mol molecular weight of PLA, while use of L-PLA and D-PLA in 1:1 ratio as a starting material gives the thermal stability of the PLA [28].

18.5.2 Ring Opening Poly Condensation

Ring opening poly condensation (RO-PC/ROP) occurs by linking single monomers through a chain reaction, and thus the derived polymer is a homo-polymer. In this polymerization process elimination of molecules is not required, further no by-product is released. The main key player in this reaction is catalyst/initiator. Catalyst or initiator has ancillary ligand, metal site (catalysis occur), an initiating group and site (propagation occur). This process facilitate in the production of high molecular weight polymer which is just double of its initial monomers. As for example, polyethylene has double molecular weight compared to its monomer ethylene. Basically, ring-opening co-polymerization reactions are involved in the production of polyesters and polycarbonates [29–31].

This polymerization process basically is a propagation process of cyclic monomer using different ions. Depending upon the initiators this process is further classified into anionic ROP, cationic ROP, and radical ROP [32]. In PLA synthesis process, lactide is used as cyclic monomer and controlled polymerization (controlling purity of cyclic monomer and synthesis process) process gives high molecular weight PLA. In 1935, first PLA was synthesized through ROP process and after 20 years high molecular weight PLA is produced using purification method. However, several factors like temperature, rate of heating, pressure, solvent and catalyst has an immense impact on the final product. On the other hand, synthesis of lactide and its purity has a great significance as it acts as a cyclic monomer in synthesis process. Changes in above mentioned factors assists in the production of high yield lactide. There are several report of lactide synthesis with the yield of 40–77% [23]. However, using stannous octoate-toluene as a catalyst and at 220–240 °C, 80% yield of lactide was obtained [33]. Using zinc oxide-stannous octoate as a catalyst at 180–206 °C temperature, 86.4% yield was obtained [34]. After that, 95–97% yield was gained using stannous octoate when the temperature was 170–250 °C [35].

Several researchers got different yield of PLA with varying molecular weight by changing process condition [23]. Maximum yield of 97–99% with a molecular weight of 93,300 g/mol was reported by Jacobsen *et al.* [36], where stannous octoate was used as catalyst, polyethelyne glycol as solvent, temperature was 180–185 °C and time was 7 min, considered as controlling parameters. Maximum molecular weight (468,000 g/mol) with the yield of 95–96%, was reported by Korhonen *et al.* [37], where stannous octoate was used as catalyst, poly-glycerine as solvent, temperature at 160–200 °C, and time taken was 3–5 min. Higher hydroxyl groups of co-initiators led to get high molecular weight (>400,000 g/mol) PLA in faster way without hampering its thermal property [37]. At low pressure (0.001 kPa), at 140 °C for 10 h PLA with high molecular weight (100,000 g/mol) was achieved by Kaihara *et al.* [38]. After that, better quality PLA with molecular weight 160,000 g/mol was achieved at 200 °C for 1 h, without solvent [37].

Use of positively charged intermediate in ROP is known as cationic ROP (CROP). CROP facilitate in the production of high molecular range of polymers with a broad range of physicochemical properties. As CROP is implemented in all types of heterocyclic monomers, thus produce polyethers, polyamides, polysulfides, polyesters, polyamines, polyphosphazenes, polysiloxanes, polyacetals, and polyphosphates. General process of CROP involves initiation, propagation, and termination [39]. Among the two mechanism of CROP, the first one is involved in the addition of monomer via S_N1 and S_N2 mechanism by the cationic centre at the chain end. Further the polymer structure can be controlled, where use of appropriate side group play an important role. Several industrial polymer such as polyacetals, 1,3,5-trioxane, 1,3-dioxolane, polytetrahydrofurans, poly 3,3-bis(chloro-methyl) oxetanes and poly-siloxanes is derived through this process [40].

In anionic ROP (AROP), heterocyclic monomers undergoes polymerization through the attack of nucleophilic initiator at the hetero-carbon atom of the adjacent ring. Breakdown of hetero-carbon atom occurred by the activities of the nucleophile. Repeated attack of anion to cyclic compound, repeats the polymerization reaction. Some of the AROP occurs through the side reactions, where hetero atoms in the polymer chain attack the nucleophile at the hetero-carbon atom bond. However, attack on the same polymer resulted in de-polymerisation. There are limited numbers of heterocyclic monomers that can undergo polymerization. However, there are several advantages of AROP over CROP like AROP produce high molecular weight of polymer. In AROP, molecular weight can be controlled, block co-polymers, and polymers with reactive-end can be synthesized and in some cases facilitates formation of stereo-regular polymers. Nylon-6, polysiloxanes, poly-lactide, polypropylene oxide, and poly ethylene oxide are the some examples of AROP process [41].

Via radical ROP (RROP) process molecular weight of PLA can be controlled. RROP are advantageous over AROP and CROP because in RROP contamination of water do not cause any hindrance, emulsion polymerization can occur in water and causes no contamination with initiator derived intermediates. This process is beneficial where particular molecular weight should be maintained such as in tooth filling materials, molding, and coating of electrical and electronic constituents [32].

18.6 Applications of PLA

With advancement in polymer science, although several synthetic polymers like poly trimethylene terepthalate (PTT) came into the market but unique characteristics of poly-lactide have gained the attention of researchers worldwide. Biodegradability, thermal plasticity, biocompatibility and scaffolding properties of poly-lactide, recommended its application in medical science, textile, and packaging industry. Production of such polymers from monomers of lactic acid and further its degradation via enzymatic hydrolysis by soil microbes not only makes the entire system sustainable but also reduces the production of any waste. A wide range of application of PLA in medical industry is considered as one of the notable invention of the century [4, 42, 43].

Over the last few years, rigorous research in the concerned field had led to several successful events. As a result, poly-anhydrides of PLA gained huge approbation for its application in surgical implants and drug delivery systems. Biopolymers based on PLA serves several advantages over metallic stents or non-biodegradable artificially fabricated stents. PLA can be easily moulded and is easily acceptable by the human body system, therefore no chance of any complication due to rejection by immunological system. Scaffold of PLA can act as a support for tissue regeneration in wounded part of the human body system naturally. And after regeneration, there is no pain of removal because they are easily removed by the body itself after a passage of time interval. As a result, the process eliminates the need of any donor. It has been seen that surface properties of any biomaterial plays a crucial role for its application. The dynamic property of PLA made it an ideal polymer to be modified by physical, chemical, radiation, and plasma induction methods [44–46].

While considering packaging industry, application of conventional packaging material and techniques limits the ability to prolong the shelf life of food material. To overcome such bottlenecks, the concept of active packaging came into the market. By definition, it is an intelligent system in which the packaging components interact with internal environment system or food directly to meet consumer demands like freshness of the product, quality, and enhanced organoleptic properties. Although, considerably very little research has been performed for using PLA as a packaging material but several attempt have been performed for enhancing this versatile polymer. To prevent pro-oxidative action, several anti-oxidants were added intentionally so that they migrate inside the food material. Researchers suggested that the use of this kind of polymers serve the purpose of sustained release of anti-oxidants into the food material and as a result limits the oxidative action of dairy products rich in high fat. Based on several research reports, it can be hypothesized that successive incorporation of active agents like organic acids, bacteriocins and enzymes can further add unique property to the polymer [42].

18.7 Conclusion

From the above discussion, it is possible to conclude that production of lactic acid using cost effective agro-residues not only reduces the cost of lactic acid and PLA production but also contribute in reducing global waste production. On the contrary, production of PLA as a packaging material also simultaneously aid to reduce the usage of non-biodegradable petroleum based polymers. Application of PLA as a medical aid in tissue regeneration opens a new avenue of research and hope for patients to get cured without complications. Extreme versatility of PLA and its tailor-made surface changeable properties bear immense potential to be used commercially in different forms as an active packaging material in food industry for the upcoming generation.

References

1. Datta, R., Henry, M., Lactic acid: Recent advances in products, processes and technologies – a review. *J. Chem. Technol. Biotechnol.*, 81, 1119–1129, 2006.
2. Jamshidian, M., Tehrany, E.A., Imran, M., Jacquot, M., Desobry, S., Poly-lactic acid: Production, applications, nanocomposites, and release studies. *Comp. Rev. Food Sci. Food Saf.*, 9, 552–571, 2010.
3. Garlotta, D., A literature review of poly-lactic acid. *J. Polym. Environ.*, 9(2), 63–84, 2001.
4. Bishai, M., De, S., Adhikari, B., Banerjee, R., A comprehensive study on enhanced characteristics of modified polylactic acid based versatile biopolymer. *Eur. Polym. J.* 54, 52–61, 2014.
5. Oh, H., Wee, Y. J., Wun, J.S., Han, S.H., Jung, S., Ryu, H.W., Lactic acid production from agricultural resources as cheap raw materials. *Bioresour. Technol.*, 96, 1492–1498, 2005.
6. Gao, C., Ma, C., Xu, P., Biotechnological routes based on lactic acid production from biomass. *Biotechnol. Adv.*, 29, 930–939, 2011.
7. Wang, Y., Tashiro, Y., Sonomoto, K., Fermentative production of lactic acid from renewable materials: Recent achievements, prospects, and limits. *J. Biosci. Bioeng.*, 119(1), 10–18, 2015.
8. Panesar, P.S., Kennedy, J.F., Gandhi, D.N., Bunko, K., Bioutilisation of whey for lactic acid production. *Food Chem.*, 105, 1–14, 2007.
9. Abdel-Rahman, M.A., Tashiro, Y., Sonomoto, K., Lactic acid production from lignocellulose-derived sugars using lactic acid bacteria: Overview and limits. *J. Biotechnol.*, 156, 286– 301, 2011.
10. Abdel-Rahman, M.A., Tashiro, Y., Sonomoto, K., Recent advances in lactic acid production by microbial fermentation processes. *Biotechnol. Adv.*, 31, 877–902, 2013.

11. Abdel-Rahman, M.A., Sonomoto, K., Opportunities to overcome the current limitations and challenges for efficient microbial production of optically pure lactic acid. *J. Biotechnol.*, 236, 176–192, 2016.
12. Wasewar, K.L., Heesink, A.B.M., Versteeg, G.F., Pangarkar, V.G., Reactive extraction of lactic acid using alamine 336 in MIBK: euilibria and kinetics. *J. Biotechnol.*, 97(1), 59–68, 2002.
13. Ryu, H.W., Kim, Y.M., Wee, Y.J., Influence of operating parameters on concentration and purification of L-lactic acid using electro dialysis. *Biotechnol. Bioproc. Eng.*, 17(6), 1261–1269, 2012.
14. Li, Y., Shahbazi, A., Willliams, K., Wan, C., Seperate and concentrate lactic acid using combination of nanofiltration and reverse osmosis membranes. *Appl. Biochem. Biotech.*, 147(1–3), 1–9, 2008.
15. Gonzalez, M.I., Alvarez, S., Riera, F.A., Alvarez, R., Lactic acid recovery from whey ultrafiltrate fermentation broths and artificial solutions by nano filtration. *Desalination*, 228, 84–96, 2008.
16. Wasewar, K.L., Seperation of lactic acid: Recent advances. *Chem. Biochem. Eng. Q*, 19(2), 159–172, 2005.
17. Kumar, S., Babu, B.V., Process intensification for separation of carboxylic acids from fermentation broths using reactive extraction. *J. Fut. Eng. Technol.*, 3(3), 19–26, 2008.
18. Martinez, F.A.C., Balciunas, E.M., Salgado, J.M., Gonzalez, D.J.M., Converti, A., Oliveira, R.P.S., Lactic acid properties, applications and production: A review. *Trends Food Sci. Technol.*, 30, 70–83, 2013.
19. Pal, P., Dey, P., Developing a sustainable technology for clean production of lactic acid. International conference on chemical, ecology and environmental sciences. Bangkok, pp. 166–170, March 17–18, 2012.
20. Evangelista, R.L., Mangold, A.J., Nikolov, Z.L., Recovery of lactic acid by sorption: Resin evaluation. *Appl. Biochem. Biotechnol.*, 45/46, 131–144, 1994.
21. Cao, X.J., Yun, H.S., Koo, Y.M., Recovery of (+) lactic acid ion exchange resin Amberlite IRA-400. *Biochem. Eng. J.*, 11, 189–196, 2002.
22. Lee, H.J., Xie, Y., Koo, Y.M., Wang, N.H.L., Seperation of lactic acid from acetic acid using a four-zone SMB. *Biotechnol. Progr.*, 20, 179–192, 2004.
23. Hu, Y., Daoud, W.A., Cheuk, K.K.L., Lin, C.S.K., Newly developed techniques on polycondensation, ring-opening polymerization and polymer modification: Focus on poly (lactic acid). *Materials*, 9(133), 1–14, 2016.
24. Gupta, A., Kumar, V., New emerging trends in synthetic biodegradable polymers–polylactide: A critique. *Eur. Polym. J.*, 43, 4053–4074, 2007.
25. Steinborn-Rogulska, I., Rokicki, G., Solid-state polycondensation (SSP) as a method to obtain high molecular weight polymers. *Polimery*, 58, 1–13, 2013.
26. Achmad, F., Yamane, K., Quan, S., Kokugan, T., Synthesis of polylactic acid by direct polycondensation under vacuum without catalysts, solvents and initiators. *Chem. Eng. J.*, 151, 342–350, 2009.
27. Kim, K.W., Woo, S.I., Synthesis of high-molecular-weight poly (L-lactic acid) by direct polycondensation. *Macromol. Chem. Phys.*, 203, 2245–2250, 2002.

28. Fukushima, K., Kimura, Y., An efficient solid-state polycondensation method for synthesizing stereo complexed poly (lactic acid)s with high molecular weight. *J. Polym. Sci. A Polym. Chem.*, 46, 3714-3722, 2008.
29. Piringer, O.G., Baner A.L., *Plastic packaging: Interaction with food and pharmaceutical.* 2nd ed., p. 632, Wiley-VCH Verlag GmbH & Co., KGaA, Weinheim, 2008.
30. Kricheldorf, H., Ring-opening polycondensation. In: Polycondensation. Springer, Berlin, Heidelberg, pp. 135–145, 2014.
31. Paul, S., Zhu, Y., Romain, C., Brooks, R., Saini, P.K., Williams, C.K., Ring-opening copolymerization (ROCOP): Synthesis and properties of polyesters and polycarbonates. *Chem. Commun.*, 51, 6459–6479, 2015.
32. Nuyken, O., Pask, S.D., Ring-opening polymerization-an introductory review. *Polymers*, 5, 361–403, 2013.
33. Liang, S., Wang, H., Chen, X., Li, F., Synthesis and purification of intermediate lactide. *New Chem. Mater.*, 33, 66–70, 2005.
34. Chen, J., Cheng, C., Wang, Y., The synthesis of lactidecatalyzed by ZnO-Sn(Oct)2. *Polym. Mater. Sci. Eng.*, 23, 74–76, 2007.
35. Sanglard, P., Adamo, V., Bourgeois, J., Chappuis, T., Vanoli, E., Poly(lactic acid) synthesis and characterization. *Chim. Int. J. Chem.*, 66, 951–954, 2012.
36. Jacobsen, S., Fritz, H.G., Degée, P., Dubois, P., Jérôme, R., New developments on the ring opening polymerisation of polylactide. *Ind. Crops Prod.*, 11, 265–275, 2000.
37. Korhonen, H., Helminen, A., Seppälä, J.V., Synthesis of polylactides in the presence of co-initiators with different numbers of hydroxyl groups. *Polymer*, 42, 7541–7549, 2001.
38. Kaihara, S., Matsumura, S., Mikos, A.G., and Fisher, J.P., Synthesis of poly(L-lactide) and polyglycolide by ring-opening polymerization. *Nat. Protoc.*, 2, 2767–2771, 2007.
39. Goethals, E.J., De Clercq, R.R., Cataonic ring-opening polymerization. In: New methods for polymer synthesis, W.J. Mijs (Ed.), pp. 67–109, Springer, New York, 1992.
40. Penczek, S., Kubisa, P., Cationic ring-opening polymerization. In: *Ring-opening polymerization*, D.J. Brunelle, (Ed.), pp. 13–86, Hanser Publishers, Munich, Germany, 1993.
41. Inoue, S., Aida, T., Anionic ring-opening polymerization. In: *New methods for polymer synthesis*, W.J. Mijs, (Ed.), pp. 33–65, Springer, New York, 1992.
42. Ahmed, J., Varshney, S.K., Polylactides—chemistry, properties and green packaging technology: A review. *Int. J. Food Prop.*, 14, 37–58, 2011.
43. Jamshidian, M., Tehrany, E.A., Imran, M., Jacquot, M., Desorby, S., Poly-lactic acid: Production, applications, nanocomposites, and release studies. *Compr. Rev. Food Sci. Food Saf.*, 9, 552–571, 2010.
44. Lasprilla, A.J.R., Martinez, G.A.R., Lunelli, B.H., Jardini, A.L., Filho, R.M., Poly-lactic acid synthesis for application in biomedical devices — A review. *Biotechnol. Adv.*, 30, 321–328, 2012.

45. Cheung, H.Y., Lau, K.T., Lu, T.P., David, H., A critical review on polymer-based bio-engineered materials for scaffold development. *Composites: Part B.*, 38, 291–300, 2007.
46. Lopes, M.S., Jardini, A.L., Filho, R.M., Poly (lactic acid) production for tissue engineering applications. *Proc. Eng.*, 42, 1402–1413, 2012.

19

A New Perspective on Fermented Protein Rich Food and Its Health Benefits

Jagriti Singh, Akanksha Rastogi, Debajyoti Kundu, Mohan Das and Rintu Banerjee*

Agricultural and Food Engineering Department, Indian Institute of Technology, Kharagpur, West Bengal, India

Abstract

In recent times, consumer's negative perception towards synthetic preservatives and growing interest in minimally processed and nutritionally enriched food products drives the researchers to rethink and reinvent new food products to meet the demand. The consumer society is shifting towards an organic product for consumption therefore the market demand is thriving towards natural products. Thus the fermentation technology is considered to be an alternative established and reliable process. Over the generations, this pioneering practice of food fermentation has expanded and improved to preserve and fortify the available food resources, in order to fulfill the requirements of the consumers. Adequate consumption of high-quality food is necessary for optimal growth and development and metabolic regulation in humans. Proteins are the building blocks of the body and are required in large amount by all age groups. Fermented protein-rich food and beverages like soyabean, whey beverages, tempeh, soya sauce, fermented grains, red bean, etc. are few examples which are enriched with antioxidant, amino acid, bioactive peptides, and protein content. Fermented protein products are high in demand for their low cost and safety in comparison to commercially available protein hydrolysates. This book chapter covers implications and opportunities for current and emerging protein rich fermented food products.

Keywords: Fermented food, biopeptide, protein digestion, soyabean, DDGS

**Corresponding author*: rb@iitkgp.ac.in.

Arindam Kuila and Vinay Sharma (eds.) Principles and Applications of Fermentation Technology, (417–436) © 2018 Scrivener Publishing LLC

19.1 Introduction

A healthy, balanced, and nutritious diet is important for good health and wellbeing. Protein is the most fundamental component of the human body and high protein diet is a prime requirement for all age groups. They are the core macromolecules which are required for proper metabolic functioning of the human system. Beans, cheese, whey concentrate, egg, milk, cereals, sausages, yogurt, beef, fish etc. are some of the example of a protein-rich diet. In addition to it distillers dried grains with solubles (DDGS) is a novel source of protein. It has components that make it potentially valuable as an ingredient for human foods as it is rich in polyunsaturated fatty acids, antioxidants/phenolic acids, dietary fibre and xanthophylls. Generally, processing of food from raw to final product involves several stages which deprive the natural nutritional value of the food product. Thus, due to unavailability of "complete set" of proteins and other compounds (vitamin, minerals), the end product has a lesser amount of nutritional compounds. Hence, the processed food fails to provide complete nutrition required by human body. In order to overcome these limitations, researchers have given all their efforts in selecting and optimizing the most suitable natural processing methods which provide organic products without involvement of any toxic components.

Fermentation, a well-established process since medieval age, is a traditional food processing and food preservation method, has the potential to improve nutritional factor of food substance by altering the native composition [1]. Many traditional food products are fermented and are evergreen because of their high demand for people, eg. bouza, kambucha, kefir, cheese, idli, dosa, etc. Starter cultures of lactic acid bacteria (LAB) and other microorganisms are used in the production of fermented beverages, dairy, meat, and vegetable products. Fermented products have an extended shelf life, and distinct flavour profiles and textures. The preservation effect is due to the production of lactic acid and other organic acids, which reduce the pH and inhibit the growth of pathogenic and spoilage organisms [2]. Both submerged and solid state fermentation can be used for this purpose. Microbial fermentations results in degradation of various anti-nutritional factors, an increase in amount of small-sized peptides and improved content of both essential and non-essential amino acids [3]. Compared to raw food with fermented food, fermented protein-rich substance has several beneficial effects including increased average daily gain, improved growth performance, better protein digestibility, and functionality. The increasing demand for fermented food product is promoted by long set evidence on their health benefits, beyond the traditionally recognized effects on the digestive system. Fermented foods have unique functional properties imparting some health benefits to consumers due to the presence of functional microorganisms, which possess probiotics properties, antimicrobial, antioxidant, peptide production, etc. [4]. With the rising concern of consumer and development in biotechnology, proteinaceous fermented foods are in high demand as they have multiple of nutritional benefits and also increases bioavailability/bioabsorbability of peptide and amino acids in human cells [5]. Some of the common health benefits of protein-rich fermented foods are synthesis of nutrients, prevention of cardiovascular disease, prevention of cancer, gastrointestinal disorders, allergic reactions, diabetes, anti-oxidant, anti-microbial, anti-fungal, anti-inflammatory, anti-diabetic and anti-atherosclerotic activity. Several methods are developed to manipulate fermentation techniques for practical and gastronomical applications, and thousands of protein-rich fermented foods and beverages have been established across all cultures including soyabean, soya sauce, soy hydrolysates, tempeh, peanut hydrolysate, miso, yogurt, sausages and red bean [6].

A protein has no nutritional value unless it is hydrolyzed by proteases and peptidases to amino acids, dipeptides, or tripeptides in the lumen of the small intestine. Amino acids provide nitrogen, hydrocarbon (CHO), sulfur (essential components of organisms) which cannot be replaced by any other nutrients (including carbohydrate and lipids) because neither nitrogen nor sulfur is produced in the body. Amino acids are essential precursors for the synthesis of proteins, peptides, and low-molecular-weight substances (e.g., glutathione, creatine, nitric oxide, dopamine, serotonin, RNA, and DNA) with enormous physiological importance. Dietary glutamate, glutamine, and aspartate are major metabolic fuels for the mammalian small intestine whereas glutamine in the arterial blood is the exclusive source of energy for this organ in the post-absorptive state. Thus, amino acids are essential for the health, growth and development, reproduction, and survival of organisms. Protein molecules when hydrolyzed by hydrolytic enzyme (pectin, pectinases, etc.) present in small intestine break down into small peptide fragments releasing amino acids which are ultimately absorbed by cells [7]. Furthermore, amino acids have specific biological activities that can influence human health and prevent diseases. Unlike carbohydrate and lipids, protein has no storage cells and thus there is a continuous demand for high proteinaceous diet. Whey protein is a high protein diet, naturally found in milk, is considered as a complete protein and contains all the amino acids. Animal proteins like egg, meat, and fish contain all the essential amino acids in comparison with plant-based source. In this chapter, the bioactivities of fermented protein-rich products and their bioavailability and bioabsorbability in human body are discussed in the following sections.

19.2 Sources of Fermented Protein

Proteins are grouped into two categories, i.e. endogenous and exogenous. Endogenous proteins are those which are synthesized within the body whereas exogenous proteins are those which are obtained from an external source (through diet). Proteins derived from plant and animal origin contains wide range of amino acids in their structures [8,9]. Animal protein from egg, milk, meat, fish, and poultry are high in demand as they provide good percentage of protein and amino acids. Similarly, a vegetable protein (fermented rice, nuts, beans, soy, etc.) provides a full source of essential amino acids and reduces the intake of saturated fat and cholesterol. Traditional and cultural foods such as natto, miso, tofu, fermented vegetables, etc. are cholesterol-free, and contain high bioactive-peptide. Beside from these products, vegetable protein is also found in a fibrous form called as textured vegetable protein (TVP). Soy flour considered as an alternative to animal protein (meat) having low-calorie and low-fat with high phytochemicals and fibres. In the following sections fermented proteins based foods from both sources (animal and plant) are discussed in detail.

19.3 Protein in Biological System

Proteins are nitrogen-containing polymers that are made up of amino acids joined together by peptide bonds. They majorly provide structural component of muscles and tissues in the body and promotes in producing hormones, enzymes, and hemoglobin. Bones, cartilages, skin, hair, nails, and blood are all made up of proteins. All proteins are not similar; they differ in their nutritional profile, digestibility, and bioavailability [24]. Depending upon the sequence of the amino acids, protein forms different parts of the body. They perform several functions which include: an assembly of proteins forming antibodies in the body which immobilize the infectious antigens and help the body to fight against multiple diseases. Actin and myosin are muscle proteins which help in muscle contraction. Protein also helps in the production of enzymes in the body which helps in digestion of food substances. For example, the lactase helps to break down milk products and to digest them. Pepsin is another enzyme which helps in the digestion of proteins. Some hormones, made up of proteins,

Table 19.1 Fermented food products and their important health benefit.

S.No.	Food product	Microorganism	Health benefits	References
1.	Tempe	*Klebsiella pneumoniae*	Highly rich in vitamin B12 prevents oxidative stress causing diseases such as diabetes, cancer, and damage of pancreatic beta cell.	[10]
2.	Kefir	*Saccharomyces florentius*	Rich in dietary minerals, vitamins (vitamin A, vitamin B1 vitamin B2, vitamin B3, vitamin B6, vitamin B9, vitamin B12, vitamin C, vitamin D, and vitamin E), essential amino acids, and conjugated linoleic acid.	[11, 12]
3.	Fermented sausages	*Lactobacillus alimentarius*	Improves gut health and provides natural defense.	[13]
4.	Soyabean	*Lactobacillus plantarum*	Enhances amino acids like leucine, isoleucine, valine, aspartic acid and proline and decreases trypsin inhibitor content.	[14]
5.	Fermented milk products/yogurt	*Lactobacillus delbruecki, L. rhamnosus*	Rich source of calcium, vitamin B-2, vitamin B-12, potassium, and magnesium. Also Improves blood pressure, metabolism and bone health.	[15]
6.	Fermented meat/fish	*E. faecium*	Reduces cholesterol, calories, sodium content and nitrites.	[16]

(*Continued*)

Table 19.1 Cont.

S.No.	Food product	Microorganism	Health benefits	References
7.	Cheese	*Geotrichum candidum*	Contains calcium, protein, phosphorus, zinc, vitamin A and vitamin B12. Prevents cavity, cancer and improves bone strength.	[17, 18]
8.	Kimchi	*Bacillus mycoides, B. pseudomycoides, B. subtilis, L. carnosum*	Prevents cancer, detoxification of heavy metals in liver, kidney, and small intestine	[19]
9.	Ogi	*Lactobacillus* sp., *Saccharomyces* sp., *Candida* sp.	Good source of carbohydrates, vitamins B (pantothenic acid – B5, niacin – B3, riboflavin and thiamine – B1), folic acid, vitamin A and C, potassium, Chromium, selenium, zinc, phosphorous, magnesium.	[20]
10.	Miso	*Aspergillus oryzae,* Zygosaccharomyces, *Pediococcus* sp.	Good source of copper, manganese, vitamin K, protein, and zinc.	[21]
11.	Kvass	*Lactobacillus cerevesiae*	Rich in vitamin B1 and B6, magnesium, phosphorus, amino acids, and pantothenate. It is also rich in lactic acid.	[22]
12.	Chibuku	*Lactobacillus cerevesiae*	Improves bowel function, reduces heart disease and diabetes. Rich in anti-oxidant property.	[23]

are the chemical messengers which stimulate a number of chain reactions in the body. For instance, insulin is one of the important hormones formed by protein which influences the concentration of blood sugar and therefore helps to regulate the digestion of glucose. Other examples of protein-based hormones are somatotropin and oxytocin. Collagen and elastin are structural protein which provides support to ligaments and tendons while keratin provides a protective cover for the hair. Some proteins transport substances from one part of the body to another. Hemoglobin is the most important example of this type of protein. It helps to circulate oxygen throughout the body. Therefore, adequate consumption of high-quality protein diet is essential for optimal body growth and development [25].

19.4 Bioabsorbability of Protein

All proteins differ in their amino acid content, bioavailiability and the rate at which they are been absorbed. Protein absorbtion takes place in the small intestine. It is first hydrolyzed by enzymes into peptides and amino acids thereafter, these are absorbed by the cells lining the small intestine. During digestion and absorption of protein, it has to pass through many organs. Protein digestion begins in the stomach using gastric enzymes. The gastric gland also secretes a large quantity of hydrochloric acid (HCl) which initiates the digestion by gastric proteinases under optimum pH range. Pepsin, an essential peptic enzyme present in the stomach is active at a low pH (below 3), helps in breakdown of protein molecules [26]. Once the protein leaves the stomach they get exposed to the proteolytic enzymes produced by the pancreas and chemical breakdown takes place rapidly. The pancreatic proteases is divided into two categories, i.e. (i) endopeptidases, such as trypsin and chymotrypsin, which attack peptide bonds located within the amino-acid chains of proteins and polypeptides, breaking them into smaller fragments; and (ii) exopeptidases, such as carboxypeptidases A and B, which cleave the terminal bonds of proteins or peptides, splitting off amino acids [26]. Further breakdown of the oligopeptides is done by the peptidases present in the intestinal mucosa. The adsorbed pancreatic proteases in conjunction with superficially placed mucosal peptidases complete the process of protein digestion [26]. Once protein is digested, the body can utilize its nutrients to build and repair many of the cells in the body.

19.4.1 Absorption of Peptides and Amino Acids

Bioactive peptides are organic substances which have specific protein fragments that have a positive impact on body functions and on human health.

Absorption of peptide needs adenosine triphosphate (ATP), the energy source the body utilizes during protein absorption. Peptide absorption is first diffused across the mucus layer before absorption across the epithelia. The absorption of di- and tripeptides occurs in the small intestinal epithelial cell by co-transport with H+ ions via a transporter called PepT1. Once a protein crosses the monolayer of intestinal epithelial cells, it can either enter the capillaries of the portal venous system or the lymphatic lacteal [27]. Once inside the enterocyte, the vast bulk of absorbed di- and tripeptides are digested into amino acids by cytoplasmic peptidases and exported from the cell into blood [28].

The mechanism of amino acids absorption is almost same to that of absorption of monosaccharide (carbohydrates). The luminal plasma membrane of the absorptive cell bears four sodium-dependent amino acid transporters - one each for acidic (L-glutamic and n-aspartic acids), basic (L-lysine), neutral (L-leucine, L-alanine, L-methionine), and imino (L-proline) amino acids. During the absorption process the amino acids released from the peptide are utilized by the carrier protein transport system [29]. Each amino acid group has a carrier protein that is responsible for transporting it from the intestines to the mucosa cells. Sodium (Na^+) and potassium (K^+) pump are playing an important role to pass the mineral through villi via intestine and finally into the bloodstream. These transporters first binds to sodium after that they bind with amino acids. The sodium-amino acids bounded transporter then undergoes a conformational change that dumps Na^+ and the amino acid into the cytoplasm, followed by its reorientation back to the original form. Thus absorption of amino acids is dependent on the electrochemical gradient of Na^+ across the epithelium. Further absorption of amino acids, like that of monosaccharides, contributes to generating the osmotic gradient that drives water absorption. The basolateral membrane of the enterocyte contains additional transporters which export amino acids from the cell into blood. These transporters are independent of Na^+ gradient pump [29].

19.5 Fermented Protein-Rich Food Products

In the following sections some of the common fermented food products with associated heath benefits are discussed.

19.5.1 Soyabean (Gycine max)

Soya is most widely accepted vegetable protein which is considerably equivalent to wheat, barley, and rice as a nutritional crop. It belongs to the

Fabaceae family, contains approximately 40% of protein by weight [30]. Glycinin, phaseolin, and legumelin are main content of proteins in soybean. Soya protein is an attractive alternative to animal protein and also popular among people who are lactose intolerant. It has a high concentration of branched-chain amino acids. Fermentation process has degraded the anti-nutritive and allergenic compounds of soyabean, thus increasing the possibilities of utilization of various processed products of soya [31]. A wide diversity of microbes has been used to ferment soya meal for nutritional enrichment. The fermentation environment and dietary quality of the fermented soyabean thus produced can vary depending on the type of microorganism used. The fermentation process is facilitated by the use of a mold or a bacterium. In case of bacterial fermentation, various *Lactobacillus* species and *Bacillus subtilis* are preferred [32]. The fermentation process can be achieved by either solid state fermentation or by submerged fermentation. Much like submerged fermentation, the process related to solid state fermentation has been reported to upgrade the nutritional quality of soyabean [33]. The peptide content and fibrinolytic enzyme activity is increased significantly. Solid state fermentation also resulted in an increase of *in-vitro* trypsin digestibility and nitrogen solubility under alkaline conditions and improvement of the nutritional quality of soyabean [34]. Fermentation with *Aspergilli* almost completely eliminates phytate, resulting in a protein source for feed with highly available phosphorus [35,36] as well as zinc. Fermented soyabean has approximately 10% more crude protein than raw soyabean. According fermentation of soyabean with *A. oryzae* did not affect the essential amino acids concentration but increased the concentrations of glycine, glutamine, and aspartic acid [37]. Fermentation with LAB like *Lactobacillus plantarum* results in protein hydrolysis and increased liberation of free amino acids, thus the resulting fermented soyabean has significantly higher total free amino acids content as compared to raw soyabean [38]. There is an increase in concentrations of certain amino acids like histidine, serine, valine and lysine after fermentation which might increase the antioxidant property of the soya product.

The soya product can be separated into three distinct categories; flour, concentrates, and isolates. Soy flour can be further divided into natural or full-fat, defatted and lecithinated (lecithin added) forms [39]. Of the three different categories of soy protein products, soy flour is the least refined form. Soy concentrate has a high digestibility and is found in nutrition bars, cereals, and yogurts. Isolates are the most refined soy protein product containing the greatest concentration of protein, but unlike flour and concentrates, contain no dietary fibre. They are very digestible and can be easily used in foods such as sports drinks and health beverages as well as infant formulas. The health benefits associated with soy protein are related to the physiologically active components that are part of soy, such as protease inhibitors, phytosterols, saponins, and isoflavones. These components have been noted to demonstrate lipid-lowering effects, increase LDL-cholesterol oxidation and have beneficial effects on lowering blood pressure. Isoflavones are beneficial for cardiovascular health, possibly by lowering LDL concentrations increasing LDL oxidation [40] and improving vessel elasticity. Also, isoflavones are considered as phytoestrogens, i.e. they exhibit estrogen-like effects and bind to estrogen receptors and have potential to reduce the risk for breast cancer risk [41].

19.5.2 DDGS (Distillers Dried Grain with Solubles)

Distillers dried grains are the primary by-product of brewery industry. Cereals like rice, corn, wheat, maize, barley, etc. are used as a raw starting material for the production of alcohol (beer/whiskey). Grinding of the selected grains is done in order to remove the dust particle. Liquefaction process (addition of water) is carried out followed by addition of amylolytic enzymes, i.e. amylase and amylopectin. These enzymes help in breakdown of starch into monomeric form (glucose). Fermentation begins by addition of yeast, e.g. *S. cerevisiae,* a seed culture, converting sugars into alcohol. Distillers grains is the by-product obtained in equal proportion with alcohol and carbon dioxide (CO_2). These are basically a thick stillage (fermented grains) left after distillation of alcohol. As it is a fermented product, it contains a large percentage of yeasts which greatly enhance the nutritional properties of the fermenting grains. The nutritional profile of distillers dried grain depends upon the type of raw material being used. Natural, wild yeasts are excellent sources of lysine, riboflavin, niacin and thiamin, and other amino acids, and vitamins. Further, the growing yeasts reduce the phytate concentration in the grain, improving their digestibility. The yeasts ferment sugars to alcohol in acidic conditions. The acid conditions of the beer are produced by the fermentation action of LAB, primarily *Lactobacillus, Leuconstococcus, Streptococcus,* and *Pediococcus.* These bacteria add more nutritional value, in the form of protein, amino acids, and vitamins, to the food product. Fermented grain foods generally have protein content 8–20% higher than was originally in the grain. Distiller's grains also have enhanced values of thiamine, riboflavin, niacin and amino acids, all vital nutrients for good health.

19.5.3 Tempe

Tempe is a vegetable-based fermented food containing high-quality source of protein. Soyabean is the main ingredient used in the production of tempe. Composition of soyabean gets altered during the fermentation process resulting in enhanced protein content. During fermentation soyabean is hydrolyzed by the protease (proteolytic enzyme) produced by the microbes (*Aspergillus*) [42]. Production process is carried by a single-stage fermentation in which cooked soybeans are inoculated with a mould, wrapped in fresh banana leaves and left for an overnight at ambient conditions (30–35 °C, high humidity) [43]. After 24 hours, a dense web of fungal hyphae forms which looks like a cake that can be sliced without disintegrating. Banana leaves are a practically natural choice for wrapping cooked soyabean in the tropics, being fresh and readily available. Final cooking is done for consumption as a snacks or a meal. Generally, *igosporus* is used as a mould, although other members of this genus, such as *Rhizopus stolonifer, Rhizopus oryzae* and *Rhizopus arrhizus* can also be used for the good fermentation. Tempe contains high protein content with respect to the base material used, i.e. soyabean. Upon characterization of tempe it was observed that it contains significant amount of vitamin B12 [44]. The glutamic and aspartic acid content was found to be the highest whereas, leucine and lysine content is decreased.

19.5.4 Red Bean (Phaseolus Vulgaris)

Red bean is herbaceous annual plant belongs to the family leguminosae [45]. It contains high crude protein, fibres, and minerals have diverse micronutrient composition, which can be further enhanced by fermentation. Fermentation of beans enhances flavors, increases textural properties and improves digestibility [46]. Fermentation process also reduces flatulence-causing sugars and enhances protein digestibility when carried out in a natural and controlled manner. During dehulling and cooking of red beans, 90% of the tannins get eliminated from the beans as large amount of tannins are present in seed coat. In general practice, several food products prior to fermentation, removal of seed coat is done from the substrate in order to eliminate anti-nutritional factors like tannins [47]. During open fermentation of red bean maximum protein digestibility around 92.50% is noticed. It might be due to protein denaturation and inactivation of trypsin inhibitor by heat treatment. Acting microbes produces some proteolytic enzymes during fermentation which are responsible for increased protein digestibility of fermented red beans [48]. After 48 hours of controlled fermentation, the saponin contents were reduced to almost zero. During open fermentation the pH drops significantly which might be due to the production of LAB. A drop in pH is a positive indication for protection of food items from pathogenic microbes. Fermentation also helps in reduction of phytate content (25% less than the initial value), trypsin inhibitors, raffinose oligosaccharides, and saponins of red bean flour. Fermentation offers unique nutritional advantages for making the protein of coarse-grained red beans more digestible by reducing tannins content by 45%. Open fermentation had a better effect on the breakdown of the oligosaccharides, raffinose, and stachyose. In addition to this pure culture is not required for fermentation thus, making the process cost-effective. The product obtained can be marketed as dried beans powder. Moreover, the extruded products obtained from the processed beans can be further commercialized [48].

19.5.5 Fermented Peanuts (Arachis Hypogae)

Peanut press cake is fermented with species of *Neurospora* to produce 'oncom' also commonly known as 'ontjom'. De-oiled peanut is a primary raw material used for fermentation. The peanut press cake is a moderately dry substrate, so in order to make it moist it is soaked in water for 4–5 hours. Water is drained and seeds are mixed with the other ingredients such as salt, spices, etc. Cooking of the peanut is done for some time and flat cake is formed in a wide tray. Inoculum of a mould (*Neurospora*) is added and wrapped with banana leaves. After that, it was transferred to woven bamboo trays and kept at ambient temperature for 3–4 days for proper fermentation to occur. Under these conditions the mould grows through the cakes. *Neurospora* forms pink spores on the finished product. Some forms of oncom use the mould of *Rhizopus* spp. which gives a black oncom, again because of spore formation. The nutritional content of oncom is considered to be quite high, especially in enhanced production of vitamin B12, protein content and low-fat content. Black oncom has better nutritional content compared to red oncom, since it has higher protein content; however red oncom has lower fat content compared to black oncom. The amount of aflatoxin in oncom is decreased by 50%. Moreover, phytic acid existed which is major antinutritional content of peanut is also decreased by 50%. After inhibition of phytate, absorption of minerals such as calcium and magnesium is increased [49].

19.5.6 Sufu

Sufu or furu is a fermented soyabean-based product originating in China. It has a higher percentage of protein-nitrogen (10–12%) than other oriental soybean foods, such as miso and natto [50]. Nutritionally, soybean milk, tofu, and sufu have the same health benefits. It is a cheese-like product with a spreadable creamy consistency and a distinct flavour. Sufu is a popular side dish consumed mainly with breakfast rice or steamed bread. Sufu is made by fungal solid state fermentation of tofu (soybean curd) followed by aging in brine containing salt and alcohol. Several types of sufu can be distinguished, according to processing method or according to colour and flavour. Choice of processing can result in mould fermented sufu, naturally fermented sufu, bacterial fermented sufu, or enzymatically ripened sufu. Depending on the choice of dressing mixture, red, white, or grey sufu may be obtained. The stages of the process are discussed and include the preparation of tofu, the preparation of pehtze, salting, and ripening. Fungal starters include *Actinomucor spp., Mucor spp.* and *Rhizopus spp.* Raw soybeans are soaked in water, then ground with water and the liquid extract filtered off, a milky fluid results, which is colloquially called 'soy milk'. The protein in this extract can be precipitated with calcium and/or magnesium salts to give the curd called 'tofu' (or 'dofu'). For sufu production this is cut into small cubes, dried for 10 minutes in an oven at 100 °C, then inoculated with *Actinomucor elegans* and incubated until mould mycelium completely covers the tofu. The cubes are then placed in brine, which may contain rice wine, red rice, soy sauce and various other flavourings, possibly after the moulded cubes have been dry-salted for 3–4 days. The steeping in brine lasts for about 3 months, after which the product is ready for consumption. Post-fermentation was the main stage responsible for the hydrolysis of protein together with the increase in the content of amino-type nitrogen and free amino acid [50].

19.5.7 Kefir

Kefir is a fermented yogurt-like thick beverage prepared by incubating milk with "kefir grains", which contain sugars, proteins, LAB, and yeast [51]. It is characterized by its creamy texture, tart taste, and fizzy effervescence. Due to its claimed health benefits, i.e. reduction of lactose intolerance symptoms, stimulation of the immune system, lowering cholesterol, and antimutagenic and anticarcinogenic properties, kefir has become an important functional dairy food product. It can be prepared from any type of milk such as pasteurized, unpasteurized, whole fat, low fat, and no fat [52]. The production involves a complex microbial system that has not only been found to be nutritionally beneficial, but has also been proven to inhibit a number of food-borne pathogens and spoilage microorganisms [53]. The microbiological and chemical compositions of kefir indicate that it is a much more complex probiotic. Since yeasts and bacteria present in kefir grains have undergone a long association, the resultant microbial population exhibits many similar characteristics, making isolation and identification of individual species difficult [54]. Kefir has an antibacterial effect against many pathogenic organisms due to the inherent formation of organic acids, hydrogen peroxide, acetaldehyde, carbon dioxide, and bacteriocins. Kefir consumption reduces serum cholesterol and phospholipids. They also have anti-carcinogenic properties and inhibition of tumor growth. Pregnant and nursing women can safety consumes kefir as it promotes the absorption of nutrients, increases immunity, helps the body adjust to hormonal changes and prevents infections such as yeast overgrowth [55].

19.5.8 Fermented Whey Beverage

Whey protein is a by-product obtained from processing of milk product, soyabean etc. as a valuable food component with important nutritional and functional properties. Due to its high protein content, whey protein has gained acceptance as a functional food component. Whey protein is found mainly in 3 forms i.e. whey concentrates, whey isolates and whey hydrolysates [57]. Whey proteins are better source of protein due to their high content of essential amino acids thus possessing high biological value [58]. It is a dairy-based non-conventional beverage which has gained wide acceptance from consumers because of its high nutritional properties [59]. The initial protein contents of whey are less than 5%. The protein content of fermented whey beverage showed almost twice higher than that of other yogurt products. LAB used as a starter culture increases the viscosity of the fermented whey produced. It produces different organic acids by degrading some component in the raw material [60]. Therefore, the rich organic acid profile of fermented whey product is an indicator of the metabolic activity of added bacterial cultures. These acids act as natural preservatives and contribute to the distinguishing sensory properties [61]. Lactic acid contents greatly increased as much as approximately 4 folds during fermentation. Whey based beverage provides an abundant source of branched-chain amino acids. These amino acids are used to build and repair muscles. Whey protein is especially high in leucine which plays a very important role in initiating the genetic transcription pathways that boost protein synthesis. Elevated leucine consumption improves the stimulation of protein synthesis. This improves tissue healing, exercise recovery, strength production and adaptation to stress. Whey protein is also very rich in cysteine which is a critical agent used to make the super intracellular anti-oxidant glutathione. Ingesting large amounts of whey protein increased cellular glutathione levels. Research states that whey protein can be considered as "whole food" for boosting glutathione levels. Due to this anti-oxidant capacity, whey protein is being researched for its ability to prevent degenerative diseases like heart disease, cancer, and neurodegenerative disorders like Parkinson`s [61].

19.5.9 Salami

Salami, made up of animal meat, is fermented and air-dried food product. The production of salami involves three major steps: preparation of raw materials, fermentation, and ripening and drying [62]. Raw meat, usually of pork, beef or chicken depending on the type of salami that is produced, is milled and mixed with ingredients such as salt, sugar, spices, and yeast. Meat is sliced into thin strips and their casting is done according to the desired size and shape. Fermentation enhances flavor and texture of meat. Direct exposure of meat into acidic solution might cause protein denaturation and form coagulant therefore causing uneven texture. In order to avoid this manufactures hang the salami in humid atmosphere for 3–4 days for proper bacterial exposure. The bacteria produce lactic acid which lowers the pH and coagulates the proteins, reducing the meat's water-holding capacity [63]. The bacteria-produced acid gives tangy flavor and also provides smooth texture. Salami flavor relies as much on how these bacteria are cultivated as it does on the quality and variety of the other ingredients. Wine is also added as an ingredient to enhance the growth of other beneficial bacteria. Starter cultures such as LAB and coagulase-negative cocci (CNC) like specific strains of Staphylococcus xylosus or Micrococcus are most commonly used in salami production [64]. More species of LAB and CNC were discovered during the last decades and they were found to have different fermentation temperatures with variable rates of acidification. Drying process is carried out happens after fermentation [65]. This stage causes the main physical and microbial changes through the large amount of water loss. Salami is a source of complete protein thus containing all the amino acids. It is also rich in minerals like sodium, calcium, potasium, and iron which help in improving the immune system and increasing RBC's. It is a good source of Vitamin B12 which is helpful in making brain healthy [66].

19.6 Conclusion

Fermented foods which are rich in proteins had gained its importance in recent past and attracted the interest of consumers which drastically raised the market size. An appropriate mixture of animal- and plant-based foods is a practical way to ensure balanced provision of dietary protein for the young and the adult. Consumption of fermented soyameal, whey based beverage, tempe, fermented peanut cake, etc. shows an outstanding bioactivities such as antihypertensive, antioxidant, immunomodulating, enrichment of vitamins, and minerals. Distillers grains is a novel protein source in terms of its potential to sustainably deliver protein for the future, considering drivers and challenges relating to nutritional, environmental, and market domains. Due to its health promoting-disease-preventing nature, fermented foods had become an important functional dietary product around the world. The production of bioactive peptides in fermented foods increases the bioabsorbability of peptides and amino acids. Under the impression of such positive effects on human body, scientific researchers and industries are putting joint effort in developing health-promoting products, thereby creating a wider acceptability of fermented-based protein-rich products.

References

1. Ghosh, J.S., Solid state fermentation and food processing: a short review. *J. Nutr. Food. Sci.* 6, 453, 2015.
2. Leroy, F., De Vuyst, L., Lactic acid bacteria as functional starter cultures for the food fermentation industry. *Trends Food Sci. Technol.*, 15(2), 67–78, 2004.
3. Tamang, J.P., Shin, D.H., Jung, S.J., Chae, S.W., Functional properties of microorganisms in fermented foods. *Front. Microbiol.*, 7, 578. 2016.
4. Thapa, N., Tamang, J.P., Functionality and therapeutic values of fermented foods, in: *Health benefits of fermented foods* Tamang, J.P. (Ed.), pp. 111–168, New York: CRC Press, 2015.
5. Nout, M.J.R., Fermented foods and food safety. *Food Res. Int.* 27, 291–298, 1994.
6. Farhad, M., Kailasapathy, K., Tamang, J.P., Health aspects of fermented foods, in: *Fermented foods and beverages of the world,* J.P. Tamang, K. Kailasapathy (eds.), pp 391–414, CRC Press, New York, NY, 2010.
7. Sánchez, A., Vázquez, A., Bioactive peptides: a review, *Food Quality Safety,* 1(1), 29–46, 2017.
8. Carrasco-Castilla. J., Hernández-Álvarez, A.J., Jiménez-Martínez, C., Gutiérrez-López, G.F., Dávila-Ortiz, G., Use of proteomics and peptidomics methods in food bioactive peptide science and engineering. *Food Eng. Rev.*, 4, 224–243, 2012.
9. Bhat, Z.F., Kumar, S., Bhat, H.F., Bioactive peptides of animal origin: a review. *J. Food Sci. Technology-Mysore*, 52, 5377–5392, 2015.
10. Astuti, M., Health benefits of tempe, in: *Health benefits of fermented foods,* Tamang J.P., (Ed.), pp. 371–394, CRC Press, New York, NY, 2015.

11. Guzel-Seydim, Z.B., Kok-Tas, T., Greene, A.K., Seydim, A.C., Review: functional properties of kefir. *Crit Rev Food Sci Nutr.* 51(3), 261–268, 2011.
12. Ahmed, Z., Wang, Y., Ahmad, A., Khan, S.T., Nisa, M., Ahmad, H., Afreen, A., Kefir and health: a contemporary perspective. *Crit Rev Food Sci Nutr.* 53(5), 422–434, 2013.
13. Askild, H., Lars, A., Anette, M., Tone, M.R., Even Heir. Health and safety considerations of fermented sausages. *J. Food Qual.*, 2017.
14. Amadou, I., Tidjani, A., Foh, M.B.K., Kamara, M.T., Le, G.W., Influence of *Lactobacillus plantarum* Lp6 fermentation on the functional properties of soybean protein meal. *Emir. J. Food Agric.*, 22, 456–465, 2010a.
15. Toldrá,F., Improving the sensory quality of cured and fermented meat products. Processed Meats Improving Safety, Nutrition and Quality. A volume in Woodhead Publishing Series in *Food Sci. Technol. Nutri.*, pp. 508–526, 2014.
16. Keizo, A., Strategies for designing novel functional meat products. *Meat Sci.*,74(1), 219–229, 2006.
17. Huth, P.J., Park, Keigan, M., Influence of dairy product and milk fat consumption on cardiovascular disease risk: a review of the evidence. *Adv.Nutr.*, 3(3), 266–285, 2012.
18. Barbara, E., *The pocket guide to cheese*, Lansdowne Press/Quarto Marketing Ltd., 1981.
19. Han, E.S., Kim, H.J., Choi, H.K., Health benefits of Kimchi, in: *Health benefits of fermented foods*, Tamang J.P. (Ed.), 343–370, CRC Press, New York, 2015.
20. FAO, Cereal fermentation in African countries. *Fermented Cereals – A Global Perspective*, United Nations FAO, 2006.
21. Shurtleff, W., Aoyagi, A., *The book of miso. savory, high-protein seasoning*, 2nd ed., pp. 232–237, Berkeley: Ten Speed Press, 2001.
22. Alexander, A., *The poetic outlook of slavs about nature*, 1865–1869, pp. 260, reprinted 2014.
23. Gadaga, T.H., Mutukumira, A.N., Narvhus, J.A., Feresu, S.B., A review of traditional fermented foods and beverages of Zimbabwe. *Int. J. Food Microbiol.*, 53, 1–11, 1999.
24. Guoyao, W., Dietary protein intake and human health. *Food Funct.*, 7, 1251–1265, 2016.
25. Henchion, M., Hayes, M., Mullen, A.M., Fenelon, M., Tiwari, B., Future protein supply and demand: strategies and factors influencing a sustainable equilibrium. *Foods*, 6(7), 53, 2017.
26. Porter, C.J., Drug delivery to the lymphatic system. *Crit. Rev. Ther. Drug Carrier Syst.*, 14, 333–393, 1997.
27. Langguth, P., Bohner, V., Heizmann, J., The challenge of proteolytic enzymes in intestinal peptide delivery. *J. Controlled Release*, 46, 39–57, 1997.
28. Johnson, L.R., Digestion and absorption in: *Gastrointestinal physiology*, L.R. Johnson (Ed.), pp. 120, St. Louis, MO, USA, 2001.
29. Vahdatpour, S., Mamaghani, A.P., Goloujeh, M.S., Maheri-Sis, N., Mahmoodpour, Hamid, T.V., The systematic review of proteins digestion and

new strategies for delivery of small peptides. *Electronic J. Biol.*, 12(3), 265–275, 2016.

30. Tamang J.P., Naturally fermented ethnic soybean foods of India. *J. Ethnic Foods.*, 2, 8–17, 2015.
31. Kishida, T., Ataki, H., Takebe, M., Ebihara, K., Soybean meal fermented by *Aspergillus awamori* increases the cytochrome p-450 content of the liver microsomes of mice. *J. Agric. Food Chem.*, 48, 1367–1372, 2000.
32. Yang, Y.X., Kim, Y.G., Lohakare, J.D., Yun, J.H., Lee, J.K., Kwon, M.S., Park, J.K., Choi, J.Y., Chae, B.J., Comparative efficacy of different soy protein sources on growth performance, nutrient digestibility, and intestinal morphology in weaned pigs. *Asian Australas. J. Anim. Sci.*, 20, 775–783, 2007.
33. Singh, K., Linden, C.J., Johnson, E.J., Tengerdy, P.R., Bioconversion of wheat straw to animal feed by solid substrate fermentation or ensiling. *Indian J. Microbiol.*, 30, 201–208, 1990.
34. Song, Y.S., Frias, J., Martinez-Villaluenga, C., Vidal-Valdeverde, C., de Mejia, E.G., Immunoreactivity reduction of soybean meal by fermentation, effect on amino acid composition and antigenicity of commercial soy products. *Food Chem.*, 108, 571–581, 2008.
35. Ilyas, A., Hirabayashi, M., Matsui, T., Yano, H., Yano, F., Kikushima, T., Takebe, M., Hayakawa, K., A note on the removal of phytate in soybean meal using *Aspergillus usami*. Asian Australas. *J. Anim. Sci.*, 8, 135–138, 1995.
36. Golbitz, P., Traditional soyfoods - processing and products. *J. Nutrition,* U5 (Suppl. 3), 570–572, 1995.
37. Hirabayashi, M., Matsui, T., Yano, H., Nakajima, T., Fermentation of soybean meal with *Aspergillus usamii* reduces phosphorus excretion in chicks. *Poult. Sci.*, 77,552–556, 1998.
38. Hong, K.J., Lee, C.H., Kim, S.W., *Aspergillus oryzae* 3.042GB-107 fermentation improves nutritional quality of food soybeans and feed soybean meals. *J. Med. Food.*, 7,430–434, 2004.
39. Lena, D.G., Patroni, E., Quaglia, G.B., Improving the nutritional value of wheat bran by a white rot fungus. *Int. J. Food Sci. Technol.*, 32:513–519, 1997.
40. Frias, J., Song, Y.S., Martínez-Villaluenga, C., De Mejia, E.G., Vidal-Valverde, C., Immunoreactivity and amino acid content of fermented soybean products. *J. Agric. Food Chem.*, 56:99–105, 2008.
41. Ikeda, R., Ohta, N., Watanabe, T., Changes of Isoflavones at various stages of fermentation in defatted soybeans. *J. Jpn. Soc. Food Sci. Technol.*, 42, 322–327, 1995.
42. Ashenafi, M., Microbiological evaluation of tofu and tempeh during processing and storage. *Plant Foods Hum. Nutr.*, 45, 183–189, 1994.
43. Piskorz, B.J., Nowak, J., Szebiotko, K., Fermented tempeh-like products as useful way of utilizing legume seeds. *Acta Biotechnologica,* 14, 105–110, 1994.
44. Truesdell, D.D., Green, N.R., Acosta, P.B., Vitamin-Bl2 activity in miso and tempeh. *J. Food Sci.*, 52, 493–494, 1987.
45. Choy, S.Y., Prasad, K.M.N., Wu, T.Y., Ramanan, R.N., A review on common vegetables and legumes as promising plant-based natural coagulants in water clarification. *Int. J. Environ. Sci. Technol.*, 12(1), 367–390, 2015.

46. Diana, M., Quílez, J., Rafecas, M., Gamma-aminobutyric acid as a bioactive compound in foods: a review. *J. Funct. Foods*, 10, 407–420, 2014.
47. Marathe, S.A., Deshpande, R., Khamesra, A., Ibrahim, G., Jamdar, S.N., Effect of radiation processing on nutritional, functional: sensory and antioxidant properties of red kidney beans Radiat. *Phys. Chem.*, 125, 1–8, 2016.
48. Hayat, I., Ahmad, A., Masud, T., Ahmed, A., Bashir, S., Nutritional and health perspectives of beans (*Phaseolus vulgaris* L.): an overview. *Crit. Rev. Food Sci. Nutr.*, 54 (5), 580–592, 2014.
49. Beuchat, L.R., Fungal fermentation of peanut press cake economic botany. *Econ. Botany*, 30(3), 227–234, 1976.
50. Cheng, Y.Q., Hu, Q., Li, Li-Te., Saito, M., Yin, L.J., Production of sufu, a traditional chinese fermented soybean food, by fermentation with mucor flavus at low temperature. *Food Sci. Technol. Res.*, 15(4), 347–352, 2009.
51. Angulo, L., Lopez, E., Lema, C., Microflora present in kefir grains of the Galician region (North-West of Spain). *J. Dairy Res.*, 60(2), 263–267, 1993.
52. Semih, O.E., Cagindi, C., Kefir: a probiotic dairy-composition, nutritional and therapeutic aspects. *Pak. J. Nutr.*, 2(2), 54–59, 2003.
53. Paucean, A., Carmen, S., Probiotic activity of mixed cultures of kefir's lactobacilli and non-lactose fermenting yeasts. *Bulletin UASVM, Agriculture*, 65(2), 2008.
54. Edward, R.F., Kefir a complex probiotic. *Food Sci. Technol. Bulletin: Funct. Foods*, 2(1), 2006.
55. Sandra, E., No fear of kefir. Benefits, love stories about kefir. Available from: http://www.benefitsofkefir.com, 2013.
56. Kwon, C.S., Park, M.Y., Cho, J.S., Choi, S.T., Chang, D.S., Identification of effective microorganisms from kefir fermented milk. *Food Sci. Biotechnol.*, 12, 476–479, 2003.
57. Pescuma, M., Habert, E.M., Mozzi, F., de Valdez, G.F., Functional fermented whey-based beverage using lactic acid bacteria. *Int. J. Food Microbiol.*, 141(1-2):73–81, 2010.
58. Hugunin, A., US whey ingredients in yogurt and yogurt beverages US Dairy Export Council, 2008.
59. Shiby, V. K., Radhakrishna, K., Bawa, A.S., Development of whey-fruit-based energy drink mixes using Doptimal mixture design. *Int. J. Food Sci. Technol.*, 48 742–748, 2013.
60. Wu, S.C., Su, Y.S., Cheng, H.Y., Antioxidant properties of Lactobacillus-fermented and non-fermented Graptopetalum paraguayense E. Walther at different stages of maturity. *Food Chem.*, 129 804–809, 2011.
61. Adhikari K., Grûn I.U., Mustapha A., Fernando L.N. Changes in the profile of organic acids in plain set and stirred yogurts during manufacture and refrigerated storage. *J. Food Qual.*, 25 435–451, 2002.
62. Fonseca, G.G., Silva, P., Paulo S., Almeida P., Luiz, A., Reduction of drying and ripening times during the Italian type salami production. *Trends Appl. Sci. Res.*, 1(5), 504–510, 2006.
63. Barbut, S., Color development during natural fermentation and chemical acidification of salami-type products. *J. Muscle Food*, 21(3), 499–580, 2010.

64. Aquilanti, L., Garofalo, C., Osimani, A., and Clementi, F., Ecology of lactic acid bacteria and coagulase negative cocci in fermented dry sausages manufactured in Italy and other Mediterranean countries: an overview. *Int. Food Res. J.*, 23(2), 429–445, 2016.
65. Aquilanti, L., Santarellia, S., Silvestria, G., Osimania, A; Petruzzellib, A., Clementia, F., The microbial ecology of a typical Italian salami during its natural fermentation. *Int. J. Food Microbiol.* 120(1–2), 136–145, 2007.
66. Cevolia, C., Fabbria, A., Tabanellib, G., Montanarib, C., Gardinia, F., Lanciottia, R., Guarnieria, A., Finite element model of salami ripening process and successive storage in package. *J. Food Eng.*, 132, 14–20, Retrieved 19 March 2016.

20

An Understanding of Bacterial Cellulose and Its Potential Impact on Industrial Applications

Akanksha Rastogi, Jagriti Singh, Mohan Das, Debajyoti Kundu and Rintu Banerjee*

Agricultural and Food Engineering Department, Indian Institute of Technology, Kharagpur West Bengal, India

Abstract

Bacterial cellulose is a biopolymer whose interest in the global market is increasing because of its advanced properties concerning the high purity, fibril structure, high degree of polymerization, and high crystallinity. To enhance the production of bacterial cellulose at pilot plant scale, new approaches are tried upon with respect to culture condition, agitation in bioreactors, supplementation of nutrition, and its cost effectiveness. This chapter also discusses on the genetically modified strain for enhanced cellulose production. In addition, bacterial cellulose has immense potential in many industrial sectors such as paper industries, textile industries, electronics, food industries, and biomedical devices. It plays a major role in tissue engineering with application in wound care and revival of the damaged organs. Moreover, it is having multifunctional properties in food industries and is engaged in maintaining shelf life, as thickening and gelling agent, and as a packaging material due to its water binding capacity. Bacterial cellulose shows good barrier properties in preventing food rancidity and also improves the food rheology. Due to the unique properties of bacterial cellulose, it is attractive innovative products have been developed and commercialized making it an attractive biopolymer for industrial application.

Keywords: Bacterial cellulose, genetically modified bacteria, crystallinity, bioreactor

**Corresponding author*: rb@iitkgp.ac.in

Arindam Kuila and Vinay Sharma (eds.) Principles and Applications of Fermentation Technology, (437–458) © 2018 Scrivener Publishing LLC

20.1 Introduction

Biopolymers are those polymers that are produced from the living beings. Amongst all, cellulose is one of the most important, affordable polymers in the earth which is extracted from the plants and their wastes. This biopolymer mixed with different another glycans molecules like hemicelluloses and lignin. To obtain only cellulose, it is treated with alkali and acid following the chemical process. Cellulose is also procured from the *in vitro* synthesis by using enzymatic pathway instead of chemical pathways. On the basis of morphological localization these polysaccharides are located intracellular or extracellular region. Two forms of extracellular polysaccharides are found into the culture like as loose slime that is non-attachable and sticky substance to the cell wall or viscosity increased into the liquid medium, microcapsules and capsules are separated by centrifugation method.

As the demand of pure cellulose is getting high, the different microorganisms such as algae, fungi, bacteria are used for the production of extracellular natural form of cellulose. Since 1886, it has been discovered by Brown that pure cellulose can also be produced from bacteria due to its special properties like crystalline structure, purity, high mechanical, strong water holding capacity, biodegradability, high polymerization degree and biocompatibility [6]. Because of this reason it can be catching the research bug for different application sectors like biomedical and biotechnology field. These exopolysaccharides are repeating unit of monosaccharides glucose sugar molecules. Several current developments accounted for the cellulose-producing bacteria, that contain the genera of bacterial cellulose makers include *Gluconacetobacter*, *Acetobacter*, *Burkholderia*, *Achromobacter*, *Enterobacter*, *Salmonella*, *Aerobacter*, *Rhizobium*, *Pseudomonas*, *Agrobacterium*, *Alcaligenes*, *Azotobacter*, *Escherichia*, *Sarcina* and *Dickeya* [37,39]. Amid the bacteria, a gram-negative purple bacterium is one of the most popular types of vinegar bacteria named as *Acetobacter*.

It is secreted from acetic acid producing bacteria where pellicles are appeared on the top of the liquid surface. This non-photosynthetic microorganism can acquire glucose, sugar alcohol, glycerol substrates and responsible to convert it into pure form of cellulose. Both have different physical and chemical properties but they are sharing the same molecular formula $(C_6H_{10}O_5)_n$. According to Oh *et al.* [34], it is a linear and homopolysaccharides that consisting β-D 1–4 linkage glucopyranose unit with C4-OH and C1-OH group present at every end. It has the tendency to clump and form the tight and highly ordered structural units due to their conformation and chemical nature [34].

Usually, bacterial cellulose properties, composition, and its morphology are affected by the many factors like temperature, nutrients, pH, additives, reactor design and availability of oxygen into the culture medium [27]. This optimal design supports the cell growth of microorganism and improves the bacterial cellulose production. Another key factor is a bioreactor which is also responsible to reduce the cost for the high yield of cellulose and use cheap nutrients like by-products from the previous production process. Due to the requirement of large space and time, static culture reactors are replaced by the other type of supportive reactor which increases the productivity of cellulose [50].

In addition, it is also produced by the members of the genus *Leifsonia* belonging to the class of *Actinobacteria* which is gram-positive in nature and accountable to give high yield of bacterial cellulose. It has been seen that it shows less crystallinity and high amorphous region in pellicle membrane so it can be used directly for the different application without converting their crystalline nature of cellulose. It could be a novel foundation in term the production of bacterial cellulose because of their high yield and uses of cheap nutrient sources.

Nowadays, the alternative reactors used for the high yield of cellulose are stirred tank reactor, trickling bed reactor, rotating disk bioreactor, bubble column bioreactor, and biofilm reactor [29,30]. Due to the perfect structure, feasibility, and biocompatibility of bacterial cellulose, it leads to the variety of application fields like paper, food, textiles, medical, and electronics. Currently, bacterial cellulose-based materials are probably used in medical field that include artificial skin, scaffolds for tissue engineering, drug delivery, wound healing materials, and blood vessels. In addition, it also has the capability to uphold a physical barrier that helps to reduce the pain, transfer the drug into the injured region, and bacterial infection.

This chapter focuses in amalgamating the features of cellulose structure, properties, molecular biology, biosynthesis, several bioreactors for the enhancement of bacterial cellulose, and their applications in different industries by comparing the already well-published reviews and articles.

20.2 Cultivation Conditions for Production of Bacterial Cellulose

20.2.1 Fermentation Process

The high yield of bacterial cellulose from bacteria is usually dependent on the two main factors of culture conditions like the composition of media and eco factors. For instance, dissolved oxygen content, type of reactor, pH, and temperature are the major elements for the high productivity and their growth of cellulose producing bacteria.

20.2.2 Composition of Culture Media

20.2.2.1 Carbon Source

To increase the production of bacterial cellulose, one of the most effective parameter like carbon sources are used that include oligosaccharides, organic acids, sugar, and other alcohols and monosaccharides. In addition, different bacterial strain is also responsible to produce various yields of cellulose and the yield is also dependent upon the source of nutrients. In addition, it has been found that sucrose gave the maximum yield (3.83 g/L) of bacterial cellulose in Hestrin and Schramm (HS) medium [25]. Among all, D-arabitol and D-mannitol showed maximum productivity approximately 6.2- and 3.8-fold greater yield of bacterial cellulose compared to glucose [21]. In comparison to other sugars, the less suitable carbon source is galactose into the culture media that depends on the capability of bacteria to utilize those carbon sources or not. As well as, it was found that other than glucose, mannitol and fructose are also showing the same rate for the bacterial cellulose production that is transported through a cell membrane.

Utilization of sugar alcohol and other carbon sources for the production of bacterial cellulose are very costly when it goes for large-scale production. To make the process cost effective, some researchers are trying to reduce the cost of the used carbon source in to the culture medium. Nowadays, sugarcane molasses, processed rice bark, coconut water, soy flour extract, beet molasses, konjac powder hydrolysate, and fruit juices are used as a carbon source into the culture media for industrial production by using this inexpensive waste products.

20.2.2.2 pH for Bacterial Cellulose Production

Depending upon the different strains of bacteria the classified and selected range of pH is 4 to 7. But highest production of bacterial cellulose is obtained at pH 6.5 [49]. Production of gluconic, lactic, and acetic acid as secondary metabolites into culture medium leads the low production of bacterial cellulose due to the high consumption of carbon and nitrogen source. Some products of bacterial cellulose are available in the medicinal field such as Biofill and Gengiflex at low pH range 4 to 4.5 that avoids the contamination in the medium during cellulose production [21]. In addition, it has been found that to maintain the pH into the media some researchers add the corn steep liquor as a buffer which increases the viscosity of the medium. But it shows some disadvantages due to an increase in viscosity and also causes the inhomogeneous mixing of component culture.

20.2.2.3 Temperature for BC Production

It has been shown that 20 to 40 °C range of temperature is most favourable for the cellulose producing *Acetobacter* sp. While the optimum temperature is 30 °C for the bacterial cellulose production, but at the lower temperature (25 °C) the yield of bacterial cellulose is not that much affected. But variation in temperature can change the crystallinity and morphology of that bacterial cellulose. For instance, in HS media, *Acetobacter xylinum* ATCC 23769 are produced, band of cellulose II at 4 °C temperatures with phosphate buffer having pH 7 while at 28 °C temperatures they produced the ribbon of cellulose I. Hence, incubation time plays a very crucial role in morphology and crystallization of cellulose [15].

20.2.2.4 Dissolved Oxygen on BC Production

This is very important parameter for cell metabolism, the high yield of cellulose and the quality of bacterial cellulose. Production of bacterial cellulose is irreversibly related to the dissolved oxygen content. It has been reported that, if dissolved oxygen is high in culture medium then gluconic acid concentration is high and it ultimately reduces the pH, the cell viability, and the productivity of bacterial cellulose. For instances, maximum cellulose production was obtained at 10% dissolved oxygen in batch-fed culture [19].

20.3 Bioreactor System for Bacterial Cellulose

Generally, bacterial cellulose produced in static culture medium is a very common method for the production of cellulose on the top of liquid media culture. They have high water holding capacity due to their extensive interior surface area. But gradually the production of cellulose is getting moderate due to the slow growth rate of oxygen transfer and less nutrition availability in the culture medium. Pellicles are entrapped and cells are dying due to the lack of oxygen. It also depends on bacterial strain which takes 5–14 days and that is not efficient for pilot scale production. Even so, one Philippines's nata-de-coco native dessert is the first example of another form of bacterial cellulose through static culture method. On the other hand, agitated cultures are used to increase the rate of mass transfer concerning their low shear stress and it also increases the oxygen rate into the culture media [50]. In shaken culture, sometimes they show few disadvantages of this reactor, for example they endorsed mutation of cellulose producing bacteria into non-cellulose producing mutants. So to overcome this problem, some reactors are designed similarly so that it can improve the production and reduce the mutation steps in the culture media and also reduce the labour cost. For enhancement and their scale-up we are discussing some new bioreactor designs for bacterial cellulose production.

20.3.1 Stirred Tank Reactor

To address the problem of the production of non-cellulose-producing microorganism from mutant bacteria the stirred tank reactors are used. Besides that, the doubling time of microorganism was found increasing in submerged, that is approximately completed in around 4–6 hours compared to static culture where it takes 8–10 days [4].

The flow property of bacterial cellulose in liquid culture was found to be shear-thinning behaviour and non-Newtonian fluid was detected [24]. Higher stirring speed is directly related to the improved oxygen transfer with in turn increases the yield of bacterial cellulose. The reactor equipped with impeller and spin filter is made up of stainless steel.

20.3.2 Trickling Bed Reactor

It provides the greater ratio of surface to volume of liquid culture and also supplies the oxygen rate from that bacteria can easily contact with air compared to stirred tank reactor. This type of reactor gives the 3D space, surface to adhere the bacteria and grow in different semisolid and liquid–solid environment and form the pellicle on the top liquid of culture. In this tank culture broth pass from bottom to the top meanwhile, air is circulated. Obtained cellulose has high degree of polymerization, high water holding capacity, and biocompatibility [30].

20.3.3 Airlift Bioreactors

This is another bioreactor which is important for the high yield of bacterial cellulose. Due to their simple design and easy maintenance, it is broadly used for the biochemical process. From the lower part of the reactor oxygen-enriched air was supplied, which in turn drove the circulation of the culture media. Low production of bacterial cellulose was certified to the limited dissolved oxygen content in the culture broth. To reduce the bubble size add draft tube that having rectangular wire mesh to enhance the volumetric mass oxygen transfer rate due to this reason it is used modified airlift bioreactor. In comparison to the conventional bubble column reactor, it increases the bacterial cellulose production five times better than from 2.82 to 7.72 g/L and also increases mass oxygen transfer rate by 50%. To reduce the shear stress that is produced by bacteria by adding agar into the culture medium which increases the viscosity of the media and reduces the chances mutates to non-cellulose producing strains [52].

20.3.4 Aerosol Bioreactor

Due to limited supply of nutrients and their wall effect leads to dormant growth of bacterial cellulose and also impedes the elimination of the product from the active zone. To overcome this problem novel bioreactor are designed in this way so that they are able to get rid of that issue by feeding the glucose and oxygen directly to the reactor. In this type of reactor aerosol sprayed of glucose from nozzle on the living bacteria that are distributed at the surface of liquid interface and allowed to constant growth up to 8 weeks. Pellicles are formed on the top of the liquid media because of the highest oxygen concentration and nutrition diffusion is there so that bacteria grow happily on that surface [17].

20.3.5 Rotary Bioreactor

Homogeneity problem arose during bacterial cellulose production in airlift and agitation system because of the adhesion of cellulose in different parts of the reactor. To overcome this issue a new reactor designed in this way that they are exposed to culture medium first and then again through air. In this reactor, eight circular discs are there and for inoculation separately inlets are present. During the production of bacterial cellulose half of the part of rotating discs is merged with medium and other half part of disc contacts with air [42]. It has been reported that beyond eight discs in the reactor it can decrease the productivity of cellulose, increase the agglomeration of bacterial pellicles and also low the mechanical strength of the product due to the low space between the circulatory discs. At the end of this process finally got the improved mechanical properties, enhanced cell adhesion and used in paper fibre [46].

20.3.6 Horizontal Lift Reactor

All the reactors work on the principle of batch process to extract the bacterial cellulose. In horizontal lift reactor, there are long tanks with the culture medium at the end, bacterial pellicle is lifted and separated out from the culture medium without disturbing the nanofibre of the bacterial cellulose which is present within the bacterial pellicle. Height of the bacterial pellicle can be adjusted by increasing the tank length to allow the bacteria to grow for longer time.

20.3.7 Other Type of Bioreactor

There are few other important bioreactors which are also responsible to produce the high yield of bacterial cellulose. In case of membrane bioreactor, cellulose are produced within static condition to exploit the high surface area of the membrane. In addition, membranes are used which are made up of hydrophilic polyether sulfone and have the pore size of 0.45 mm. *Gluconacetobacter xylinus* are passed through one side of the membrane and other nutrients which are required for the bacterial growth are passed through the other side of the membrane [16] and also reduced the downstream cost.

Another is cell recycle; biofilm (immobilized cell) and hollow fibre bioreactor are used to enhance the production of bacterial cellulose. It reduces the cost and increases the production yields by 2.5 times in comparison to controlled one.

20.4 Plant Cellulose vs. Bacterial Cellulose

In plant, cellulose is a structural component of the cell wall and is important for plant survival while in bacteria, cellulose is not vital for survival but does give endurance benefits. In the early 19th century, scientist discovered cellulose by treating plant tissue with acid and ammonia and are also recognized as a highly potential renewable biopolymers component of plant biomass. It produced approximately 1011–1012 tons per year by nature. It has high molecular weight polysaccharide that is composed of glucose moiety. In plants, nearly 50% of carbon mass is used to form cellulose which gives rigidity to the cell. Generally, the plant uses cellulose to make their stems and leaves strong. The molecules of cellulose are arranged in parallel and joined together by the hydrogen bond. It is a long, chain-like structure which is joined with other cellulose molecules. Besides that, there is another source that produced cellulose, named as bacterial

Table 20.1 Different strains producing different yield of bacterial cellulose by using different carbon source.

Microorganism	Carbon source	Additional nutrients	Reactor	Incubation time	Yield (g/L)	References
Komagataeibacter pasteurianus	Beet molasses	None	Static	7 days	0.03	[8]
Gluconacetobacterxylinus	Glucose	Fabrics Cellulosic	Static	14 days	10.80	[10]
Acetobacter xylinum BPR2001	Molasses	None	Fed-Batch	72 h	7.82	[1]
Gluconacetobacterhansenii PJK (KCTC 10505 BP)	Glucose	Ethanol	Static	72 h	2.50	[35]
Gluconacetobacterxylinus ATCC 10245	Glycerol	None	Static	7 days	13	[44]
Gluconacetobacterhansenii PJK	Glucose	Ethanol	Fed Batch	18 days	20.85	[45]
Leifsonia sp. CBNU-EW3	Glucose	None	Static	15 days	3.8	[53]
Acetobacter xylinum NUST4.2	Glucose	None	Stirred Tank	72 h	3.13	[56]

cellulose because it comes from microbes. It is an extracellular product of acetic acid producing bacteria which is swollen, slippery, kind of moist skin, and gelatinous in nature. However, plant cellulose was chemically equal to bacterial cellulose and structure of both of them was combined by the hydrogen bonds. But, under the microscope the structures were not the same because fibrils structures of bacterial cellulose were mostly in oriented form. The structure difference may be due to the well-defined state of components. Bacterial cellulose is contamination free, it means it is synthesized in the pure form of cellulose. Eventhough other celluloses which are found in the mixed form of sugars like lignin, hemicelluloses, numerous glycoprotein, and pectin in plants. After treated with solvent, the cellulose is precipitated and obtained into the pure form. They both are structurally different, in case of plant cellulose it looked like ribbon-shaped which is much more complicated in comparison to bacterial cellulose.

Plant cellulose synthesis is done through cellulose synthase complexes which are arranged into rosettes form. It has been found that the encoding genes which were present in plant cellulose, has a bacterial origin because it came from the cyanobacterial endosymbiosis [33]. Bacterial cellulose producers grew statically into the liquid media and obtained into two forms, either they are floating on the air–liquid interface or they form solid surface associated film at the bottom of the conical. These biofilms are multicellular in nature and contain proteins, polysaccharides, extracellular DNA in a matrix [12]. In addition, it has been seen that producers do plant–bacteria interaction for bacterial cellulose production. For instances, the nutrient-rich rhizosphere produces bacterial cellulose by establishing the contact with plant root hair. Two well-known examples of bacterial cellulose-producing microorganisms are a tumor-inducing gram-negative bacteria *Agrobacterium tumefaciens* and mutual nitrogen-fixing symbiotic bacteria *Rhizobium leguminosarum*, plays a very important role for the production of bacterial cellulose through root–bacteria interactions.

Some basic difference between plant and bacterial cellulose are represented below for better understanding of their properties.

20.4.1 Morphology

Normally, bacterial cellulose is obtained from the cell's cytoplasmic membrane and amassed into ribbon fibril. To increase the surface area for bacterial cellulose production use high aspect ratio (L/w) which plays a key role and employed as reinforcement into a composite material and for a trouble-free transfer [18]. The length of the nanofibrils of bacterial cellulose depends on the different cellulose source [26]. In cellulose there are two main regions, one is high ordered which is more crystalline and other one is low region which is composed of crystalline and non-crystalline fabric. In addition, density and toughness of bacterial cellulose are increased in static cultivation with the high oxygen ration into the culture medium [55].

20.4.2 Crystallinity

Crystal structure of bacterial cellulose is better compared to the properties of its mechanical and interfacial than plant cellulose. The nature of crystallinity of cellulose depends on the diverse culture conditions such as pH, nutrients, additives, sort of reactor, and the most important bacterial strain [18]. In plants, cell wall cellulose is present generally in type Iβ form while approximately 70–80% type of Iα cellulose found in bacterial cellulose. Bacterial cellulose has found to be Iα-rich in a proportion of around 70–80% [51].

20.4.3 Degree of Polymerization

Length and crystallization of the cellulose play the main factor in the degree of polymerization which depends on the counting of n number units of glucose in a single chain. In comparison to plant cellulose, it has long and high crystallinity due to their high degree of polymerization [48].

20.4.4 Thermal Properties

Thermal properties of bacterial cellulose are high compared to plant cellulose by observing their thermograms. Those thermograms have two degradation steps where the first stage is volatilization and the other one is rapid volatilization of cellulose. In this first stage, cellulose decomposition occurred while in the second, which is sometimes also known as carbonaceous stage, char oxidation and carbonaceous residues are formed. On comparing the bacterial cellulose with plant cellulose, it has been found that at 22 and 24 °C bacterial cellulose degradation starts which is earlier than plant one. Though, maximum weight loss occurs in plant cellulose than in bacterial cellulose. In addition, to start the degradation process very high (142 and 150 °C) activation energy in plant cellulose than in bacterial cellulose is required [6].

20.4.5 Mechanical Properties

Bacterial cellulose has been found that it is having higher tensile strength and Young's modulus mechanical properties due to their presence of high degree of polymerization, increases the surface area and their crystallinity compared to plant cellulose origin. In the two different culture conditions, it has been obtained that static culture of bacterial cellulose is showing the highest mechanical properties than agitated bacterial culture. Generally, bacterial cellulose's tensile strength and Young's modulus are in the range of 200–2000 MPa and 15–138 GPa, respectively [54].

20.4.6 Water Absorption Properties

Through hydrogen bonding it can hold more water due to their nature of porous network in bacterial cellulose. They behave as a free bulk water because in hydrogel they contained only 10% water of the 99 %(w/w) in water interaction of bacterial cellulose and remaining 89% was hopped with fibril network [13]. It has also been reported that their water holding capacity depends upon the different cultivation method. In rotating biological fermentor, bacterial cellulose can clench five times more water than in static culture method.

20.4.7 Optical Properties

These light transmittance properties are higher because of the high wavelength of visible rays in bacterial cellulose. Due to this reason, it has many applications in various fields like lenses, display devices, and coatings. In the case of plant cell, it has low transparency due to their formation of large fibril, low yields of homogenization, aggregation, and low light transmittance value [48]. Basically, transmittance is associated with the process yield. Bacterial cellulose based nanocomposites are showing high optical properties, flexibility and dimension stability with good reinforcement in resins [11].

20.5 Compositional View of Bacterial Cellulose

Previous studies described that cellulose fibres are produced from the bacteria but before that they are formed as the homogenous slimy layer within that culture. According to the X-ray crystallographic studies the bacterial cellulose having a crystal structure known as Cellulose I where the two sugar molecules are arranged in parallel fashion. This native cellulose I found in two different modifications Iα and Iβ that are popular by name is triclinic and monoclinic respectively. The ratio of Iα/Iβ is depended upon the origin of cellulose Iβ structure explains different conformations and hydrogen bonding for neighbouring chains. Inspite of cellulose I and II, still three types of structure are present where changes happened only in crystal structure but not in the chemical structure of cellulose family. These are cellulose III, cellulose IV, and cellulose X. The triclinic (Iα) phase contained cell dimensions a = 0.674nm, b = 0.593nm, c = 1.036nm, it is a chain axis, α = 117°, β = 113°, and γ = 81°and one disaccharide molecules per unit cell [14]. Likewise, the model of Sarko and Muggli the monoclinic (Iβ) phase has the cell dimensions a =0.801 nm, b = 0.817nm, c = 1.036nm (chain axis), whereas, α = β =90°, and γ = 97.3 and two disaccharide molecules per unit cell [40]. Among all, cellulose II is one of the most stable crystal structures which are formed by the alkali treatment and the regeneration of crystal structure. Major factors behind the production of different nature of cellulose are their arrangement of molecules and also it depends upon the sources. So other than crystal nature, cellulose is also having amorphous domain in different ranges and different physical and chemical properties. Amorphous cellulose is also known as cellulose II which is less ordered in nature so any reactant can easily penetrate in this region. There are many methods such as, acid, enzymatic, ball milling and mixed solvents which are used to convert crystalline cellulose into amorphous form. This regeneration of cellulose can be analysed by the X-ray diffraction pattern, solid state ^{13}C-NMR spectra with wide resonance and Raman spectra techniques [7].

20.6 Molecular Biology of Bacterial Cellulose

Biosynthesis of bacterial cellulose involves the enzymes, catalytic complexes, regulatory proteins, and also managing a multi-step process. UDP-Glc (Uridine Diphosphoglucose) behaves as a precursor for the formation of cellulose followed by the polymerization of glucose molecule. However, GDPG (guanidine diphosphoglucose) are used into the biosynthesis of green plant cellulose. Many gram-negative bacteria species have the capability for cellulose production, such as *Agrobacterium tumefaciens, Escherichia coli, G. Xylinus,* and *Rhizobium leguminosarum.* Central structural genes are the most important factor which is essential for the cellulose biosynthesis; the product of structural gene involves the enzyme and structural proteins. These multimeric enzymes are located on the cytoplasmic membrane while terminal complexes (TCs) found in the cellulose producing organisms. These terminal complexes are of two types, first is linear terminal and other one is rosette terminal. TCs is a transmembrane protein complex that extent both cytoplasm membrane and outer membrane. Linear terminal has been seen that *Dictyostelium* sp., algae, and bacteria *Acetobacter xylinum.* In *Acetobacter xylinum,* TCs is present on the outer membrane as a single row of particle and flat microfibril is produced from the TCs subunit [23], while rosette terminal complex found on the algae and land plants [9].

Likewise it includes cellulose synthase, form an operon and regulatory gene on the chromosome. Some strains having more than one operon which are responsible for the cellulose biosynthesis but *in vivo* studies tell that only one operon at a time responsible for the biosynthesis of cellulose [41]. In general, for protein binding, two genes are present named as c-di-GMP (bis-3,5 cyclic diguanylic acid) and cellulose synthase. In operon, the first gene, cellulose synthase (catalytic subunit) that is encoded by *bcsA* sometimes it is also known as *acsA/celA* which stands for *Acetobacter* cellulose synthesis bind with UDP glucose whereas, binding protein c-di-GMP is encoded by *bcsB (acsB,celB)* gene present in cytoplasm membrane and interact with different enzymes. In type 2 cellulose synthases, the two gene *bcsA* and *bcsB*a refused together in a single open reading frame which in terms shows the tight coupling functional of the two protein products. According to Matthysse [32], during cellulose biosynthesis two different lipid intermediates participated in the *Agrobacterium tumifaciens* organism where initial glucose-lipid derivative is formed by the *celDE* gene product. But in the case of *G. xylinus* there is no information available of lipid intermediates and nor the homologous *celDE* [32].

In *S. typhimurium*, AgfD is a response controller of the LuxR domain which was noticed under the specific environmental conditions to regulate the transcriptionally AdrA and also regulate the thin aggregative fimbriae [38]. When AgfD is deleted, AdrA activates the cellulose biosynthesis. In view of the fact that, transcription of *bcsA* and *bcsC* are not dependent on the AgfD so the activation of cellulose biosynthesis by AdrA occurred at a posttranscriptional level [57].The production of c-di-GMP by the AdrA's domain (GGDEF) or the stabilization of Bcs proteins, when it contacts with AdrA by protein–protein communications might elicit the synthesis of cellulose. The same genes are present in *E. coli*, and *Salmonella* serotypes so it could be predicted that they also have same regulatory pattern which can be useful for cellulose biosynthesis.

20.7 Importance of Genetically Modified Bacteria in Bacterial Cellulose Production

Gluconacetobacter xylinum utilizes the glucose and sucrose as a carbon source via oxidation process but the end product is ketogluconate not cellulose and that's why it could be the reason for decrement of the pH, production rate and their growth into the culture medium. To obtain desired amount of cellulose it is necessary to isolate the ketogluconate-negative *Gluconacetobacter* strains [31]. By using UV mutagenesis, from original one, isolate the non-ketogluconate-producing mutant. When it is compared with parental strain, genetically modified *Gluconacetobacter xylinum* BPR2001 produced by dgc1-disrupted mutants. This modified gene dgc1 synthesized from the c-di-GMP and plays a very crucial role for the production of bacterial cellulose [2]. It also improves the biodegradability of cellulose *in vivo* because normal cellulose is not hydrolyzed in human digestive system and also restricts the various applications like in biomedical field.

Besides the production of cellulose from bacteria, *Gluconacetobacter* also secrete viscous water-soluble polysaccharides named as acetan. It is equivalently synthesized into the media with the cellulose production. For self-synthesis acetan carbohydrate molecule used the UDP-Glc which is also necessary for the cellulose synthesis, so it may lead to decrease the production of bacterial cellulose. To acknowledge this issue some researchers are trying to isolate the EP1 non-producing acetan mutant strain from the *Gluconacetobacter xylinum* BPR2001 [20].

20.8 Applications of Bacterial Cellulose in Different Industrial Sector

Beyond, the role of natural synthesis of bacterial cellulose and its hydrogel it is having an important role in different sectors of medical, food, cosmetics, etc. In food various range of applications of bacterial cellulose like from vegetarian meat, a traditional dessert, low cholesterol diet, and as a food additive and dietary assist to novel applications, such as immobilization of enzymes and cells. By using bacterial cellulose in the manufacturing of traditional dessert named as nata de coco famous in Philippines formed by the fermentation of coconut water for the biosynthesis of BC. Little chopped pieces of BC are dipped into the sugar syrup and served as a sweet candy [36].

Due to their high purity, biocompatibility, and hydrophilicity nature of bacterial cellulose, it offers many applications towards human and veterinary medicine.

20.8.1 Skin and Wound Healing

For tentative wrapping of wounds, the membrane of bacterial cellulose is used because of their high mechanical strength and easy flowbility for liquid and gases. For instance, in the market, bacterial cellulose products are available in the name of Biofill and Gengiflex which can be used for burn and ulcers in human and also have a wide role in surgery and dental implants. It gives immediate pain relief, reduced the infection rate, faster healing, transparency, reduced the time and cost. One disadvantage of Gengiflex is having limited elasticity in high mobility area [22]. Bacterial cellulose composites used as a skin tissue repair material because of having antimicrobial activity against microorganism. Some composite like bacterial cellulose-collagen type I are bound with outer and inner surface of the bacterial cellulose membrane with hydrogen bonding between collagen and bacterial cellulose. It is showing the antioxidant property with low adsorption of few proteases and interleukins.

20.8.2 Bacterial Cellulose Composites

The other composites like Bacterial cellulose-gelatin hydrogel which is used to improve the strength of fracture and showing high elastic modulus. In addition, bacterial cellulose-poly(3-hydroxubutyrate-co-4-hydroxubutyrate) scaffold shows high proliferation of fibroblast cell of Chinese hamster lung compared to alone poly(3-hydroxubutyrate-co-4-hydroxubutyrate) [5].

20.8.3 Artificial Blood Vessels

Artificial blood vessels are prepared by synthetic material and used to restore the circulatory blood. Due to hardening of arteries it blocks the coronary artery blood vessel around the heart. Nowadays, for by-pass operation bacterial cellulose materials are used and showing low risk of blood clot compared to the use of synthetic material such as Teflon, Dacron, and vinyon. Bacterial cellulose shows good contact with blood and not allows clotting that blood in vessels. PVA-BC nanocomposite is polyvinyl alcohol bacterial cellulose which is formed using thermal processing with confined strain and also adds the small quantity of bacteria cellulose.

20.8.4 In Paper Industry

In paper industry bacterial cellulose is used as a binder in paper. A very small amount of microfibril is added to the paper that gives strength and durability to the paper. To improve the tensile strength and filler retention of paper sheet use the obtained cellulose from both processes agitated and static culture at the wet end. In addition, during paper sheet formation add

5% bacterial cellulose with wood pulp to improve the fire resistance, kaolin retention, and their strength properties [3].

20.8.5 In Food Industry

Chronic diseases are not prevented or cured by any medicines, hence to reduce this risk, sometimes doctors prefer to take the dietary fibres in their regular food. Bacterial cellulose is also included as a dietry fibre food and it is classified as a "generally recognized as safe (GRAS)"[47]. It is also used as food additives and forms a low-calorie food item into the market and also replaces the fat molecule meat product that is emulsified [28]. It is worth mentioning that it also has the role in the production of low cholesterol food products. In addition, it has been seen that it reduces the serum trigylcerides, liver and serum cholesterol level *in vivo* system.

Monascus is a red pigment which is generally used in the production of fermented foods. Nowadays, monascus combined with bacterial cellulose having meat like taste and texture looks like lean meat and liver which could replace the meat. It also contains major amount of fibre, healthy nutrients, and limited calories. Monascus mycelium plays as a base flavour added food but does not gives the flavour to any food product.

20.8.6 Applications of Bacterial Cellulose in Other Fields

It is noteworthy that, bacterial cellulose is also used as a membrane for separation and purification due to the present of their porous structure. In ion exchange, bacterial cellulose combined with acrylic acid have high absorption capabilities for heavy metals through metal ion, shows improved electrochemical properties and also does not shows any environmental problems like disposal type during preparation. In addition, molecular imprinted polymer combined with bacterial cellulose is used for the separation of required enantiomers from racemic mixture because it is having multiple numbers of binding sites and carries the further separation part.

For photonic and optoelectronic devices, bacterial cellulose behaves as a substrate by the incorporation of conductive properties and it converts into electrically conductive sheets. It is immobilized with dyes (electrochemical) and attached to electrodes. It is highly flexible, reflective and high contrast and also the responsible for the application of reversible colour changes [43]. Bacterial cellulose has many new applications in biotextiles, organ engineering, bedsores, and biological nonwoven fabric.

20.9 Conclusion

Pure bacterial cellulose has applications in different industrial sectors due to its special properties and characteristics. Improvement in bacterial cellulose production can be achieved by designing fermentors in order to increase growth rate of microorganisms without the fibrils being mechanically disrupted. Agitation in the bioreactor is another factor that helps bacterial cellulose produced in forming three-dimensional structure, high tensile strength, and branched pellicle formation of cellulose. In current scenario, cellulose also obtained from gram-positive bacteria which are gaining equal importance compared to gram-negative bacteria because of its amorphous nature which have further potential applications in various fields. To meet the demands of bacterial cellulose, the combined use of airlift and stirred tank reactor are preferred sometimes for large-scale production. In addition, it has the potential of substituting plant cellulose due to its characteristic features, such as nanofibril structure, high crystallinity and porosity, water holding capacity, antimicrobial and mechanical properties. Bacterial cellulose composite has become popular for the development of hydrogels which hold relevance in various applications. In addition, the properties of bacterial cellulose have improved which ultimately can be used in a biomedical, food, and electronic applications.

References

1. Bae, S., Shoda, M., Bacterial cellulose production by fed-batch fermentation in molasses medium. *Biotechnol. Prog.*, 20(5), 1366–1371, 2004.
2. Bae, S.O., Sugano, Y., Ohi, K., Shoda, M., Features of bacterial cellulose synthesis in a mutant generated by disruption of the diguanylate cyclase 1 gene of *Acetobacter xylinum* BPR 2001. *Appl. Microbial. Biotechnol.*, 65(3), 315–322, 2004.
3. Basta, A.H., El-Saied, H., Performance of improved bacterial cellulose application in the production of functional paper. *J. Appl. Microbiol.*, 107, 2098–2107, 2009.
4. Bielecki, S., Krystynowicz, A., Turkiewicz, M., Kalinowska, H., *Bacterial cellulose*, pp. 31 Wiley-VCH Verlag GMBH & Co. KGaA, Weinheim, Germany 2005.
5. Cai, Z.J., Yang, G., Bacterial cellulose/collagen composite: characterization and first evaluation of cytocompatibility. *J. Appl. Polym. Sci.*, 1205, 2938–2944, 2011.
6. Campano, C., Balea, A., Blanco, A., Negro, C., Enhancement of the fermentation process and properties of bacterial cellulose: a review. *Cellulose.*, 23(1), 57–91, 2016.

7. Ciolacu, D., Ciolacu, F., & Popa, V. I., Amorphous cellulose—structure and characterization. *Cell. Chem. Technol.*, 45(1), 13, 2011.
8. Coban, E.P., Biyik, H., Evaluation of different pH and temperatures for bacterial cellulose production in HS (Hestrin-Scharmm) medium and beet molasses medium. *Afr. J. Microbiol. Res.*, 5, 1037–1045, 2011.
9. Emons, A.M.C, Role of particle rosettes and terminal globules in cellulose synthesis, in: *Biosynthesis and biodegradation of cellulose,* C. H. Haigler and P. J. Weimer (Eds.), pp. 71–98, Marcel Dekker, Inc., New York, N.Y., 1991.
10. Feng, Y., Zhang, X., Shen, Y., Yoshino, K., Feng, W., A mechanically strong, flexible and conductive film based on bacterial cellulose/graphene nanocomposite. *Carbohydr. Polym.*, 87(1), 644–649, 2012.
11. Fernandes, S.C.M., Oliveira, L., Freire, C.S.R., Silvestre, A.J.D., Neto, C.P., Gandini, A., Desbri′eres, J., Novel transparent nanocomposite films based on chitosan and bacterial cellulose. *Green Chem.,* 11, 2023–2029, 2009.
12. Flemming, H.C., Wingender, J., The biofilm matrix. *Nat. Rev. Microbiol.,* 8, 623–633, 2010.
13. Gelin, K., Bodin, A., Gatenholm, P., Mihranyan, A., Edwards, K., Stromme, M., Characterization of water in bacterial cellulose using dielectric spectroscopy and electron microscopy. *Polymer.,* 48, 7623–7631, 2007.
14. 8, A.P., Teleman, O., Interface between monoclinic crystalline cellulose and water: breakdown of the odd/even duplicity. *Langmuir,* 13(3), 511–518, 1997.
15. Hirai, A., Tsuji, M., Horii, F., Communication: culture conditions producing structure entities composed of Cellulose I and II in bacterial cellulose. *Cellulose,* 4(3), 239–245, 1997.
16. Hofinger, M., Bertholdt, G., Weuster-Botz, D. Microbial production of homogeneously layered cellulose pellicles in a membrane bioreactor. *Biotechnol. Bioeng.*, 108(9), 2237–2240, 2011.
17. Hornung, M., Ludwig, M., Gerrard, A.M., Schmauder, H.P., Optimizing the production of bacterial cellulose in surface culture: evaluation of substrate mass transfer influences on the bioreaction (Part 1). *Eng. Life Sci.*, 6(6), 537–545, 2006.
18. Huang, Y., Zhu, C.L., Yang, J.Z., Nie, Y., Chen, C.T., Sun, D.P., Recent advances in bacterial cellulose. *Cellulose.*, 21, 1–30, 2014b.
19. Hwang, J.W., Yang, Y.K., Hwang, J.K., Pyun, Y.R., Kim, Y.S., Effects of pH and dissolved oxygen on cellulose production by *Acetobacter xylinum* BRC5 in agitated culture. *J. Biosci. Bioeng.*, 88(2), 183–188, 1999.
20. Ishida, T., Sugano, Y., Nakai, T., SHODA, M., Effects of acetan on production of bacterial cellulose by *Acetobacter xylinum. Biosci. Biotechnol. Biochem.*, 66(8), 1677–1681, 2002.
21. Jonas, R., Farah, L.F., Production and application of microbial cellulose. *Polym. Degrad. Stabil.*, 59, 101–106, 1998.
22. Kawecki, M., Krystynowicz, A., Wysota, K., Czaja, W., Sakiel, S., *et al, Bacterial cellulose biosynthesis, properties and applications,* International Review Conference Biotechnology, Vienna, Austria, 14–18, 2004.

23. Kimura, S., Chen, H.P., Saxena, I.M., Brown, R.M., Itoh, T., Localization of c-di-GMP-binding protein with the linear terminal complexes of *Acetobacter xylinum*. *J. Bacteriol.*, 183(19), 5668–5674, 2001.
24. Kouda, T., Yano, H., Yoshinaga, F., Kaminoyama, M., Kamiwano, M., Characterization of non-Newtonian behavior during mixing of bacterial cellulose in a bioreactor. *J. Ferment. Bioengineer.*, 82(4), 382–386, 1996.
25. Kurosumi, A., Sasaki, C., Yamashita, Y., Nakamura, Y., Utilization of various fruit juice as carbon source for production of bacterial cellulose by *Acetobacter xylinum* NBRC 13693. *Carbohyd. Polym.*, 76, 333–335, 2009.
26. Lavoine, N., Desloges, I., Dufresne, A., Bras, J., Microfibrillated cellulose its barrier properties and applications in cellulosic materials: a review. *Carbohyd. Polym.*, 90, 735–764, 2012.
27. Lee, K.Y., Bismarck, A., Susceptibility of never-dried and freeze-dried bacterial cellulose towards esterification with organic acid. *Cellulose*, 19, 891–900, 2012.
28. Lin, K.W., Lin, H.Y., Quality characteristics of Chinese-style meatball containing bacterial cellulose (Nata). *J. Food Sci.*, 69, 107–111, 2004.
29. Lin, S.P., Hsieh, S.C., Chen, K.I., Demirci, A., Cheng, K.C., Semicontinuous bacterial cellulose production in a rotating disk bioreactor and its materials properties analysis. *Cellulose*, 21, 835–844, 2014.
30. Lu, H.M., Jiang, X.L., Structure and properties of bacterial cellulose produced using a trickling bed reactor. *Appl. Biochem. Biotechnol.*, 172, 3844–3861, 2014.
31. Masaoka, S., Ohe, T., Sakota, N., Production of cellulose from glucose by *Acetobacter xylinum*. *J. ferment. Bioengineer.*, 75(1), 18–22, 1993.
32. Matthysse, A.G., Thomas, D.L., White, A.R., Mechanism of cellulose synthesis in *Agrobacterium tumefaciens*. *J. Bacterial.*, 177(4), 1076–1081, 1995.
33. Nobles, D.R., Brown, R.M., The pivotal role of cyanobacteria in the evolution of cellulose synthases and cellulose synthase-like proteins. *Cellulose*, 11, 437–448, 2004.
34. Oh, S.Y., Yoo, D.Il., Shin, Y., Kim, H.C., Kim, H.Y., Chung, Y.S., Park, W.H., Youk, J.H., Crystalline structure analysis of cellulose treated with sodium hydroxide and carbon dioxide by means of X-ray diffraction and FTIR spectroscopy. *Carbohyd. Res.*, 340, 2376–2391, 2005.
35. Park, J.K., Jung, J.Y., Park, Y.H., Cellulose production by *Gluconacetobacter hansenii* in a medium containing ethanol. *Biotechnol. lett.*, 25(24), 2055–2059, 2003.
36. Phisalaphong, M., Chiaoprakobkij, N., Applications and products nata de coco. in: *Bacterial nanocellulose: a sophisticated multifunctional material,* Gama, M., Gatenholm, P., Klemm, D. (Eds.), pp. 143–156, CRC Press, Boca Raton, 2012.
37. Römling, U., Galperin, M.Y., Bacterial cellulose biosynthesis: diversity of operons, subunits, products, and functions. *Trends Microbiol.*, 9, 545–557, 2015.

38. Römling, U., Sierralta, W.D., Eriksson, K., Normark, S., Multicellular and aggregative behaviour of Salmonella *typhimurium* strains is controlled by mutations in the agfD promoter. *Mol. Microbial.*, 28(2), 249–264, 1998.
39. Ross, P., Mayer, R., Benziman, M., Cellulose biosynthesis and function in bacteria. *Microbiol. Rev.,* 55, 35–58, 1991.
40. Sarko, A., Muggli, R., Packing analysis of carbohydrates and polysaccharides. III. Valonia cellulose and cellulose II. *Macromolecules.*, 7(4), 486–494, 1974.
41. Saxena, I.M., Brown, R.M., Identification of a second cellulose synthase gene (acsAII) in *Acetobacter xylinum. J. Bacteriol.*, 177(18), 5276–5283, 1995.
42. Serafica, G., Mormino, R., Bungay, H., Inclusion of solid particles in bacterial cellulose. *Appl. Microbiol. Biotechnol.*, 58(6), 756–760, 2002.
43. Shah, J., Brown, R.M., Towards electronic paper displays made from microbial cellulose. *Appl. Microbial. Biotechnol.*, 66(4), 352–355, 2005.
44. Sherif, M.A.S., Evaluation of different carbon sources for bacterial cellulose production. *Afr. J. Biotechnol.*, 4(6), 478, 2005.
45. Shezad, O., Khan, S., Khan, T., Park, J.K. Production of bacterial cellulose in static conditions by a simple fed-batch cultivation strategy. *Korean J. Chem. Eng.*, 26(6), 1689–1692, 2009.
46. Shezad, O., Khan, S., Khan, T., Park, J.K., Physicochemical and mechanical characterization of bacterial cellulose produced with an excellent productivity in static conditions using a simple fed-batch cultivation strategy. *Carbohydr. Polym.* 82(1), 173–180, 2010.
47. Shi, Z., Zhang, Y., Phillips, G.O., Yang, G., Utilization of bacterial cellulose in food. *Food Hydrocoll.,* 35, 539–545, 2014.
48. Siro, I., Plackett, D., Microfibrillated cellulose and new nanocomposite materials: a review. *Cellulose,* 17, 459–494, 2010.
49. Son, H.J., Heo, M.S., Kim, Y.G., Lee, S.J., Optimization of fermentation conditions for the production of bacterial cellulose by a newly isolated Acetobacter. *Biotechnol. Appl., Biochem.*, 33(1), 1–5, 2001.
50. Song, H.J., Li, H.X., Seo, J.H., Kim, M.J., Kim, S.J., Pilot-scale production of bacterial cellulose by a spherical type bubble column bioreactor using saccharified food wastes. *Korean J. Chem. Eng.*, 26, 141–146, 2009.
51. Sun, D.P., Zhou, L.L., Wu, Q.H., Yang, S.L., Preliminary research on structure and properties of nano-cellulose. *J. Wuhan. Univ. Technol. Mat. Sci. Ed., 22*, 677–680, 2007.
52. Valla, S., Kjosbakken, J., Cellulose-negative mutants of *Acetobacter xylinum. Microbiology*, 128(7), 1401–1408, 1982.
53. Velmurugan, P., Myung, H., Govarthanan, M., Yi, Y. J., Seo, S. K., Cho, K. M., *et al*, Production and characterization of bacterial cellulose by Leifsonia. *Biotechnol. Bioprocess Eng.*, 20(3), 410–416, 2015.
54. Vitta, S., Thiruvengadam, V., Multifunctional bacterial cellulose and nanoparticle-embedded composites. *Curr. Sci.*, 102, 1398–1405, 2012.

55. Watanabe, K., Yamanaka, S., Effects of oxygen tension in the gaseous phase on production and physical properties of bacterial cellulose formed under static culture conditions. *Biosci. Biotechnol. Biochem.*, 59, 65–68, 1995.
56. Zhou, L.L., Sun, D.P., Wu, Q.H., Yang, J.Z., Yang, S.L., Influence of culture mode on bacterial cellulose production and its structure and property. *Wei sheng wu xue bao= Acta microbiologica Sinica*, 47(5), 914–917, 2007.
57. Zogaj, X., Nimtz, M., Rohde, M., Bokranz, W., Römling, U., The multicellular morphotypes of *Salmonella typhimurium* and *Escherichia coli* produce cellulose as the second component of the extracellular matrix. *Mol. Microbial.*, 39(6), 1452–1463, 2001.

Index

MPM 070319
Printed in Singapore